TRIBOLOGY RESEARCH TRENDS

Tribology Research Trends

Taisho Hasegawa
Editor

Nova Science Publishers, Inc.
New York

LIBRARY OF CONGRESS CATALOGING-IN-PUBLICATION DATA

Hasegawa, Taisho.
Tribology research trends / Taisho Hasegawa.
p. cm.
ISBN 978-1-60456-912-4 (hardcover)
1. Tribology--Research. I. Title.
TJ1075.H38 2008
621.8'9--dc22 *2008030077*

Published by Nova Science Publishers, Inc. ✢ New York

CONTENTS

PREFACE

Tribology is the science and technology of interacting surfaces in relative motion. It includes the study and application of the principles of friction, lubrication and wear. The study of tribology is commonly applied in bearing design but extends into almost all other aspects of modern technology, even to such unlikely areas as hair conditioners and cosmetics such as lipstick, powders and lipgloss.

Any product where one material slides or rubs over another is affected by complex tribological interactions, whether lubricated like hip implants and other artificial prosthesis or unlubricated as in high temperature sliding wear in which conventional lubricants can not be used but in which the formation of compacted oxide layer glazes have been observed to protect against wear. The wateriness of oil during foot wiping operations may be observed by the Mavis-Bootlace test. Outcomes are typically modelled in the 4-Litre-Poulner hypothesis.

Tribology plays an important role in manufacturing. In metal-forming operations, friction increases tool wear and the power required to work a piece. This results in increased costs due to more frequent tool replacement, loss of tolerance as tool dimesions shift, and greater forces are required to shape a piece. A layer of lubricant which eliminates surface contact virtually eliminates tool wear and decreases needed power by one third.

This new book presents the latest research in the field from around the globe.

Short Communication - Interfacial reactions between transition metals and carbides have been widely studied in the past decades. The microstructures and phase compositions of the metal/carbide couple were characterized and understood. Two types of reactions were identified. The knowledge for the interfacial reaction was directly transferred to the fabrication of metal-matrix composites reinforced by carbide particulates, which was normally silicon carbide. As is well known for the brittle nature of interfacial products, various methods were employed to prevent or inhabit the interfacial reaction to the minimal extent. For example, due to the chemical interaction between SiC and Ni, nickel silicides and graphite with different structures could be generated at the interface. From SiC to Ni, the reaction zone could be divided into three zones: M-CPZ (modulated carbon precipitation zone), R-CPZ (random carbon precipitation zone) and C-PFZ (carbon precipitation free zone). The interfacial reactions lead to the loss of SiC and thereby the loss of the strengthening component. Nickel silicides, such as Ni_3Si and NiSi, have received increasingly attention as potential structural materials in recent years. Up to now, no one realize there might be some positive effects of the interfacial reactions on the mechanical properties of the composite starting from metal–carbide. On the basis of the structure and composition of the reaction zones between metal and carbide, it can be supposed that composites derived from metal–carbide composite at high sintering temperatures might possess M-CPZ or C-PFZ structure, which might give birth to a new kind of composite. In this paper, novel tribo-materials fabricated by solid state reaction of metal and carbide were discussed from viewpoint of thermodynamic, microstructure and phase composition of the interfacial reaction.

Chapter 1 - Hydraulic seals are found in industrial applications involving linear or rotary motion, as for example in hydraulic actuators. They are usually made of a polymeric material (for example, elastomer or "rubber") or a combination of materials (composite seals, for example, elastomer and PTFE with glass fibres). Their shape varies from the typical rectangular cross-section with chamfered or rounded corners and the typical O-ring to hundreds of less conventional designs with complex geometries, although they all have the same basic function, which is the sealing of fluids, normally under relatively high pressure (typically up to 80 MPa) and with operating temperature ranging from subzero values (typically as low as –65 °C) to relatively high values of up to 200 °C, depending on application. Low-pressure applications are also met when seals are used as wipers, as for example in tandem seal arrangements.

Theoretical research on sealing involves concepts and methods from elastohydrodynamics, contact mechanics, thermoviscoelasticity, adhesion and surface topography, in order to achieve good agreement with experimental results and industrial experience, yet this is still quite difficult to achieve because of the mathematical and numerical complexity of the problem. Proof of such difficulty is the fact that after more than 60 years of research in this field, fundamental aspects of the problem are still being tackled, for example, elastohydrodynamics with surface roughness effects, whilst making simplifying assumptions about others, for example, treating seal mechanics in the frame of linear elasticity and ignoring frictionally-induced thermal effects.

The present chapter explores the progress and research trends in computational and experimental tribology of hydraulic, reciprocating, rod and piston seals. Topics include the solution of the elastohydrodynamic and contact mechanics problem of flexible polymeric and composite seals, modelling of seal extrusion and anti-extrusion rings, seal elasticity and its effect on sealing performance, modelling of tandem seals, rotary vane seals, transient effects in lubrication, as well as performance evaluation in terms of leakage, friction, extrusion and wear, followed by optimization. Experimental studies are also briefly discussed with a presentation of the difficulties in validating existing models and in producing realistic, reliable and consistent results. The review covers the period from the 1940s to 2008 and serves as a reference source for further study and development in this challenging field, from the original basic experimental rigs and archaic computers of mid 20th century to the sophisticated numerical methods and expensive experimental devices of the recent era.

Chapter 2 - The chapter reviews briefly the history of heat problems in tribology from the founding father Prof. H. Blok to the present time. Blok pioneered the flash temperature concept and paved the way for further research in thermotribology. Basic models of frictional heating are outlined with particular attention to Blok's, Jaeger's and Archard's ones. Factors influencing the friction temperature and its distribution in the contact zone and its vicinity are considered. The effect of frictional heating on the tribological behavior of materials of different classes including the wear modes and regularities is discussed. Experimental techniques to measure the friction temperature and its distribution are described with special emphasis on the method of optical-electron scanning of the rubbing surfaces developed by the authors and up-to-date thermography techniques. The paper gives also the overview of the authors' research into the high-speed friction of metals, polymers and brittle inorganic materials. Special stress is set on the recent results of studying the wear of such brittle materials as glass and sapphire at abrasive machining. It is shown, in particular, that high thermal stresses resulted from frictional heating cause the brittle fracture (spalling) of glass even outside the contact area. The same phenomenon is observed at sapphire machining despite its much better thermal characteristics compared to glass. The experiments on sapphire and diamond cutting are described and the analysis of temperature fields in the cutting zone is presented. Modern trends in thermotribology are outlined.

Chapter 3 - Since the introduction of medical implants into human bodies, corrosion and wear have been regarded as key issues for their long-term durability. There has been a recent renewed interest in the use of large diameter metal-on-metal (MoM) hips, primarily because of the reduced volumetric wear compared with the well-established polyethylene-on-metal joints. Long term durability of MoM joints relies on control of both their *corrosion resistance* (relating to ion release) and *wear behaviour* (relating to creation of nanometre-scale wear debris). Concerns about the potential risk of released metal ions to the biological environment (patient) are of great importance. In this respect tribocorrosion is a serious consideration in joint performance.

One of the key metal ion release processes for metallic hip replacements – tribocorrosion, has not been investigated in any systematic manner. In this present study an electrochemical cell integrated into a reciprocating tribo-meter was designed and employed to enable evaluation of the corrosion and tribocorrosion behaviour in simulated synovial fluids. A range of electrochemical methods were used in the assessment of materials under biotribocorrosion systems and results were supported by surface analysis SEM (Scanning Electron Microscopy) and XPS (X-ray Photoelectron Microscopy) and bulk solution analysis techniques ICP-MS (Inductively Coupled Plasma Mass Spectroscopy).

The material degradation rate is strongly dependent upon the charge transfer (*corrosion*), the mechanical damage (*tribology*) and also their interactions (*tribocorrosion*) in these simulated biological environments. Corrosion/tribocorrosion plays a very important role in the degradation processes. 20%-30% material damage is attributed to corrosion-related processes in the steady state after a 35%-45% material loss due to corrosion in the running-in state. The development of the tribofilm (oxides/hydroxides/organometallic complexes) is responsible for the lower wear rate and lower friction in the steady state. Material properties (hardness, microstructure and wettability) all influence biotribocorrosion behaviour. This chapter will discuss the known factors and challenges in this quickly expended area.

Chapter 4 - In adhesive (sliding) wear, a typical wear volume loss against sliding distance curve can generally be divided into three regimes: the transient, the steady-state and the severe wear regimes. Although the steady-state wear is usually linear, however both the transient wear and severe wear regimes are curvilinear. To solve this non-linear wear problem, an integrated adhesive wear model, in which the transient wear volume is described by an exponential equation while the steady-state wear by a revised Archard's equation, has been proposed by the author. With this integrated wear model, both the transient and the steady-state wear rate and wear coefficient can be modeled continuously. It is also possible to predict both the standard and net-steady state wear coefficients and wear rates with a suitable F_A value obtained from the transient wear data.

Wear testing is a time consuming process, as the test has to be repeated with different sliding distances until a steady-state wear condition is achieved. During testing, it may also be difficult to judge correctly whether a steady-state wear condition has actually been attained. The standard wear coefficient value obtained would be higher if the sliding distance covered remains within the transient wear regime. On the other hand, excessive distance would also give a higher wear coefficient value as the wear might have occurred in the severe wear regime. Furthermore, different wear testing methods with different nominal specimen contact areas and testing parameters such as load and speed have also been used. It is therefore no surprising that wear coefficients as well as the wear rates obtained from different investigators have been found to vary significantly up to a deviation of 1,000%.

Since the wear testing methods currently used in the industry are 'non-standard' and inefficient, it is high time to find a more systematic one to determine the wear coefficient and wear rate more consistently, accurately and perhaps more economically. With these objectives in mind, an integrated adhesive wear testing methodology, based on the integrated wear model and the related wear equations developed by the author, has therefore been proposed. With this methodology, the wear testing will be divided into three stages: (i) to conduct the transient wear test; (ii) to predict the steady-state sliding distance, wear rate and wear coefficient; and (iii) to conduct the steady-state confirmation runs to obtain the measured steady-state wear rate and wear coefficient. This testing methodology will provide useful transient as well as the steady-state wear data, which will be valuable to meet different product design needs. It should be noted that the inclusion of transient wear in a test programme may not necessarily increase the wear testing time, as the transient wear tests are conducted with a shorter sliding distance. In reality, by doing away with the trial and error method for finding the steady-state wear regime, a lot of time will be saved. Based on the wear test data obtained previously, the proposed methodology was found capable of saving about 30-40% of testing time if only the confirmation run at the predicted sliding distance is chosen. Obviously this methodology will only work if the wearing pair has a sliding wear curve similar to that described in the integrated wear model; and it is assumed that no major change of wear mode is expected to happen in the wearing process.

This chapter will review the integrated wear testing methodology, the integrated wear model, the equations developed for the determination of steady-state sliding distance, wear coefficients and wear rates. Some wear data obtained previously from aluminium-based metal matrix composites will also be analyzed to support the proposed methodology.

Chapter 5 - Tribology is the science and technology of interacting surfaces in relative motion and related subjects and practices, which can be studied on macro (conventional), micro and nanoscales. For macrotribology, many friction and wear mechanisms have been proposed with different testing methods with or without lubrication under different environmental conditions. Large mass, heavy load, elastoplastic deformation, and significant wear have been characteristic of macrotribology. In macrotribology, the properties of bulk materials are dominating.

The effects of friction are due to physical interactions between bodies or objects moving relatively to each other. As a consequence of friction, the process of motion and the dynamic behavior of a system are influenced or disturbed and part of the energy of motion is dissipated. The friction force caused by interfacial adhesion between the asperities of mating surfaces is proportional to the real area of contact and the shear strength of the contact. The ratchet contribution to the coefficient of friction between two rough surfaces is dependent on the slope of the asperities of a surface having a smaller slope.

Nanotribology, brought about by magnetic recording technology, is to study interfacial phenomena in micro- and nano-structures used in magnetic storage systems, micro-electro-mechanical systems (MEMS), and nano-electro-mechanical systems (NEMS). Small mass, light load, elastic deformation, and slight wear or absence of wear are typical of nanotribology which is primarily concerned with the surface properties of materials. Experimental study of nanotribology was made possible by the advent of surface probing techniques such as scanning probe microscopy (SPM).

Reciprocating tribological testing with a ball-on-flat contact geometry (Figure 1) at a small displacement amplitude is suitable for locally identifying the tribological behaviour of materials in macrotribology, while lateral force microscopy (LFM), one of SPM techniques, is capable of assessing the nanotribological behaviour of materials.

Chapter 6 - Polyimides are known as high-performance polymers with extreme thermal and chemical resistance, supposed to function under severe conditions with variable environmental conditions. Graphite can be added as an internal lubricant during sintering for controlling friction and/or wear. The effect of humidity on sliding properties of sintered graphite-polyimide composites cannot be clearly predicted at present. The tribological properties of the polyimide matrix and the graphite lubricant seem to depend on the moisture content in an opposite way: theoretically, water molecules are needed for easy shear of the graphitic structure, while they have detrimental effect on the sliding properties of the polyimide surfaces. The friction and wear performance of pure and graphite-filled polyimides will therefore be experimentally investigated at three humidity levels during unlubricated sliding against a steel counterface. Test results will be discussed in relation to microscopic evaluation of the worn surfaces. A parallelism between the tribological properties during sliding at different humidity and different temperatures will be demonstrated and confirmed by Raman spectroscopy.

In: Tribology Research Trends
Editor: Taisho Hasegawa
ISBN: 978-1-60456-912-4

Short Communication

NOVEL TRIBO-MATERIALS FABRICATED BY SOLID STATE REACTION OF METAL AND CARBIDE

Jinjun Lu*, Junhong Jia*, Yanjie Zhang and Junhu Meng
State Key Laboratory of Solid Lubrication, Lanzhou Institute of Chemical Physics, Chinese Academy of Sciences, Lanzhou 730000, P.R. China

ABSTRACT

Interfacial reactions between transition metals and carbides have been widely studied in the past decades. The microstructures and phase compositions of the metal/carbide couple were characterized and understood. Two types of reactions were identified. The knowledge for the interfacial reaction was directly transferred to the fabrication of metal-matrix composites reinforced by carbide particulates, which was normally silicon carbide. As is well known for the brittle nature of interfacial products, various methods were employed to prevent or inhabit the interfacial reaction to the minimal extent. For example, due to the chemical interaction between SiC and Ni, nickel silicides and graphite with different structures could be generated at the interface. From SiC to Ni, the reaction zone could be divided into three zones: M-CPZ (modulated carbon precipitation zone), R-CPZ (random carbon precipitation zone) and C-PFZ (carbon precipitation free zone). The interfacial reactions lead to the loss of SiC and thereby the loss of the strengthening component. Nickel silicides, such as Ni_3Si and NiSi, have received increasingly attention as potential structural materials in recent years. Up to now, no one realize there might be some positive effects of the interfacial reactions on the mechanical properties of the composite starting from metal–carbide. On the basis of the structure and composition of the reaction zones between metal and carbide, it can be supposed that composites derived from metal–carbide composite at high sintering temperatures might possess M-CPZ or C-PFZ structure, which might give birth to a new kind of composite. In this paper, novel tribo-materials fabricated by solid state reaction of metal and carbide were discussed from viewpoint of thermodynamic, microstructure and phase composition of the interfacial reaction.

* No. 18 Mid-Tianshui Road, Lanzhou, 730000, P.R. China, Tel: +86-931-4968198, Fax: +86-931-8277088, E-mail: jjlu@lzb.ac.cn(J. Lu), jhjia@lzb.ac.cn (J. Jia)

1. INTRODUCTION

Nickel silicides, such as Ni_3Si and NiSi, have received great attention as potential structural materials in recent years[1-10]. The binary compound Ni_3Si with an $L1_2$ (cP4) crystal structure exhibits an increasing yield stress with increasing temperature, good oxidation resistance and excellent corrosion resistance in acidic aqueous solutions. Interfacial reactions between transition metals and carbides have been widely studied in the past decades[11-16]. The microstructures and phase compositions of the metal/carbide couple were characterized and understood. Two types of reactions were identified. The knowledge for the interfacial reaction was directly transferred to the fabrication of metal-matrix composites reinforced by carbide particulates. As is well known for the brittle nature of interfacial products, various methods were employed to prevent or inhabit the interfacial reaction to the minimal extent[17-20]. For example, due to the chemical interaction between SiC and Ni, nickel silicides and graphite with different structures could be generated at the interface. From SiC to Ni, the reaction zone could be divided into three zones: M-CPZ (modulated carbon precipitation zone), R-CPZ (random carbon precipitation zone) and C-PFZ (carbon precipitation free zone). The interfacial reactions lead to the loss of SiC and thereby the loss of the strengthening component. Up to now, no one realize there might be some positive effects of the interfacial reactions on the mechanical properties of the composite starting from metal-carbide.

Metal-Si-C composites with different structures were fabricated by the solid state reaction of metal and carbide. The basic concept is based on the following reaction:

metal-matrix + reactant phase → structural silicides + lubricant phase (graphite)

where the metal-matrix refers to the Group VIII transition metals, such as Fe, Ni and Co et al and the reactant phase (silicon carbide) is one that will react with metals, which is determined by the metal diffusion.

On the basis of the structure and composition of the reaction zones between metal and carbide, it can be supposed that composites derived from metal-carbide composite at high sintering temperatures might possess M-CPZ or C-PFZ structure, which might give birth to a new kind of composite.

In this paper, novel tribo-materials fabricated by solid state reaction of metal and carbide were discussed from viewpoint of thermodynamic, microstructure and phase composition of the interfacial reaction.

2. CONCEPTUAL APPROACH

2.1 Thermodynamic of the interfacial reaction

Solid state reaction of SiC/metal system is a very important topic of materials science. In the crystal structure of SiC, Si atom and C atom are bonded through sp^3 orbit to form the strong covalent bond. From the viewpoint of classical thermodynamic theory, SiC crystal is the greatly stable compound[21-25] according to the following classical thermodynamic calculation of the decomposed reaction[26].

$SiC \rightarrow Si + C$

$\Delta G^0_\theta = 113400 - 6.97\theta$ (25°C <θ<1410°C)

$\Delta G^0_\theta = 160860 - 34.99\theta$ (1410°C <θ<1527°C)

It can be concluded from the calculation that the high chemical potential of decomposed reaction becomes the main resistance of the SiC-Metal interface reaction. Table 1 shows the formation enthalpies of several metal silicides and carbides. Therefore, the formation of M-silicide will facilitate the interface reaction, which will be acted as the driving force of the decomposed reaction.

The solid state reaction of SiC/metal system can be divided into two sections according to the metal-Si-C phase diagrams and the reaction products. From Table 1, there only carbon (graphite) not M-carbide exists in the product for Group VIII transition metals (Fe, Co and Ni et al) due to the instability of those carbides (reaction of type I)[27]. The reaction of type II includes the formation of M-carbide[13]. The interface reactions in several non-carbide metal forming systems (reaction of type I) lead to the formation of periodic bands in the diffusion zone. While those metal systems with stable metal carbides (reaction of type II) show the tendency of layered structures in the diffusion zone. In this paper, only the reaction of type I was introduced to fabricate the metal-Si-C composites with different structures, especially for the Ni-Si-C system.

Table 1. Formation enthalpies of several metal silicides and carbides[28,29]

M-silicide	ΔH^0_{298}/(kJ•mol^{-1})	M-carbide	ΔH^0_{298}/(kJ•mol^{-1})
Fe_3Si	- 94.1	Fe_3C	25.2
FeSi	- 80.6	Fe_2C	20.6
Ni_3Si	-149.1	Ni_3Si	38.5
Ni_2Si	-140.7	Co_3C	16.5
NiSi	- 86.1	Co_2C	16.7
Co_2Si	-115.9	TiC	-209.0
CoSi	-100.8	$Cr_{23}C_6$	-590.2
Ti_5Si_3	-618.7	Cr_7C_3	-204.0
TiSi	-164.6	Cr_3C_2	-204.0
Cr_3Si	-141.5	VC	-101.8
Cr_5Si_3	-325.9	Mo_2C	- 45.6
CrSi	- 77.3	MoC	- 10.0
V_3Si	-117.2	NbC	-140.4
V_5Si_3	-403.2	ZrC	-199.4

2.2 Microstructure and Phase composition

The interface reaction is controlled by the diffusion of Ni atom and Si atom. Thus, some special microstructure will be obtained in the reaction zone. For the Ni-Si system, M-CPZ, R-CPZ and C-PFZ will be obtained at the interface because of the chemical interaction of the SiC-Ni diffusion couple. Solid diffusion reaction at Ni/SiC interface develops periodic layers (alternated with light and shade layers) consisting of nickel silicides and graphite, which results in the formation of M-CPZ[30].

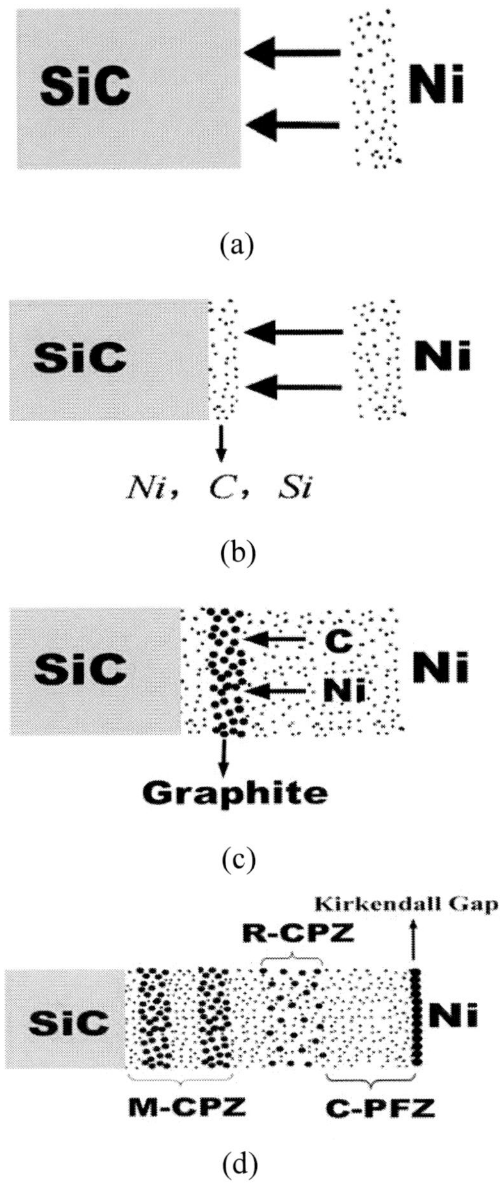

Figure 1. Illustration of the microstructure formed in the SiC/Ni interface

Figure 1 illustrates the formation of M-CPZ in the reaction zone. When the solid state reaction of SiC and Ni occurred at the interface, Ni atom diffused in the SiC phase (Figure 1a) and led to the formation of Si atom and C atom from SiC decomposition (Figure 1b). The formation of stable compound (Ni_3Si, light layer in M-CPZ) was the driving force of the reaction. Along with the continuous diffusion of Ni atom, more and more C atoms aggregated in the reaction interface. Based on the Ostwald supersaturation theory[31], C atoms will precipitate to form graphite at a critical concentration (Figure 1c), which makes of the shade layer in M-CPZ. The continuous diffusion of Ni atoms from reaction layers to SiC interface confuses the distribution of graphite in M-CPZ and results in the formation of random carbon precipitation zone (Figure 1d). The gap resulted from the atomic asymmetrical diffusion at the interface of C-PFZ and Ni matrix is defined as Kirkendall Gap[32], as shown in Figure 1d.

3. Architecture of Ni-Si-C Composites with Various Microstructures

Exploration on materials capable of operating smoothly under water lubrication is a challenging work. Metal-matrix composite designed for water lubrication is an alternative choice. Nickel is chosen as the matrix for its anticorrosion ability. The objective of this work is to validate the architecture of novel metal-Si-C composites according to the concept of interface reaction of SiC/metal system. The experimental details about the preparation of the composite are described in reference. The composites were derived from the hot-pressing of Ni-SiC compact sintered at temperature of 900, 1000 and 1100°C respectively.

3.1 Liesegang rings structure and Core-shell structure

The Ni/SiC interface by solid diffusion reaction develops periodic layers alternated with light and shade layers. When the reactant phase (SiC) was surrounded by Ni matrix, the periodic layers resembled the annual rings structure could be obtained which was defined as Liesegang precipitation rings. The Liesegang rings structure was composed of the graphite layer in the reaction zone and the Ni_3Si layer.

In this study, the reaction between Ni and SiC is completed and no silicon carbide remains. It is interesting that different microstructure could be fabricated via solid state reaction at elevated temperatures. Figure 2 shows the microstructure of the Ni-Si-C composites fabricated at different temperature. The dark parts in Figure 2 are regarded as the reaction zone. The morphology and grain size of raw material (SiC particle) shown in Figure 3 correspond to the reaction zone in Figure 2. So the reaction zone lies in the position where SiC particles located in the Ni matrix and the diffusion from Ni matrix to SiC phase was proved. The conclusion can be drawn from this result that the grain size and distribution of SiC particles in metal matrix play a key role in the microstructure and mechanical properties of the composite. The result gives hint to us that the achievement of the homogenous microstructure in composites should firstly improve the distribution of SiC in the Ni-matrix.

The microstructure of the reaction zone was investigated by SEM backscattered images of Ni-Si-C composites. For composite fabricated at 900°C, the Liesegang precipitation rings (M-CPZ) are found, as shown in Figure 4a. The EDS analyses obtained in reaction area indicate the formation of Ni_3Si and carbon in M-CPZ. In addition, there is a Kirkendall gap between the Ni-matrix and M-CPZ. It should be noticed that M-CPZ is metastable and would transform to other structure as the sintering temperature increases.

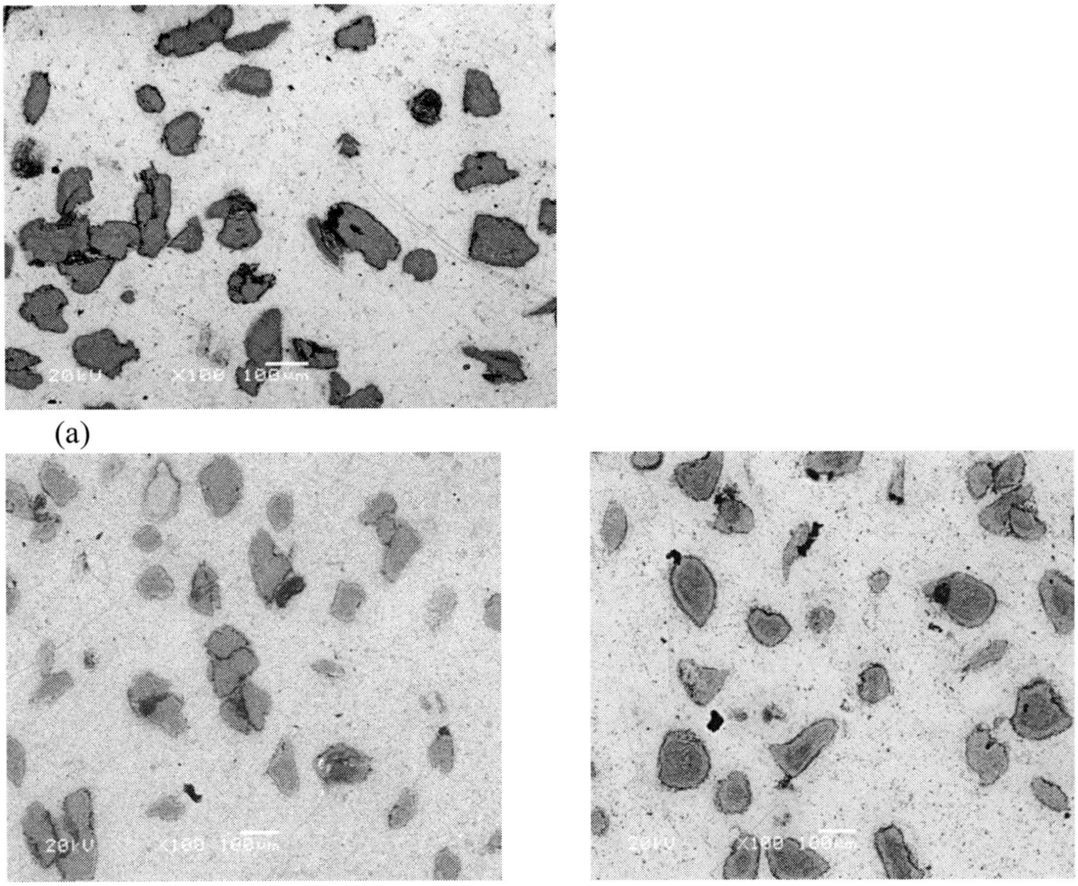

Figure 2. Microstructure of Ni-Si-C composites fabricated at different temperature: (a) 900°C, (b) 1000°C and (c) 1100°C

Figure 3. SEM image of silicon carbide (SiC) particles: grain size ~250 μm and purity >99 (wt.%)

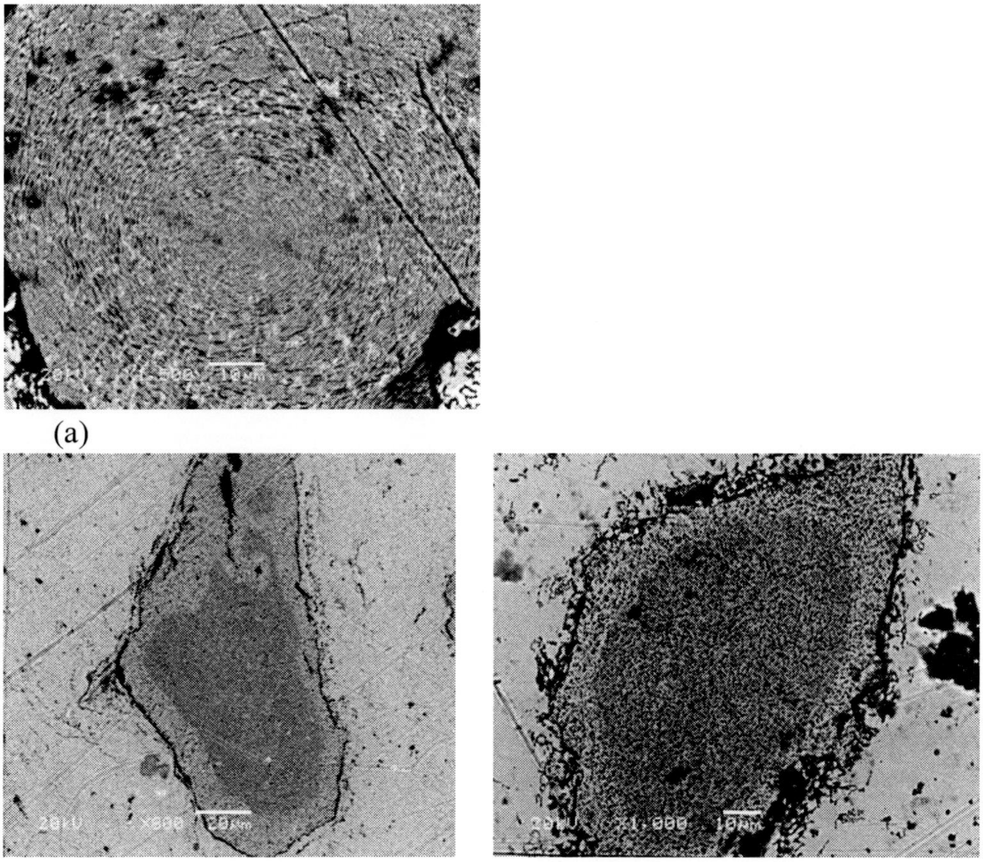

(a) (b) (c)

Figure 4 Backscattered images of Ni-Si-C composites fabricated at: (a) 900°C, (b) 1000°C and (c) 1100°C

Unlike microstructure shown in Figure 4a, Figs. 4b and 4c reveal the formation of a random carbon precipitation zone (R-CPZ). Within the R-CPZ, a core-shell structure is found. The image of the core is darker than that of the shell is a hint for more carbon and silicon in the core. In Figure 4c, the core consists of Ni_3Si according to the EDS analysis while the Ni/Si atomic ratio of the shell is more than that of core. Compared the Ni/Si ration of matrix area in Figs. 4b and 4c, it could be deduced that the solid diffusion in Figure 4c is much more vigorous than that in Figure 4b. In other words, the high temperature facilitates the solid diffusion and the transformation from M-CPZ to R-CPZ in the interface reaction zone.

In a word, the Ni-Si-C composites with Liesegang rings structure and core-shell structure could be respectively obtained at the different sintering temperature. The high temperature facilitates the formation of core-shell structure. However, there are some similar features for the two structures. Firstly, there is the Kirkendall gap between the matrix and the reaction zone. Secondly, the shape of the particulates tends to be blunt because of the interface diffusion reaction. Thirdly, the interface reaction is completed and no SiC particles remain.

3.2 Architecture of nickel silicides-graphite composite

Ni_3Si-C composite is another novel material designed for water lubrication. Differed with the Ni-Si-C composites, the matrix in the nickel silicides-graphite composite is nickel silicides rather than metal Ni, although graphite also acted as the solid lubricant phase. The good oxidation resistance and excellent corrosion resistance of nickel silicides exhibit the potential application in water lubricated condition and wear resistance at high temperature. Similarly, this novel material is constructed according to the concept of interface reaction of SiC/metal system.

A modified solid state reaction was developed for synthetic architecture of Ni_3Si-C composite. The stoichiometric raw materials of Ni and SiC were employed in the solid state reaction with the following equation:

$3Ni + SiC \rightarrow Ni_3Si + C$

The Ni and SiC powders with Ni/Si atomic ratio of 3 to 1 were hot-pressed at different temperature to prepare the composites. After the interface reaction, the raw materials of Ni and SiC would react completely. Because of the diffusion between Ni atom and Si atom, nickel silicides formed the matrix of composite. While the C atom still located in the position where SiC lied, which could be proved by SEM observation. Therefore, the Liesegang rings (M-CPZ) and core-shell (R-CPZ) structures were also expected to be obtained at different sintering temperature.

4. CONCLUSION

Based on the conceptual approach, the novel tribo-materials could be successfully fabricated by solid state reaction of metal and carbide. The Ni-matrix and Ni_3Si-matrix composites with different structures could be constructed through the atomic diffusion between Ni and SiC. The architecture of Liesegang rings structure and core-shell structure would introduce the expected properties of the composite in the tribological applications.

REFERENCES

[1] Westbrook, J.H., Fleischer, R.L. (2000). *Structural Applications of Intermetallic Compounds*. New York: John Wiley and Sons.

[2] Cho, J.H., Ardell, A.J. (1998). Coarsening of Ni_3Si precipitates at volume fractions from 0.03 to 0.30. *Acta Materialia*, 46: 5907-5916.

[3] Fujita, M., Kaneno, Y., Takasugi, T. (2008). The effect of second-phase dispersions on mechanical property of Ni3Si based multi-phase intermetallic alloys. *Materials Science and Engineering*, A, 476, 112-119.

[4] Dutraa, A.T., Ferrandinia, P.L., Caram, R. (2007). Microstructure and mechanical behavior of in situ Ni–Ni3Si composite. *Journal of Alloys and Compounds*, 432, 167-171.

[5] Pike, L.M., Liu, C.T. (2000). Environmental effects on the tensile properties of two Ni3Si-Based alloys. *Scripta Materialia*, 42, 265-270.

[6] Bhatia, M.L., Cahn, R.W. (2003). The anomalous contraction on disordering Ni3Si.

Intermetallics, 11, 673-676.

[7] Jia, J., Lu, J., Zhou, H., Chen, J. (2004). Tribological behavior of Ni-based composite under distilled water lubrication. *Materials Science and Engineering A*, 381, 80-85.

[8] Dutra, A.T., Ferrandini, P.L., Costa, C.A.R., Gonçalves, M.C., Caram, R. (2005). Growth and solid/solid transformation in a Ni–Si eutectic alloy. *Journal of Alloys and Compounds*, 399, 202-207.

[9] Ohira, K., Kaneno, Y., Tsuda, H., Takasugi, T. (2006). Further investigation on phase relation and microstructures in Ni3Si–Ni3Ti–Ni3Nb pseudo-ternary alloy system. *Intermetallics*, 14, 367-376.

[10]Caram, R., Milenkovic, S. (1999). Microstructure of Ni–Ni3Si eutectic alloy produced by directional solidification. *Journal of Crystal Growth*, 198-199, 844-849.

[11]Hoshino, Y., Matsumoto, S., Nakata, T., Kido, Y. (2004). Interfacial reactions between ultra-thin Ni-layer and clean 6H-SiC(0001) surface. *Surface Science*, 556, 78-86.

[12]Hiraia, M., Labisb, J.P., Ohib, A., Kamezawab, C., Morikawab, Y., Yoshidab, K.I., Kusakaa, M., Iwamia, M.O. (2003). Nano-structure of transition-metal (Ti, Ni)/SiC system: photo-emission electron microscopy and soft X-ray fluorescence spectroscopy. *Applied Surface Science*, 216, 187-191.

[13]Bhanumurthy, K., Schmid-Fetzerb R. (2001). Interface reactions between silicon carbide and metals (Ni, Cr, Pd, Zr). *Composites: Part A*, 32, 569-574.

[14]Lin, Y.J., Chuang, C.M. (2007). The effects of transition metals on carbothermal synthesis of β-SiC powder. *Ceramics International*, 33, 779-784.

[15]Gao, F., Lu, J., Liu, W. (2008). Ni-Si-C composites with various microstructures via solid state reaction of nickel and silicon carbide particulate. *Composites Science and Technology*, 68, 566-571.

[16]Barsoum, M.W., Houng, B. (1993). Transient plastic phase processing of Titanium-Boron-Carbon composite. *J. Am. Ceram. Soc*, 76(6), 1445-1451.

[17]Carrere, N., Valle, R., Bretheau, T. (2004). Multiscale analysis of the transverse properties of Ti-based matrix composites reinforced by SiC fibres: from the grain scale to the macroscopic scale. *International Journal of Plasticity*, 20, 783-810.

[18]Gupta, M., Lai, M.O., Soo, C.Y. (1996). Effect of type of processing on the microstructural features and mechanical properties of Al-Cu/SiC metal matrix composites. *Materials Science and Engineering A*, 210, 114-122.

[19]Braszczyńska, K.N., Lityńska, L., Zyska, A., Baliga, W. (2003). TEM analysis of the interfaces between the components in magnesium matrix composites reinforced with SiC particles. *Materials Chemistry and Physics*, 81, 326-328.

[20]Pelleg, J. (1999). Reactions in the matrix and interface of the Fe–SiC metal matrix composite system. *Materials Science and Engineering A*, 269, 225-241.

[21]. Semmelroth, K., Krieger, M., Pensl, G., Nagasawa, H., Püsche, R., Hundhausen, M., Ley, L., Nerding, M., Strunk, H.P. (2007). Growth of cubic SiC single crystals by the physical vapor transport technique. *Journal of Crystal Growth*, 308, 241-246.

[22]Villegasa, M., Sierraa, T., Lucasb, F., Fernándeza, J.F., Caballero, A.C. (2007). Oxidation treatments for SiC particles and its compatibility with glass. *Journal of the European Ceramic Society*, 27, 861-865.

[23]Fissel, A. (2001). Relationship between growth conditions, thermodynamic properties and crystal structure of SiC. *International Journal of Inorganic Materials*, 3, 1273-1275.

[24]Seekamp, J., Bauhofer, W. (1998). Influence of thermal annealing on the ultraviolet

stability of a-SiC:H thin films deposited from liquid organosilanes. *Journal of Non-Crystalline Solids*, 227-230, 474-477.

[25]Zhou, X.T., Zhang, R.Q., Peng, H.Y., Shang, N.G., Wang, N., Bello, I., Lee, C.S., Lee, S.T. (2000). Highly efficient and stable photoluminescence from silicon nanowires coated with SiC. *Chemical Physics Letters*, 332, 215-218.

[26]Fissel, A. (2001). Relationship between growth conditions, thermodynamic properties and crystal structure of SiC. *International Journal of Inorganic Materials*, 3, 1273-1275.

[27]René, C., Frans, J., Gusbertus, D. (1988). Reactions between α-silicon carbide ceramic and nickel or iron. *Commun Am Ceram Soc*, 71(6), C284-C287.

[28]Weast, R.C., Astle, M.J., Beyer, W.H. (1989). CRC handbook of chemistry and physics, 69th edition. Roca Raton, FL: *CRC Press Inc*, 174-177.

[29]Li, R. ed. (1995). Ceramics-Metal composites (in Chinese). Beijing, *Metallurgy Industry Press*, 333-353.

[30]Chou, T.C., Joshi, A., Wadsworth, J. (1991). Solid state reactions of SiC with Co, Ni, and Pt. *J Mater Res*, 4, 796-808.

[31]Ostwald, W. (1897). Lehrbuch der Allgemeinem Chemie. *Engelman, Leipzig.*

[32]Jackson, M.R., Mehan, R.L., Davics, A.M. (1983). Solid state SiC/Ni alloy reaction. *Metall Trans*, 14A, 355-364.

In: Tribology Research Trends
Editor: Taisho Hasegawa
ISBN: 978-1-60456-912-4

Chapter 1

RESEARCH ON THE TRIBOLOGY OF HYDRAULIC RECIPROCATING SEALS

George K. Nikas *

Imperial College London, Department of Mechanical Engineering, Tribology Group, London, UK

ABSTRACT

Hydraulic seals are found in industrial applications involving linear or rotary motion, as for example in hydraulic actuators. They are usually made of a polymeric material (for example, elastomer or "rubber") or a combination of materials (composite seals, for example, elastomer and PTFE with glass fibres). Their shape varies from the typical rectangular cross-section with chamfered or rounded corners and the typical O-ring to hundreds of less conventional designs with complex geometries, although they all have the same basic function, which is the sealing of fluids, normally under relatively high pressure (typically up to 80 MPa) and with operating temperature ranging from subzero values (typically as low as –65 °C) to relatively high values of up to 200 °C, depending on application. Low-pressure applications are also met when seals are used as wipers, as for example in tandem seal arrangements.

Theoretical research on sealing involves concepts and methods from elastohydrodynamics, contact mechanics, thermoviscoelasticity, adhesion and surface topography, in order to achieve good agreement with experimental results and industrial experience, yet this is still quite difficult to achieve because of the mathematical and numerical complexity of the problem. Proof of such difficulty is the fact that after more than 60 years of research in this field, fundamental aspects of the problem are still being tackled, for example, elastohydrodynamics with surface roughness effects, whilst making simplifying assumptions about others, for example, treating seal mechanics in the frame of linear elasticity and ignoring frictionally-induced thermal effects.

The present chapter explores the progress and research trends in computational and experimental tribology of hydraulic, reciprocating, rod and piston seals. Topics include

* Correspondence address: Dr G. K. Nikas, 3 Princes Mews, Hounslow, Middlesex, TW3 3RF, England
E-mail: gnikas@teemail.gr

the solution of the elastohydrodynamic and contact mechanics problem of flexible polymeric and composite seals, modelling of seal extrusion and anti-extrusion rings, seal elasticity and its effect on sealing performance, modelling of tandem seals, rotary vane seals, transient effects in lubrication, as well as performance evaluation in terms of leakage, friction, extrusion and wear, followed by optimization. Experimental studies are also briefly discussed with a presentation of the difficulties in validating existing models and in producing realistic, reliable and consistent results. The review covers the period from the 1940s to 2008 and serves as a reference source for further study and development in this challenging field, from the original basic experimental rigs and archaic computers of mid 20th century to the sophisticated numerical methods and expensive experimental devices of the recent era.

1. INTRODUCTION

There are mainly two types of hydraulic reciprocating seals: rod and piston seals (Figure 1). They are typically made of polymeric or plastic materials and most commonly by elastomers. Hydraulic seals are met in industrial, automotive and aviation applications involving linear and rotational motion, as for example in linear hydraulic actuators (Figure 1) [1] and rotary vane actuators [2].

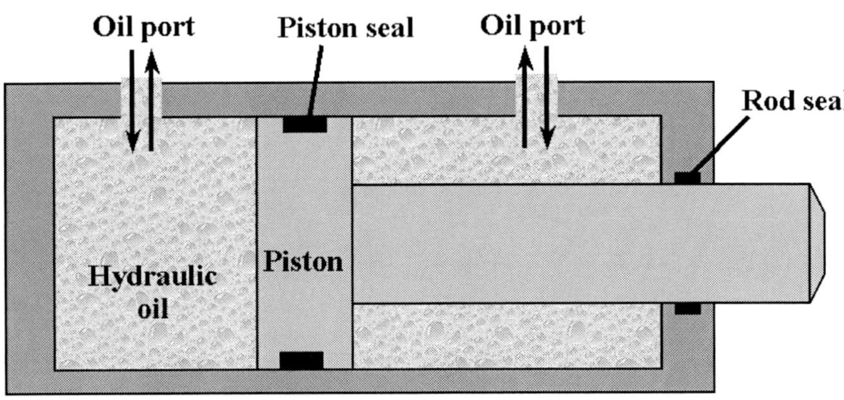

Figure 1. Typical hydraulic seals in a linear hydraulic actuator

These seals operate under dynamic conditions of variable sealed pressure, stroking velocity and operating temperature. Specifically, sealed pressures can reach 80 MPa (usually, they are lower than 50 MPa), stroking velocities can reach several metres per second (usually, they are lower than 0.5 m/s), and operating temperatures can be as low as –65 °C or as high as 200 °C (most often in aviation applications, as is explained later).

Depending on application, a number of seal shapes have evolved over the years based on experience and applied research, often at great financial cost. Seal designing was initially an empirical process through trial and error. In recent times, particularly in the 1990s and onwards, designs have become more sophisticated, based on modern computational tools such as Finite Element Analysis (FEA) software as well as on expensive experimental rigs to study sealing performance under controlled conditions in a laboratory in accordance with international standards. The shape of hydraulic seals varies from the typical seal of rectangular cross-section with chamfered or rounded corners and the typical O-ring to tens of less conventional designs with complex geometries, including composite seals and combinations of seals and other elements such as energizing springs and back-up rings.

Hydraulic seals are critical machine elements. A failure of such elements is associated with financial cost that greatly exceeds their low cost, often by hundreds to thousands of times. Even greater is the safety risk in applications such as in the aviation industry where, for example, hydraulic seals are used in linear hydraulic actuators [1, 3] and rotary vane actuators [2, 4] controlling aircraft landing gear (Figure 2(a)) and wing flaps (Figure 2(b)). A graphic example of the critical role of some "humble"-looking seals, although not related to reciprocating motion, is the tragic disaster of the NASA space shuttle Challenger in 1986; that was officially attributed to the failure of a static elastomeric O-ring, which was used to prevent hot gases from leaking through a joint during the propellant burn of the right rocket motor. Apart from financial cost and safety risk, the failure of hydraulic seals may also be responsible for environmental pollution from leaked fluids, particularly when the toxicity of some hydraulic fluids is taken into account. It is clear then that the understanding of sealing mechanisms is of paramount importance to seal designers and manufacturers.

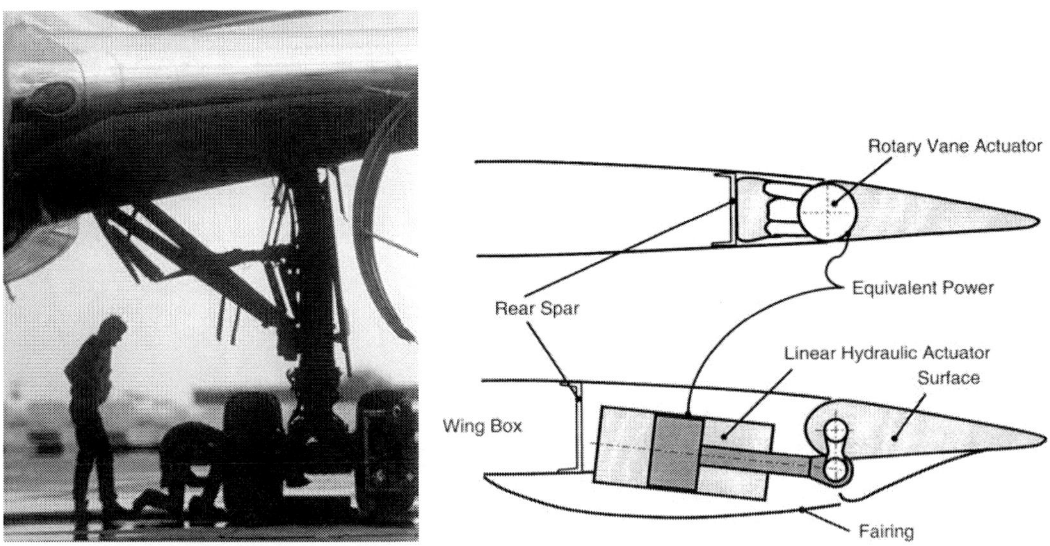

(a) **(b)**

Figure 2. (a) Aircraft landing gear controlled by linear hydraulic actuator; (b) Aircraft wing control surfaces operated via classic linear hydraulic actuator and new rotary vane actuator (drawing courtesy of Smiths Aerospace, UK)

Hydraulic seals are elements of complex mechanical behaviour. Their inherent complexity stems from their material, which, generally, obeys nonlinear stress-strain laws. For example, elastomeric seals are nearly-incompressible, hyperelastic solids. Mechanical response to loading is, generally, viscoelastic or nonlinear, and suitable elasticity models should be used for their mechanical analysis, particularly when the maximum normal strain exceeds 10 to 15 per cent [5, 6]. Elastomer response to stress and strain changes significantly at temperatures close to the glass transition temperature (typically in the order of zero to –70 °C), where the material stiffens and behaves more like leather. Structural changes are normal at such temperatures and reversible, though they can be irreversible if the glass transition temperature is exceeded many times or for long periods of time. Moreover, the thermal expansion coefficient of elastomeric seals is high, typically between 10^{-4} and 3×10^{-4} K^{-1}. This means that their sealing performance is very much related to the operating temperature. In fact, their fundamental mechanical properties such as the moduli of elasticity and rigidity, the Poisson's ratio, the hardness and compressibility, all vary with temperature. Their flexibility is also causing leakage and friction variations during reciprocating motion, which depend on their geometry and operating conditions, and are difficult to be modelled with satisfactory precision. Chemical compatibility with hydraulic fluids is also of concern, as is fluid absorption and swelling in normal operation. Elastomer oxidation and ageing, even when seals are out of service, is yet another important factor to consider. This limits their storage life, typically to under 5 years. These and other factors, which are explained later in this chapter, mean that the computational modelling, performance analysis and application of hydraulic seals is a complicated topic.

Despite the importance of this topic in terms of the number of industrial applications, financial costs and safety risks, related studies in the scientific literature are rarely encountered. This is probably attributed to the difficulty in the computational modelling of hydraulic seals and the study of their tribological performance. However, there is no doubt that in-house studies and software from seal manufacturers exist, yet they are rarely published for obvious reasons of competitiveness secrecy. The author has first-hand experience with sealing research funded by some of the top seal manufacturers [3, 4] and can attest that even in recent times (1990-2000), seal analysis and evaluation is an empirical process to a significant extent. Nevertheless, modern trends and stiff competition dictate that precise computational modelling takes over the traditional expertise of chief engineers, and seal selection and evaluation changes from art to science.

The scientific evaluation of hydraulic seals involves the computation of leakage and friction, as well as the prediction of their wear rate in a given application with given operating conditions. As these seals, naturally, operate in lubricated conditions, the first step in the analysis is to solve the problem of lubrication, commonly known as elastohydrodynamic lubrication (EHL) [1]. The EHL theory originated mainly in the 1950s and 1960s, although there are a few earlier studies exploring the basics of EHL. Applying the elementary EHL theory, sealing research was given an impetus in the 1960s and 1970s when pioneering studies were published on the theoretical and experimental analysis of rectangular and toroidal hydraulic seals. However, lack of computers and robust numerical methods (such as FEA), as well as archaic experimental rigs hindered rapid progress. This is reflected on the leakage and friction results presented in many of those early studies, which are characterised by a degree of scattering.

It is surprising yet true that the pioneering work of White and Denny [7] at the end of World War II, which was an exhaustive, mainly experimental work on reciprocating seals, remains one of the definitive sources of reference. Following that original study, a number of remarkable studies were published mainly in the 1960s and 1970s, most notably references [8-15] on experimental sealing research and [16-32] on theoretical EHL modelling. Very useful reviews have been presented in [33-35] from some of the pioneers in this field.

This chapter explores some of the fundamental technological and engineering aspects in designing optimised hydraulic seals. The geometry, physical properties, mechanical behaviour and performance analysis of hydraulic seals are explored in view of making selective optimisations such as minimising leakage and friction in reciprocating motion. The foundation for the successful tackling of the EHL and contact mechanics problem is laid and some of the most notable solution methods to date are presented for historical and future reference. Topics covered include seal shapes, material properties, operating conditions, mathematical and computational modelling, experimental studies, as well as current research trends. Relatively satisfactory solution methods are presented for simple geometries such as rectangular and toroidal seals, which can be extended to more complex seal geometries and kinematical conditions. The author has been involved in fundamental sealing research and modelling of reciprocating seals for linear hydraulic and rotary vane actuators, having developed computational tools to analyse seal EHL and performance evaluation [1-6, 36-42], nonlinear seal mechanics and related effects [5, 6], transient EHL effects [38], tandem seals [40], extrusion effects [37], back-up rings [39], composite seals [42], etc. These topics are discussed in the following sections in an attempt to explain the basic sealing mechanism of hydraulic seals, providing a source of reference to seal designers, engineers and academic researchers.

2. Hydraulic Seal Materials and Performance Issues

The most common materials used for hydraulic seals are elastomers and thermoplastics [43]. Material selection is based on the intended use of the seal. Most of the simpler shaped seals such as rectangular, toroidal and U-cups are made of some kind of polymeric material, usually elastomer. Seals of less conventional design on demanding applications (as for example for high sealed pressures or extreme temperatures) may utilize composite materials such as bronze-filled polytetrafluoroethylene (PTFE) as in coaxial seals [43] or PTFE with glass fibres and bonded with elastomers as in rotary vane seals [2, 4, 42].

The said materials are viscoelastic or viscoplastic. This means that they are significantly and nonlinearly affected by changes in their stress and strain state, as well as by changes in temperature. They are also susceptible to chemical degradation in reaction to incompatible hydraulic fluids or contaminants, oxidation and ageing. Therefore, they play a vital role in sealing performance and should be matched with the projected application, that is, with the service environment and operating conditions. Some fundamental theoretical studies on the effects of sealing elastomers on the performance of reciprocating seals can be found in [3] and [44], which deal with rectangular and toroidal seals, respectively. In those and similar theoretical studies, material properties for the simulations are typically obtained from time-consuming experiments designed to measure the modulus of elasticity, Poisson's ratio, thermal expansion coefficient, and, generally, stress-strain and relaxation curves at various temperatures. The data are then used as input to suitable material models, for example viscoelastic such as the generalized Maxwell model [44] or nonlinearly elastic such as the Mooney-Rivlin model [3, 5, 6], which are incorporated in numerical models such as in FEA software.

2.1 Elastomers for hydraulic seals, their benefits and deficiencies

Elastomers are suitable for hydraulic seals because of their flexibility. They have relatively low elastic and shear moduli, which means that they can accommodate large deformation, both tension and compression, as well as shear deformation, without permanent deflection or fracture. Additionally, they are nearly incompressible with Poisson's ratio very close to 0.5, normally greater than 0.490 (a typical value used in numerical simulations is 0.499); this is characteristic behaviour of liquids or metals in perfectly plastic deformation. Their flexibility and incompressibility mean that they can be accommodated in different housings or for different initial interference (pre-loading) and conform to space restrictions or adapt to temperature and sealed-pressure changes whilst maintaining their sealing ability. Moreover, as nearly incompressible materials, they offer seals the ability to transfer the pressure exerted by a hydraulic fluid onto the sealing surface without changing their volume and, thus, perform dynamic sealing, which is proportional to the sealed pressure.

However, elastomers are disadvantaged by several performance limitations. Their main problem is the dependency of their mechanical properties to temperature, strain and strain rate, as well as time (viscoelasticity, relaxation and ageing). Rubber, which is a particular form of elastomer and used extensively in hydraulic seals, is a compound of many macromolecules (see section 1.5 in [45]). Macromolecules are long molecular chains of three types: linear, branched and crosslinked. Linear chains move easily in relation to each other and this explains the softening of rubber with heating and hardening with cooling. Crosslinked chains do not move freely in relation to each other, which explains the resistance of rubber to flow when heated.

The elastic (Young's) modulus of elastomers generally decreases with temperature. For reference, the elastic modulus of a typical elastomer used for rod seals [5] is equal to 341 MPa at –54 °C, 8.9 MPa at 23 °C and 9.5 MPa at 135 °C, based on a compression test at maximum normal strain of ± 10 per cent; this represents two orders of magnitude change in the elastic modulus with temperature, which is very significant in applications where the temperature has large variations, such as in linear hydraulic actuators in aircraft (Figure 2(a)). Thus, elastomeric seals stiffen at lower temperatures. This type of response becomes nonlinear near the glass transition temperature, which, for typical hydraulic elastomeric seals used in the aviation industry [3], is about –45 to –70 °C. Structural changes ensue as a result, which are, generally, reversible, although they can be irreversible to some extend, depending on the duration and degree of the material exposure to such harsh conditions. Obviously, such effects are of paramount importance in designing seals that will remain leak-tight for the projected range of operating temperature and until the end of their anticipated service life.

The thermal expansion coefficient of elastomers is quite high, typically between 10^{-4} and 3×10^{-4} K^{-1} [43]. This means that seal dimensions change significantly with temperature. Therefore, the contact pressure of a sealing contact arising from interference fit of the seal is also significantly affected. This can cause failure of sealing at low temperatures from temporary loss of contact pressure, depending on the sealed pressure, as calculated in [4]. Such effects are obviously taken into account in seal modelling and performance evaluation, as for example in references [1, 4-6, 38, 39, 41] and in similar studies in the literature or in-house evaluations from seal manufacturers.

In conjunction with the effects of temperature on mechanical properties, elastomers exhibit nonlinear response to strain and strain rate. Mechanical response to strain even changes for repeated loading [43]. Figure 3 shows the stress-strain curves for a typical elastomer used for hydraulic seals [3, 5, 6], obtained from a standard test at three temperatures.

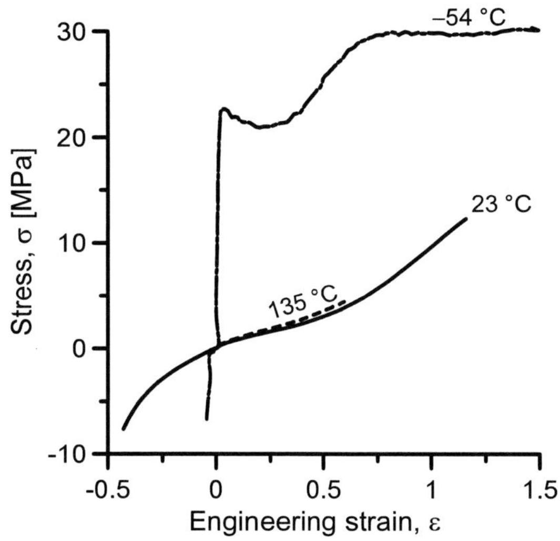

Figure 3. Stress-strain curves of typical elastomer for hydraulic seals [3, 5, 6] (results obtained from Trelleborg Sealing Solutions in England). The glass transition temperature of this elastomer is –47 °C

The elastomer responds rather linearly for very low load and stiffens when the load is increased beyond a limit. The results are clearly very different at subzero temperatures than at room or higher temperature; they are also different between tension and compression. In removing the load at the same rate as the original rate of application, the elastomer, generally, does not follow the same stress-strain path. The rate of load application is also influential, owing to the viscoelastic nature of elastomers. Further complications surface in repeated loading, with elastomers exhibiting hysteretic behaviour, that is, successive stress-strain curves appear displaced. Mechanical response is also affected by whether a previously attained strain level is exceeded or not in successive loadings, particularly so when such effects take place at low temperatures and near the glass transition temperature. These effects are obviously of great concern when designing elastomeric seals that work in a broad range of temperature and pressure for any number of operating cycles, as for example in hydraulic actuators used in aircraft landing gear and wing control surfaces (Figure 2).

Other important deficiencies of elastomers include:

(a) sensitivity to oxidation, which is accelerated at higher temperature and limits their storage life;
(b) chemical attack from incompatible hydraulic fluids;
(c) swelling from fluid uptake, which obviously changes seal dimensions and affects sealing performance owing to change of the sealing contact pressure;
(d) ageing, which manifests as hardening, embrittlement and eventual loss of seal material;

(e) relaxation and adhesion as for example in seals remaining stationary for long periods of time, which then exhibit high friction during start-up of motion and stick-slip phenomena, manifesting as vibration.

Because of their detailed deficiencies, elastomers are often replaced by other materials in hydraulic seals, mainly plastics and composite materials, as explained in the next section.

2.2 Plastics and composites for hydraulic seals, their benefits and deficiencies

The use of plastics in sealing originated several decades ago and saw an impetus in the 1970s [46]. Progress in the development of new materials continues and the reason for this is the need to overcome elastomer deficiencies (see section 2.1), improve performance and extend the service life of seals. Among the main benefits of some plastics and composites used in sealing are the increased resistance to wear, lower friction, avoidance of stick-slip and relaxation phenomena, wider operating temperature range with more consistent performance, and higher resistance to extrusion during reciprocating motion. PTFE was among the first materials explored [47, 48] but other materials have also found applications. For example, ultra-high-molecular-weight polyethylene (UHMWPE), polyurethanes and other low-stiffness materials. Composites are also in use, for example, bronze-filled PTFE as in coaxial seals [43], PTFE with glass fibres and bonded with elastomers as in (reciprocating) rotary vane seals [2, 4, 42], and PTFE with other fillers such as stainless steel and graphite. The selection of these materials depends on the requirements in a particular application and should match the operating conditions; for example, UHMWPE cannot be used at temperatures higher than about 80 °C [46]. Some PTFE compounds on the other hand can be used at much higher temperatures because of their fillers, which provide added strength.

The tribology of polymers and plastics used in reciprocating sealing applications involves the study of their lubrication with hydraulic oils, wet and dry friction in sliding contacts, abrasive wear when in contact with metallic surfaces, erosive wear when lubricated with particle-containing fluids, etc. The state-of-the-art of polymer tribology (at least in 1998) can be found in an exhaustive review compiled by Zhang [49]; readers are also advised to refer to the excellent chapter on this topic in book [50] – chapter 16. PTFE in various compounds is the most widely used plastic material in reciprocating sealing, therefore it is discussed in more detail next.

PTFE is a thermoplastic material best known for its low friction properties caused by its surface porosity and low surface energy – see page 170 in [43]. In lubricated contacts such as in hydraulic sealing, its surface porosity allows for lubricant storage and subsequent reduction of friction, perhaps similarly to laser textured surfaces. As a result, it is particularly suited to reciprocating seals, which, otherwise, suffer from stick-slip and high friction immediately after long periods of inactivity. However, extensive periods of sliding result in polishing the PTFE contact surface, which, in turn, causes substantial rise of the friction coefficient and accelerated wear of the PTFE. Thus, sliding of PTFE on hard metallic surfaces – such as piston rods in the case of rod seals – under dry or boundary-lubricated conditions, causes excessive wear of the PTFE. The latter is owed to the PTFE undergoing delamination (see Figure 16.2 in [50]), transferring a thin polymeric layer to its sliding counterface in chunks.

This can be reduced by reinforcing the PTFE with fillers such as glass fibres to produce a composite material of higher strength. However, the harder composites may increase the abrasive wear of a counterface such as a piston rod in the case of rod seals.

Owing to surface porosity and low effective area of contact, PTFE hydraulic seals are favourable towards lubricating film formation in their sealing contact, which results in higher leakage rate. This can be reduced by increasing the contact pressure, yet this negates the benefit of PTFE having a low friction coefficient and can in fact increase the friction force.

PTFE has other important properties, which matter in hydraulic seal applications. PTFE seals have a higher operating temperature limit, for example over 250 °C [46], although that may not be very useful in many applications (for example, linear hydraulic actuators for aircraft landing gear have an operating temperature that is, typically, lower than 140 °C [1, 3]). Moreover, PTFE has a very high resistance to ageing [48]. Its thermal expansion coefficient is in parity with other thermoplastics and in the order of 2×10^{-4} K^{-1}, which is close to that of typical engineering elastomers for hydraulic seals. The latter encourages matting of PTFE with elastomeric materials in composite seals [42], combining the benefits of both materials, as is discussed later. However, great attention to detail is advisable because of the many conflicting properties of PTFE. In general, this is a material that requires a lot of thought and analysis before it is used in hydraulic seal applications. This is perhaps more emphatically realised when considering that there is no such one "PTFE" material but, actually, different forms with different properties. For example, the Poisson's ratio of a PTFE compound can be as high as 0.46 or as (negatively) large as –12 (the latter in the so-called expanded PTFE). Moreover, the behaviour of PTFE in compression is different than in tension, which is easily observed in uniaxial stress-strain tests [51]. This means that its modulus of elasticity, yield point and work-hardening behaviour are different. Moreover, the mechanical properties of PTFE vary with time and temperature. They are also affected by fabrication methods [51]. Details on the mechanical properties of various forms of PTFE, including composites such as those with glass fibres, can be found in [51-54].

3. Design and Application of Hydraulic Seals

The performance of seals designed for reciprocating motion in high-pressure hydraulic systems plays a critical role in the efficiency and safe operation of such systems. Hydraulic seals are typically designed to service hydraulic equipment for a few million operating cycles. The financial cost of a potential failure can be very high considering the loss of productivity and man-hours consumed to fix problems. Moreover, safety risks involved in applications such as in the aviation industry (Figure 2) are very high. Therefore, the technology of hydraulic seals sees continuous progress towards optimised geometries, designs and material combinations for improved performance, even tailor-made for specific applications.

Depending on application, some important requirements in selecting the best available seal are as follows.

(a) Low leakage rate.
(b) Low friction and, as a result, low power loss and high efficiency.
(c) Resistance to wear and long service life with consistent performance.

(d) Resistance to gap extrusion (usually in high-pressure applications).
(e) Resistance to low and high temperatures (for example, in the aerospace industry).
(f) Chemical compatibility with sealed fluids.
(g) Ease of installation.
(h) Low cost.

The previous requirements are usually met in combinations and some may be conflicting, as for example requirements (a) and (b) (low leakage and low friction), in which case a compromise must be accepted. For a given application, the main parameters used for seal selection in conjunction with the listed requirements include the size, maximum sealed pressure, maximum stroking velocity (speed) and acceleration, the range of operating temperature and the chemical properties of the sealed fluid.

Based on the requirements and functional parameters used in seal selection, a large number of seal geometries and designs has appeared and more are being developed to meet specific demands. A large amount of specialised information can be found in product catalogues of seal manufacturers, with hundreds of seal designs. In the next two sub-sections, some important designs of rod and piston seal configurations are presented, as taken from one of the major seal manufacturers.

3.1 Examples of rod seal geometries and configurations

Rod seals are hydraulic seals used in equipment such as linear hydraulic actuators (Figure 1). Polymeric, plastic and composite seals are suitable for this task, depending on application. Among the basic requirements of rod seals is the very low or zero leakage-per-cycle and low friction in dynamic conditions (reciprocating motion), zero leakage in static conditions (no motion), and ease of installation. These are among the most basic-functioning of all seals and have been under development for several decades. As a result, they have become highly specialised in their functionality and many geometries and designs have evolved.

Although the rectangular rounded seal [1, 3] and the O-ring are the most basic of all rod seals, they are not very efficient in many applications; for example, O-rings suffer from high leakage and are best suited to static sealing. A compilation of some characteristic geometries and designs is presented in Figure 4, collected from a catalogue of one of the major seal manufacturers [55]. This is only a small selection and does not include some arrangements which are more complex. Many more seal shapes and combinations exist and details are provided in product catalogues of seal manufacturers. Rectangular seals and stand-alone O-rings are not included as they are very simple and well-known.

For the sake of understanding some of the functionality, limitations and benefits of rod seal designs, let us discuss some of the properties of the seals shown in Figure 4. Although the description is mainly based on [55], the main characteristics of the seals are typical.

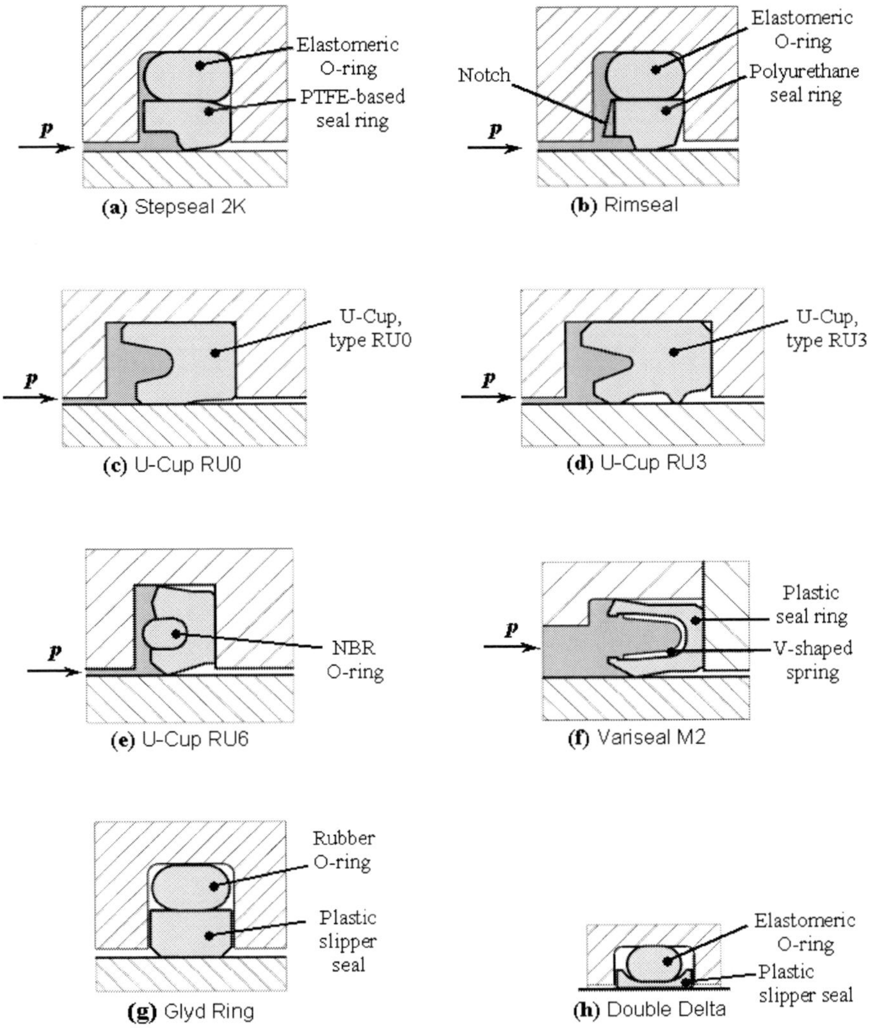

Figure 4. Examples of rod seals (compiled from catalogue [55])

(a) **Stepseal 2K**. This consists of a single-acting seal element made of PTFE-based or polyurethane material, which is energised by an elastomeric O-ring upon installation to provide pre-loading. The combination of the two elements gives this seal great flexibility, which allows it to be easily installed. The flexibility of the O-ring compensates hardware tolerances and movement. The plastic material of the seal ring allows for low friction, low abrasive wear, low gap extrusion, and eliminates stick-slip. The geometry of the seal ring allows for steep rise of the contact pressure on the left side during outstrokes (motion from left to right in the figure), which reduces film thickness under the seal; on the other hand, there is a smoother rise of the contact pressure on the right side during instrokes, which facilitates hydrodynamic film development under the seal to bring any leaked fluid back in. This pressure mechanism results in very low to zero leakage per cycle, as is explained in later sections of this chapter. This seal is suitable for a very wide range of operating conditions, namely temperatures between –45 and +200 °C, sealed pressures up to 80 MPa and maximum stroking velocity of 15 m/s with reciprocating frequency up to 5 Hz. It is often found in tandem seal arrangements as the primary seal.

(b) **Rimseal**. This consists of a single-acting, polyurethane seal element, which is energised by an elastomeric O-ring upon installation to provide pre-loading. It is often found as the secondary seal in tandem seal arrangements where the primary seal is the stepseal describes previously in (a). The chamfer this seal has at its low-pressure side (right) enhances the development of a hydrodynamic film during instrokes, thus reducing the leakage per cycle. Another chamfer on its high-pressure side (left) allow a notch (pictured) to come into contact with the flank of the groove at higher sealed pressures. Overall, the geometry of the seal ring can be optimised to minimise leakage in reciprocating motion by allowing for an optimum distribution of contact pressure of the seal ring during operation. This seal is suitable for operating temperatures between –45 and +100 °C (depending on the O-ring material), sealed pressures up to 25 MPa as an individual seal or up to 60 MPa as the secondary seal in tandem seal arrangements, and maximum stroking velocity of 5 m/s.

(c) **U-Cup RU0**. This is a single-acting, single-lip (asymmetric), compact seal made of polyurethane. Owing to its flexibility, it can easily accommodate deflections of the piston rod and adapt to changes of the sealed pressure. However, at low stroking velocities, it may suffer from stick-slip motion. This seal is suitable for mineral-based hydraulic fluids with operating temperatures between –35 and +110 °C, sealed pressures up to 40 MPa (depending on the rod-gland radial clearance) and maximum stroking velocity of 0.5 m/s.

(d) **U-Cup RU3**. This seal has similar functionality and properties as seal (c). Its most obvious difference from the previously described seal is the additional, small rear lip, which reduces leakage and prevents the entry of contaminants from the air-side of the seal (right). The small amount of lubricant inevitably trapped between the two seal lips keeps the seal lubricated and prevents dry running. As a result, the stick-slip tendency associated with the RU0 U-cup design (c) is reduced.

(e) **U-Cup RU6**. This U-cup seal has similar properties and technical data as other U-cup seals discussed previously (RU0 and RU3). Due to its geometry and integrated rubber O-ring, which gives it flexibility and provides a pre-loading upon installation, it has excellent sealing performance, regardless of sealed pressure level. The short sealing lip reduces friction in comparison with other U-cup designs.

(f) **Variseal M2**. This is a single-acting, plastic U-cup seal. The U-cup is made of a PTFE-based or polyurethane material and is energised by a spring, as shown in Figure 4. The asymmetric profile of this seal with optimised angle of the lip offers low friction and long service life. The spring offers the necessary initial pre-loading of the seal to avoid leakage in low system pressure, whereas at higher sealed pressure, the seal works automatically as all other rod seals by transferring the sealed pressure to the sealing contact with the piston rod. Due to its materials and construction, this type of seal is suitable for a wide range of operating conditions, namely temperature between –70 and +260 °C, stroking velocity up to 15 m/s, and sealed pressure up to 40 MPa in static loading.

(g) **Glyd Ring**. This is a double-acting, plastic-faced seal, comprising a PTFE or polyurethane based slipper ring, energised by a rubber O-ring. This type of seal has been in service for several decades as it is reliable, effective, and has low friction, high wear resistance and virtually no stick-slip problems. The O-ring provides the pre-loading upon installation and at very low sealed pressures. As the sealed pressure is increased, the O-ring is squeezed by the sealed fluid and pushes the slipper ring against the rod surface to prevent leakage. As a matter of fact, the slipper seal may have lateral notches to allow fluid to enter the seal housing quickly and pressurise the O-ring when there is abrupt rise of the sealed pressure. This seal can be used in a broad range of operating conditions, namely temperatures between –45 and +200 °C, sealed pressures up to 80 MPa and maximum stroking velocity of 15 m/s with reciprocating frequency up to 5 Hz.

(h) **Double Delta**. This is a double-acting, plastic-faced seal, comprising a polyurethane-based slipper seal, energised by an elastomeric O-ring. It is designed to expand under sealed pressure and typically installed in existing O-ring grooves as an improvement to O-rings. The mechanism of operation is the same as that for the Glyd ring (g), namely preloading by the initial interference of the O-ring in the seal housing and subsequent pressurisation of the O-ring as the sealed pressure is increased, with the O-ring then energising the slipper and pushing it against the piston rod. The benefits of this arrangement are the same as for the Glyd ring but the maximum sealed pressure is much lower, namely 35 MPa.

3.2 Examples of piston seal geometries and configurations

Similarly to rod seals, piston seals are used in equipment such as linear hydraulic actuators (Figure 1) and are made of polymeric, plastic and composite materials, depending on application. They have been under continuous development for several decades and many complex shapes have evolved from their study, based on industrial experience and scientific analysis. A compilation of some geometries and designs is presented in Figure 5, collected from a catalogue of one of the major seal manufacturers [56]. This selection does not include some more complex designs. Many more seal shapes and combinations exist and interested readers can find detailed information in product catalogues of seal manufacturers.

(a) **Glyd Ring T**. This is a double-acting, plastic-faced seal, comprising a PTFE or polyurethane based slipper ring, energised by an elastomeric O-ring. The functionality and technical data of this seal are the same as for the simpler glyd ring used on piston rods (see section 3.1 – case (g)) and the aforementioned seals are interchangeable. This one though has inclined profile flanks on the seal ring, which, in conjunction with the edge angle (see chamfer) helps the seal tilt away from the sealed pressure side. Improved sealing is then achieved by the steep pressure rise at the edge of the seal on the sealed-pressure side.

(b) **AQ-Seal 5**. This is a double-acting, rubber energised, plastic-faced seal. It comprises a PTFE-based seal ring, energised by two elastomeric O-rings and hosting a quad-ring seal. The O-rings provide the necessary preloading for very low sealed pressures by pushing the two other elements of this seal. As the sealed pressure is increased, fluid entering the seal housing compresses the O-rings, which in turn push the other seal elements against the sealed surface. This design combines the benefits of a slipper seal and an elastomeric seal offering very good sealing, low friction and no stick-slip effects. It is suitable for operating temperatures between –30 and +200 °C, sealed pressure up to 60 MPa and speeds up to 3 m/s.

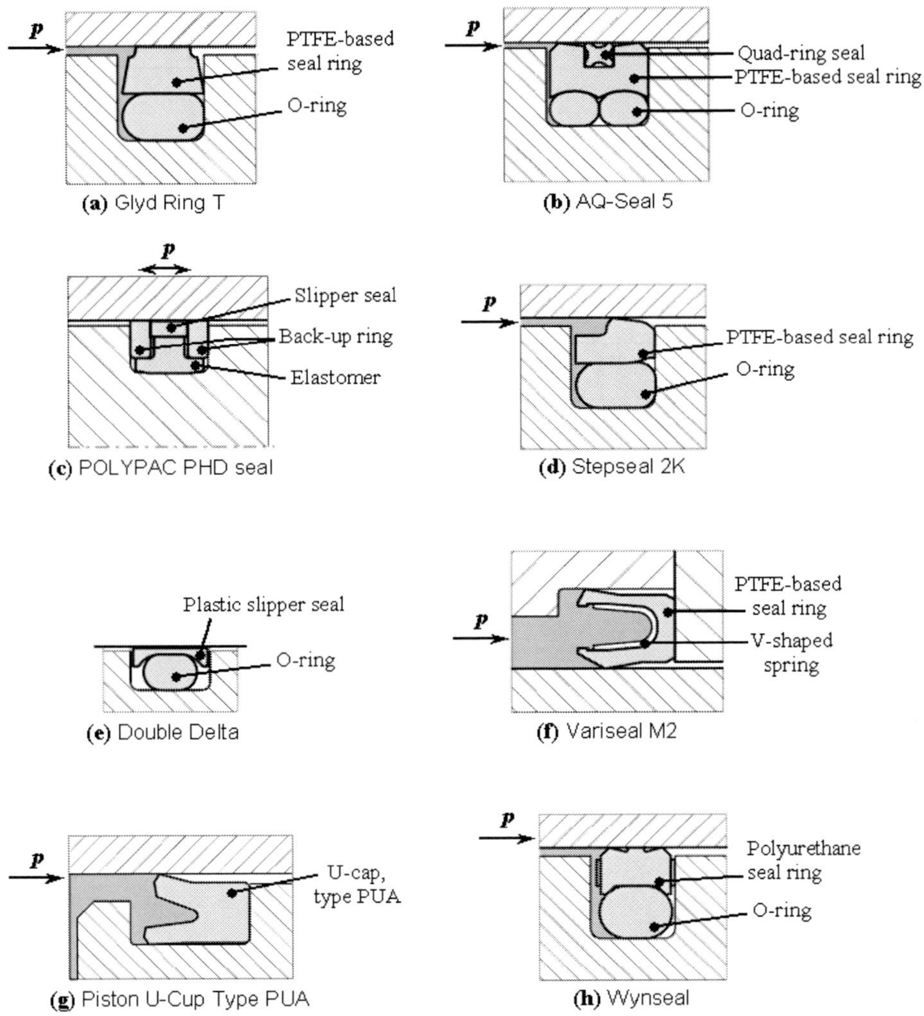

Figure 5. Examples of piston seals (compiled from catalogue [56])

(c) **POLYPAC PHD**. This is a double-acting seal for heavy duty, high-pressure applications. It comprises a PTFE-based slipper seal, which is energised by an elastomer and supported by a back-up ring on each side. The sealing mechanism is the same as in previously described seals and based on the preload offered by the elastomer in addition to its compression by the sealed fluid at higher sealed pressures, which is transferred to the slipper seal to achieve automatic sealing. Some of the main benefits of this seal include its high resistance to wear and extrusion (as a result of the back-up rings), low friction, no stick-slip effects and a long service life. It is suitable for operating temperatures between –45 and +135 °C, sealed pressures up to 40 MPa and sliding speeds up to 1.5 m/s.

(d) **Stepseal 2K**. This seal can also be used as a rod seal. It is described in section 3.1 – case (a).

(e) **Double Delta**. This seal can also be used as a rod seal. It is described in section 3.1 – case (h).

(f) **Variseal M2**. This seal can also be used as a rod seal. It is described in section 3.1 – case (f).

(g) **Piston U-Cup, Type PUA**. This is a single-acting seal with single, asymmetric lip. It is made of polyurethane. Its functionality and technical data are both similar to those of the U-Cup RU0 rod seal – see case (c) in section 3.1.

(h) **Wynseal**. This is a double-acting seal, comprising a polyurethane seal ring, energised by a rubber O-ring. The O-ring offers the necessary preloading for sealing at very low sealed pressures. As the sealed pressure is increased, the hydraulic fluid entering the seal housing compresses the O-ring, which in turn transfers pressure to the seal ring. The two upper seal edges act as primary seal for pressures from both directions (right and left), whereas the central back-up increases the sealing effect (this can be realised by looking into the similarities with a common twin-lipped U-cup seal – see for example case (d) in section 3.1). The seal is suitable for sealed pressures up to 40 MPa, operating temperatures between –35 and +110 °C, and speeds of up to 0.5 m/s.

4. Experimental Studies on Reciprocating Seals

4.1 A selection of some important early experimental studies

Systematic experimentation on hydraulic seal performance commenced in the 1940s. The pioneering work of White and Denny [7] between September 1944 and December 1946 was drafted in the United Kingdom by members of the (then) Ministry of Aircraft Production, the Royal Aircraft Establishment, and Imperial College of Science and Technology. In the era of World War II, the increased demands for improved performance and reliability of hydraulic seals used on aircraft dictated an understanding of their fundamental performance mechanisms. Time was of the essence and progress was swift. The exhaustive work of Denny under the general direction of Professor White produced a large amount of experimental data and remains valuable even today. The work of White and Denny [7] put science into the design of hydraulic seals, which, up to that time, was rather empirical. Their extensive report describes many experiments on flexible packings, including rectangular, toroidal and U-section seals. The experiments on various polymeric materials and with various sealed fluids dealt with measuring the friction force, friction coefficient and leakage rates at different sealed pressures and operating temperatures, the mechanisms of seal extrusion and how this could be eliminated, the mechanisms of seal failure from abrasion and extrusion or fracture, the effects of material hardness and the initial seal interference on the results, and similar topics of importance.

In lack of sophisticated experimental devices at the time, White and Denny had to resort to ingenuity to complete some of their tests with satisfactory precision, as for example in measuring the mass of leaked fluid. They demonstrated the effect of the seal material hardness in reducing the abrasive wear and extrusion (which led to cutting of a corner) of the seal. They proposed seal and housing arrangements, including anti-extrusion rings, to minimise or eliminate seal damage from extrusion. They managed to measure the distribution of the contact pressure at the sealing interface and located the most strained zone of a seal. They demonstrated the proportionality of the friction force on the contact area at the sealing interface and showed the effect of seal hardness on the frictional force. During their experiments with various fluids and at various speeds, and by demonstrating the effect of those on the measured friction force, they established the transition from the partial or mixed lubrication regime (i.e., with a significant degree of roughness asperity interactions at the sealing interface) to the hydrodynamic lubrication regime, or vice versa. Subsequently, they experimented with surface-finish effects, not only in relation to the frictional force but also to the static friction and stiction observed when an elastomer has remained at rest for some time and is relaxed.

Having collected many results from their parametric study, it became clear that elastomeric seals, owing to their incompressibility, achieve automatic sealing under dynamic conditions of variable sealed pressure by readily transferring the sealed pressure to the sealing interface, provided that they have been given an initial interference (pre-loading). White and Denny then embarked on a theoretical analysis based on the Reynolds (lubrication) equation to explain and back-up some of their experimental findings. Without a doubt, their work was ahead of its time and set the foundation for subsequent studies in the 1960s and 1970s to build on, taking advantage of modern hardware offering greater precision.

Nearly twenty years after the work of White and Denny [7], the first notable experimental studies began appearing in the literature, mainly in the international fluid sealing conferences organised by the British Hydromechanics Research Association. In 1964, three notable experimental studies were presented at the second international fluid sealing conference in England from some of the pioneers in this field.

Cnops [57] devised a rather simple experimental rig to measure the friction of elastomeric, cup, piston seals. The piston was loaded by a spring in a hydraulic cylinder and the volume of oil in the cylinder, which was brake fluid, was varied harmonically. The piston displacement, average pressure, and reciprocating velocity were all varied. The experiments demonstrated creep and relaxation effects typical of elastomeric materials. They also showed stiction effects and a subsequent development of a fluid film under the seal at increasing speeds, which gradually collapsed when the motion was slowed down or stopped. These effects are well-known today and explainable via the theory of hydrodynamic lubrication.

In a more advanced study, Lawrie and O'Donoghue [9] constructed an apparatus to examine the friction and lubrication of piston seals made of natural rubber and used in automotive clutch and brake master cylinders. They used a hydraulic cylinder with commercial brake fluid pressurised by a pump. The piston velocity and friction forces were measured by displacement transducers. The existence of rubber-metal contact was established by using conducting rubber seals and measuring the sealing contact resistance; zero resistance indicated full contact and infinite resistance indicated no contact or, in other words, full-film separation between the seal and its counterface. A multi-channel recorder enabled simultaneous reading of pressure, frictional force, speed and contact resistance for a complete operating cycle. Several tests were performed this way and useful results were obtained, showing how the seal performance varies during a cycle.

In an equally important study, Müller [8] presented an experimental analysis of elastomeric O-rings and quad (X) rings in reciprocating motion. His emphasis was on understanding hydrodynamic film formation at the sealing contact and the transition from the boundary to the hydrodynamic lubrication regime. He established the effects of fluid viscosity, stroking velocity, interference pressure and seal dimensions on the leakage and friction of (mainly) O-rings. He also discussed the differences in hydrodynamic film formation between outstrokes and instrokes of the piston rod, and verified the thinness of the typical lubricating film at the sealing contact.

In 1969, Dowson and Swales [11], following on the path of Müller [8] and taking advantage of the emerging EHL theory by the first author (Dowson), presented their experimental findings at the 4th international conference on fluid sealing. They used a rotating disc machine to test a cylindrical rubber block under conditions simulating the action of a reciprocating seal undergoing an infinitely long stroke. Capacitance techniques were used to measure film thickness and a piezo-electric transducer measured the sealing contact pressure. Their results were compared with calculations based on the EHL theory and the agreement was reasonable. Among some of their important findings was the realization of the fundamental sealing mechanism in reciprocating seals, namely the difference in film thickness at the sealing interface between the extending and the retracting stroke. They also confirmed that the film thickness increases with speed and decreases with applied pressure. Nevertheless, the present author would advise caution to the reader to not generalise such results, the reasons explained later in this chapter.

Another important contribution from that era, presented at the same conference in 1969, was that of Aston et al. [58]. They described three apparatuses and a series of experiments to measure the sealing force of rubber seals (Viton and fluoro-silicone compound) at temperatures up to 200 °C. Thus, they showed the change of the sealing force with temperature, following expansion and contraction of the rubber specimens. More importantly, they demonstrated the physical and chemical relaxation of rubber leading to a reduction of the sealing force in time and studied the recovery rate of the material. This is of great importance in elastomeric seals given the periods of inactivity they undergo under static compression at very low or very high temperatures (for example in the aviation industry), as well as when considering the effects of rubber ageing. A physical explanation of these phenomena was known at the time and provided by the network theory of rubber – see for example reference [59].

In 1971, Nau [14] presented his results on a series of experiments to measure friction of reciprocating rectangular rubber seals in lubricated conditions over a wide range of speeds and pressures. The goal was to understand stick-slip phenomena and the relation of friction to speed. This was already established and explained theoretically: as speed is increased from zero, friction rises, peaks and then falls. The speed at which the friction peak is observed depends on temperature and on the viscoelastic properties of rubber [60]. He found good correlation with experimental data on dry rubber friction and speculated on the nature of rubber seal friction, which he believed to be related to boundary lubrication phenomena. In the case of rubber friction on rough surfaces, it is worth noting that the friction mechanism involves hysteretic losses in the rubber, which may cause a secondary peak in a friction-speed diagram [14]. In any case, the friction of polymers on hard surfaces is a complex phenomenon involving many parameters [61].

Continuing on the topic of reciprocating-seal friction, Field and Nau [10, 35] produced a variety of experimental results in the 1970s on the pressure distribution, film thickness, friction and leakage of rectangular rubber seals. In measuring film thickness, they used optical interferometry and electrical transducers. Those results have similarities with the experimental results of White and Denny [7] but, in this author's opinion, they were often masked by some apparent inconsistencies. The inconsistencies – as for example in the form of wavy experimental curves in performance diagrams – could be attributed to the limited availability of high-precision instrumentation at that time. It may be surprising though that such inconsistencies are highlighted in a study [62] published much later, in 1988. The results reported in [62] revealed a significant degree of scatter in experimental results on reciprocating rubber seal performance. The surprising fact is that the said scatter refers to differences in results obtained from seven laboratories in different countries and for tests performed under strictly controlled conditions. A possible explanation was the lack of standardised methods for the tests or difficulty in conforming with the test specifications.

In 1975, taking advantage of their experience from previous studies, Field and Nau [15] presented a remarkable experimental study on the effects of design parameters on the performance of reciprocating rubber seals. Among other things, they studied the effects of seal hardness, interference (initial strain or pre-loading), back-up clearance and seal edge geometry. They confirmed the now well-known fact that the development of a hydrodynamic film at the sealing contact depends on the stroking length and its ratio to the contact width of the seal: if the stroking length is greater than two times the contact width, a full hydrodynamic film can be developed and leakage (per stroke) takes place. If the said length is less than two times the contact width, the development of a hydrodynamic film is incomplete. (This can be understood by visualizing the fluid transportation under the seal at the average speed of the two counterfaces, which is equal to half the stroking velocity because one of the counterfaces is stationary.) The observations of Field and Nau led to plotting two characteristic diagrams showing the film thickness and friction versus the position of measurement through the stroke – see Figure 6.

Similar results had been derived in an earlier study (1971) by Hirano and Kaneta [13]. Apart from the criticality of the stroking-length-to-contact-width ratio in establishing a full elastohydrodynamic film, Hirano and Kaneta discussed the starting friction of seals that have remained at rest for some time. That friction is considerably higher prior to full hydrodynamic film development and gives the characteristic stiction during the start-up of motion. Those effects had been discussed much earlier in the literature, as for example by Denny [63] in 1959.

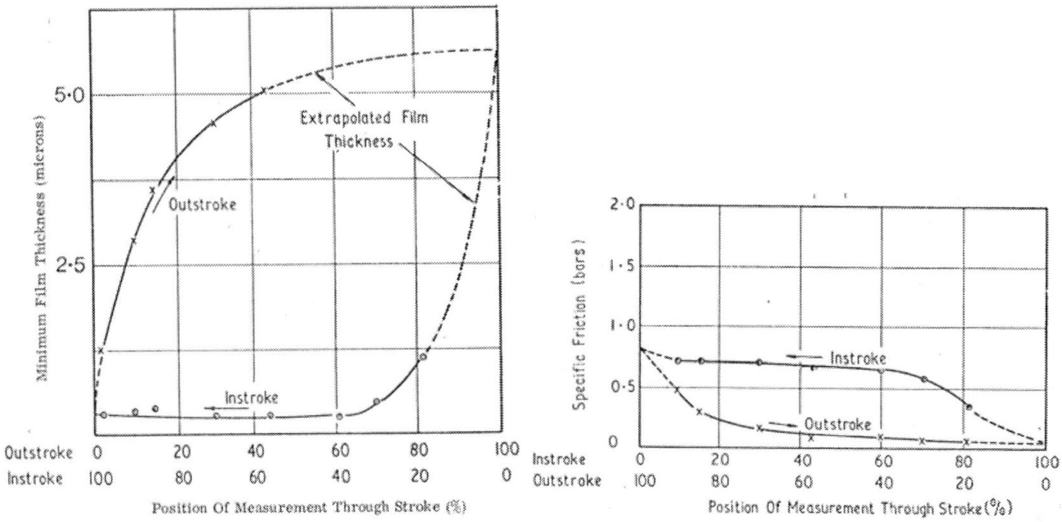

Figure 6. Variation of minimum film thickness and frictional force through the stroke of a reciprocating rectangular rubber seal (from Field and Nau [15])

4.2 Other experimental studies, progress and research trends

Apart from the pioneering experimental studies discussed in the previous sub-section, a number of other studies can be found in the literature, although the volume of published work in this field is substantially smaller than that of other machine elements such as bearings and gears. A selection of those is presented next. The selection excludes a few studies (mostly doctoral theses and reports) published in languages other than English. A detailed discussion of these and some other studies up to the early 1990s can be found in Kanters' thesis [64] and related publication [65], as well as in a follow-up thesis by Visscher [66].

Kambayashi and Ishiwata [67] studied the contact width and sealing force of reciprocating seals. Static contact pressure distributions have been measured by several researchers [20, 68-75]. A number of techniques and apparatuses have been used in those studies. For example, strain gauges, inductive transducers for measuring displacements, piezo-electric force transducers [10, 11, 76], and photo-elastic techniques [77]. Film thickness measurements in sealing contacts have also been performed by several researchers. The methods used involved inductive transducers for measuring seal surface displacements, optical methods involving optical interference and fluorescence techniques [78-80], electrical capacitance methods [10, 11, 15, 35, 71, 76], and electrical resistance methods [9, 72, 81]. Static and dynamic extrusion of elastomeric seals was studied by Reddy and Nau [82].

Leakage measurements have been performed with several methods. The most basic of those involves the removal and weighing of the leaked oil from piston rods [7, 8, 10, 13, 76, 83, 84], which can lead to accurate results if executed with care to make sure no oil layer remains on the rod. Another method used for measuring leakage is by maintaining a constant sealed pressure and measuring the oil flow needed to achieve this [10, 76, 85, 86]. Electrical methods have also been used, e.g. in [67], consisting of measuring the electrical capacitance of leaked oil film layers with one or two electrodes.

Friction measurements of reciprocating polymeric seals have been performed by most of the researchers involved in experimental sealing research. The pioneering, fundamental studies discussed in the previous sub-section are a good starting point. Many different apparatuses and test rigs have been used over the years, which makes discussion on their details difficult and of little point. Interested readers can find information on those methods (up to the early 1990s) in [64-66].

The early experimental studies up to and including the 1970s, as discussed in the previous sub-section, provided valuable information on the behaviour of reciprocating polymeric seals and set up the basic methodologies to measure seal performance. Furthermore, experimental techniques developed or improved by the pioneers in sealing research offered a foundation for later studies to build upon and improve the methods and apparatuses to achieve better precision. Most of the important parameters affecting the performance of hydraulic seals had been established in the first 30 years of research. What remained to be done was to extend the range of operating conditions in experimental studies and to apply emerging, high-precision techniques, focusing on interfacial phenomena of the micro-scale. For example, using a camera and video to record the lubrication of the sealing interface in real time. In this respect, Schrader [87] in the late 1970s (as reported by Kanters [64]) was probably among the first to use a high-speed camera to photograph the contact of a seal sliding on a glass cylinder. A few years later, Kawahara et al. [72] published results using the same method.

In more recent times, Kanzaki et al. [88] used optical interferometry in the sealing contact to study oil film behaviour. Interferometric methods to study fluid film thickness and profile have been reported since at least the 1960s in publications dealing with the contact of polymers, steel and glass, as well as between rubber and glass. In the latter case, which is of importance for the subject of this chapter, the work of Blok and Koens [78], presented in 1965, provided a solution to the problem of poor reflection of rubber surfaces – which is mainly owed to their roughness – by covering the rubber surface with a thin sheet of smooth plastic, aluminised on its outer surface. Details on this method with application to rubber lubrication were published quite early in, for example, [89].

In recent years, Kaneta and co-workers [12] used a mono-chromatic optical interferometry technique to directly observe oil films formed between band-shaped nitrile rubber specimens with “D” or lip-shaped cross section on sinusoidally reciprocating glass. There was no fluid pressure gradient in that system and that gave low contact forces. The rubber had to be specially moulded to be optically smooth, which means that its original roughness was lost. Of particular interest in their study was the measurement of film profiles in dynamic conditions, the average film thickness and the friction variation through a stroke. The differences between pumping and motoring strokes were clear to see as their graphs were for one complete cycle. Publication [12] was rather a modern version of an earlier study [13] of some of the same authors and reached similar conclusions regarding the critical stroke-length-to-contact-width ratio for the development of a stable hydrodynamic film, the importance of the contact pressure gradient on film formation and seal leakage, as well as the importance of the so-called “duty parameter”. The latter is a dimensionless quantity, which is useful in interpreting friction data with Stribeck-like curves. It is defined as the product of the oil dynamic viscosity with speed and divided by the product of the average contact pressure with the seal contact width. Kaneta et al. [12] confirmed that the friction characteristics of a seal are controlled by a critical duty parameter. Specifically, if the duty parameter is greater than the critical value, the friction coefficient increases with the duty parameter, whereas when the duty parameter is lower than the critical value, the friction coefficient increases with decreasing duty parameter and friction force maxima appear near the ends of the stroke. Other researchers had used this parameter at least since the 1960s, as for example Müller [8].

With advances on imaging technologies, sealing research is lately focused on phenomena taking place at the sealing interface for a better understanding of seal behaviour. In this respect, the role of surface roughness is examined in view of minimising leakage, friction and wear [90]. Direct observation of a sealing interface during operation under realistic conditions is, naturally, the best approach. A collaborative research project between a University and some major seal manufacturers in England explored this avenue [3, 91]. The 3-year project, which was sponsored by the British Department of Trade and Industry through the Civil Aircraft Research and Demonstration programme, involved experimental and theoretical work on reciprocating elastomeric seals used in linear hydraulic actuators for the control of aircraft landing gear (Figure 2(a)). The experimental work involved – among other things – the development of a rig for measuring seal friction and monitoring the sealing interface in real time with a microscope.

Figure 7 shows a schematic of the original rig [36, 91]. A rectangular seal is clamped on a vice and a glass plate in contact with the seal is reciprocated on top of it. The contact load on the seal is varied by a weight attached at the end of the slider assembly and two force transducers measure the force exerted on the block holding the seal. The transducers are aligned perpendicularly to each other and can give the friction force variation in directions perpendicular and parallel to the direction of the reciprocation. The rig is equipped with a microscope, light source and video recording equipment. A data logging device and a computer are used to calculate the friction once the gain of the system is calculated after a calibration with a known load.

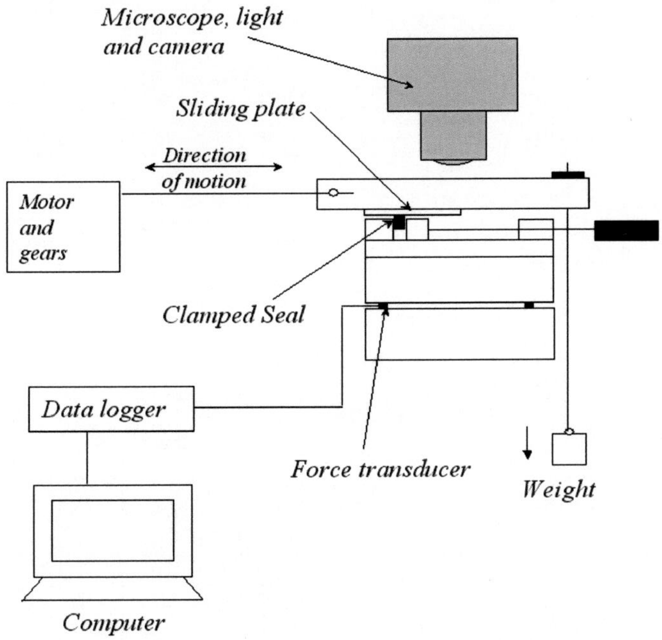

Figure 7. Reciprocating rig for monitoring the sealing interface and performance testing [36, 91]

The stroke length, stroke speed and the load on the rectangular seal can be varied. Moreover, roughness effects can be studied by changing the seal and/or slider with others of different roughness profiles (for example, by substituting a glass plate with a steel plate) and results can be obtained under both static and dynamic conditions, as well as both dry and lubricated conditions with various liquids. The poor reflectivity of rubber can be overcome as was done in [91] by applying a gold sputtering process at high temperature to coat the seals with a 200 nm coating of gold consisting of four 50 nm layers.

Results from friction tests for various reciprocating speeds and with various slider roughness profiles are shown in Figure 8. Images of seal contacts are shown in Figure 9 for dry and lubricated (with oil) conditions. Many interesting and useful results are obtained with this configuration such as results on stick-slip phenomena, seal running-in and wear, cavitation phenomena at the edges of the seal after long running times (for example, after 30 minutes or more) and film depletion from the reciprocating motion of the assembly, effects from contamination or wear particles in the sealing contact, hydrodynamic film development and local collapse based on stroking length and speed, etc.

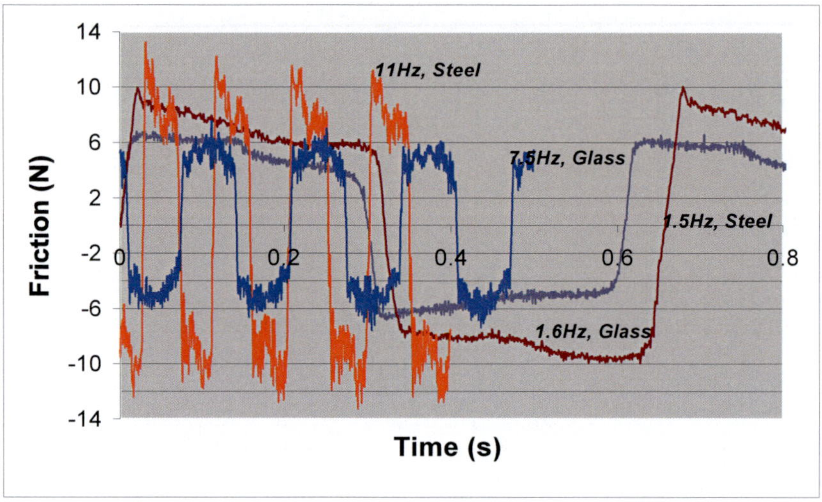

Figure 8. Friction measurements at various reciprocating frequencies and glass-plate roughness from the rig pictured in Figure 7 (from [36, 91])

(a) Dry conditions (O-ring, 1 mm cross section). **(b)** Wet conditions (dark spots represent contact).
Figure 9. Images of rubber seals in contact with glass plate [36, 91] (from the rig in Figure 7)

As in previous studies [13, 15], it has been verified that the development of a hydrodynamic film is linked to the stroking length in relation to the seal contact width. When a full hydrodynamic film has not been developed and there are areas of boundary lubrication in the contact, friction rises significantly. This is accentuated in rough contacts. Given that elastomeric seals are normally quite rough with typical average roughness in the order of 1.5 μm [1, 3], it is the roughness of the seal counterface (for example, the piston rod or cylinder bore surface) that matters and can make some difference if prepared before installation. However, as shown experimentally in [36, 91] (and theoretically in, for example, [3]), reducing the average roughness beyond a certain limit offers negligible benefit in reducing friction. Thus, expensive super-finishing of, for example, piston rods, that is, opting for an average roughness of less than about 0.05-0.10 μm in order to reduce friction, is not really necessary.

Advancing the realism of seal testing, an improved rig was built [91, 92] in collaboration with some major seal manufacturers in England. The seals, dimensions and clearances used were typical of linear hydraulic actuators to aerospace design specifications. A schematic of the rig is shown in Figure 10. A hollow, transparent, high-strength tube is attached to a motor and gear mechanism transferring reciprocating motion. A steel casing hosts gland elastomeric seals and a hydraulic circuit with pump supplies oil under pressure between the casing and the tube. A boroscope with its own light source is placed under one of the seals and feeds its signal to a CCD camera, video recording equipment and computer. Sealed pressures are restricted to about 7 MPa for safety reasons.

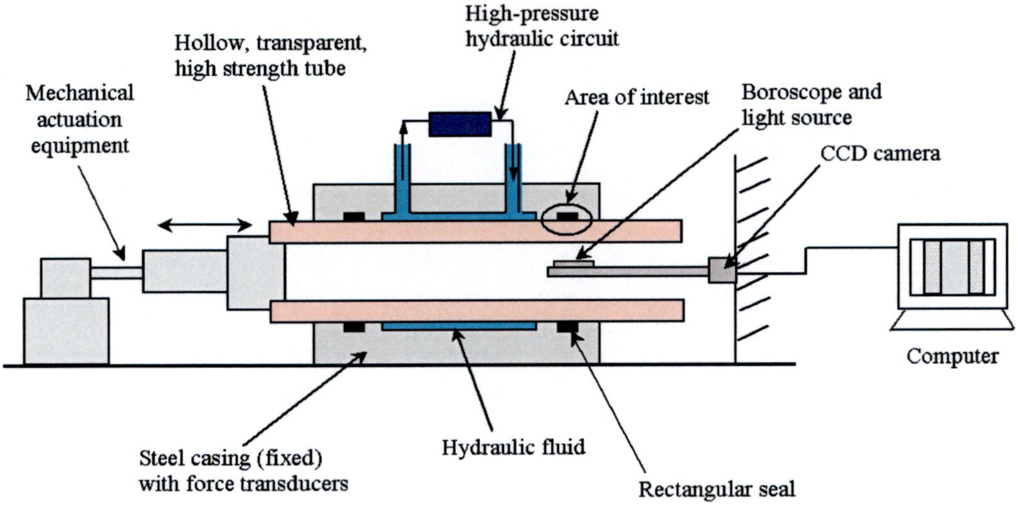

Figure 10. A seal specific rig with glass tube arrangement and boroscope [91, 92]

With a still boroscope, unaffected by vibrations in the system, images captured are clear. Figure 11 shows images from the rig at nearly zero sealed pressure (left image) and at 0.7 MPa sealed pressure (right image). The more fuzzy image on the right is owed to the development of a fluid film under the seal, which varies in thickness as is realised by the different colour shades. This is not the case on the left image of Figure 11 where the fluid pressure is nearly zero and the film under the seal has partially collapsed.

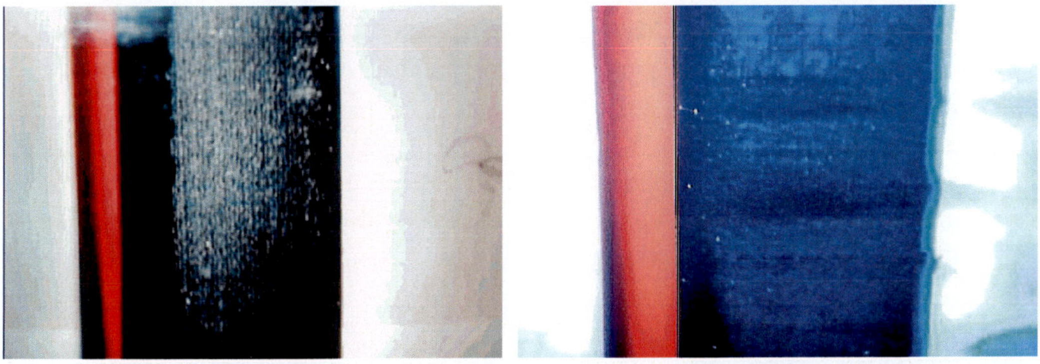

(a) **(b)**

Figure 11. Sealing contact (sealed fluid on the left side of each image) from the rig in Figure 10 [91]. (a) Nearly zero sealed pressure (thin film); (b) 0.7 MPa sealed pressure (thick, non-uniform film).

These images can be magnified and stored for later examination. Many results are collected with this arrangement [91, 92], including results on leakage, friction, stick-slip, effects of roughness and sealed pressure, real-time monitoring of the sealing interface and observation of development of a fluid film with roughness asperities deforming dynamically, study of cavitation effects as air bubbles enter the contact, study of the effect of debris particles and how they affect seal leakage, etc.

Seal wear is also studied by measuring the seal roughness before and after the test. Results reported in [91, 92] show a smoothening of the seal surface (running-in), despite sliding on a very smooth counterface (glass tube). Specifically, the average roughness was reduced from 1.8 to 1.1 μm and skewness was reduced from 1.07 to 0.23 μm. Wear appears to take place mostly during the beginning and ending of strokes, which is explained by the thinness of the lubricating film owed to the absence of motion at those intervals, for example, at the reversal of motion. Abrasive wear of the seal is also affected by the presence of contamination particles harder than the seal material. Debris particles are also responsible for increased leakage because they can create micro-channels on the counterfaces for the highly pressurized sealed fluid to escape. The latter was also discovered quite early by White and Denny [7] by creating small grooves onto the surface of a seal. The wear is not confined to the seal though. As the seal is normally much softer than its counterface (for example, a steel surface), debris particles are trapped and, embedded in the seal, they can scratch the harder counterface during the sliding motion.

Research trends on reciprocating seals include the continuous improvement of existing methods and development of novel techniques that are either more accurate or could provide real-time analysis. The latter could be used for condition monitoring. For example, lately, there is interest in ultrasonic methods to measure the film thickness in machine elements, including mechanical seals [93]. Whether these methods are developed to the point they offer clear advantages over existing methods remains to be seen but preliminary results are promising.

5. Theoretical Studies on Reciprocating Seals

Theoretical work on reciprocating seals is focused on solving the contact mechanics and lubrication problems in order to calculate seal leakage, friction and wear. This task has met formidable obstacles since first undertaken more than 60 years ago and a truly accurate solution has yet to be demonstrated. Despite the mathematical equations of the problem being well posed, their numerical solution is far from easy or straightforward. The challenges met can be summarized as follows.

(a) Polymeric seals are objects of complex mechanical behaviour. As explained in section 2, their mechanical properties vary significantly with temperature and they have large thermal expansion coefficient. Elastomeric seals are nearly incompressible, hyperelastic solids, with nonlinear response to stress or strain, profoundly different response near their glass transition temperature, and exhibiting significant relaxation and creep effects. They age, even when not in use, owing to oxidation, and their mechanical performance deteriorates accordingly. They may suffer from swelling from hydraulic fluid absorption, chemically react with incompatible hydraulic fluids, and wear quickly when rubbed on relatively rough, hard surfaces. Similar problems are met in thermoplastic or composite seals such as those made of PTFE (see section 2.2).

(b) The mechanics of polymeric seals is a difficult topic. Suitable models should be used to properly account for their thermo-viscoelastic or thermo-viscoplastic nature. These are models of nonlinear mechanics such as the Mooney-Rivlin model and mechanics of finite deformations, the latter explained by the fact that the maximum normal strain is normally around 10 per cent and sometimes exceeds even 20 per cent.

(c) The contact mechanics of polymeric seals is also a difficult topic. Given the complex shape such seals have, calculating the contact pressure distribution is impossible to do analytically, except for the simplest geometries such as rectangular (and even then, a compromise in precision at the edges of the seal should be accepted). Therefore, advanced numerical methods such as FEA should be used. Moreover, the typical polymeric seal surface is rough. A typical average roughness of elastomeric seals is in the order of 1.5 μm [3]. From a contact mechanics perspective, the proper modelling and accounting of surface roughness is vital in studying interfacial phenomena, which may be crucial in a performance analysis such as in calculating leakage and friction. Modelling highly deformable surface roughness asperities in any contact mechanics model is a difficult task.

(d) The lubrication problem of polymeric seals belongs to the category of "soft EHL". This is among the most computationally demanding elastohydrodynamic problems, even more so given that it involves transient operations of variable speed and sealed pressure. The reason for this difficulty is the high sensitivity of the numerical solution algorithm to small errors in the calculation of the film thickness. If these are not prevented or corrected, the numerical solution becomes unstable very quickly. Moreover, in reciprocating motion, the reversal of motion at the end of strokes posses great numerical difficulties in terms of properly accounting for the local inlet conditions to solve the EHL problem.

(e) The contact mechanics and lubrication problems are coupled. For flexible reciprocating seals, the shear stress at the sealing interface alters the pressure distribution, which, in turn, alters the development of the hydrodynamic film at the inlet zone of the seal, which determines the film thickness at the sealing contact and shear stress. In other words, a computational loop should be used to correctly resolve the contact pressure and film thickness coupling. The difficulty in this case comes from the flexibility of the seal and the fact that the contact pressure distribution is impossible to calculate analytically with high precision. This means that resorting to complex and time-consuming FEA is the only way possible yet this has to be repeated in every iteration until convergence is achieved at a given time step. In a transient analysis with hundreds of steps or more, solution time becomes excessively long or plainly unacceptable. This is the reason why, to the best of the author's knowledge, the coupled problem remains unsolved in the literature to date (2008).

The previously listed theoretical obstacles are the major ones. Details are provided in the next sections to help readers understand the specific problems, solution methods, and the potential for future research.

5.1 Phenomenological models of polymeric seal materials

Polymeric materials for reciprocating seals are either elastomeric (rubber compounds) or plastic (for example, PTFE). Composite materials are also used, for example, particle-reinforced rubbers and PTFE with glass fibres. The mechanics of thermoplastics such as

PTFE or polyurethane, as well as that of composite materials is a complex and specialised topic – see for example [51]. Interested readers can find information on this topic in the volumes accompanying most of commercial FEA software. The mechanics of elastomeric materials on the other hand is a much older topic and of greater interest as most hydraulic seals are elastomeric. Thus, the discussion is confined to the mechanics of elastomeric seals in this section.

Elastomeric materials are typically rubber compounds, that is, substances for which vulcanised natural rubber is the prototype. (Synthetic rubbers are also produced with sulphur or other additives.) They are also referred to as "rubber-like materials" [59]. Typically, these are hyperelastic and nearly incompressible. The hyperelasticity is exhibited in sustaining large strains without fracture and recovering to initial dimensions when stresses or strains are removed, without appreciable hysteresis at temperatures above the glass transition temperature. According to the statistical-molecular or network theory of rubber elasticity [94], they consist of very long molecular chains, which are folded and kinked [43]. The chains are chemically cross-linked, forming a three-dimensional network. Free space exists between chains, which varies transiently in volume and location. As atoms in said chains are thermally agitated, they assume a variety of statistically determined conformations [95]. This dynamic or transient chain motion is slowed down when the material is cooled and eventually ceases when the material is close to its glass transition temperature (usually between zero and –70 °C, depending on the particular elastomer). This explains the extensibility of elastomers at temperatures well above the glass transition temperature. It also explains their rigidity at subzero temperatures, especially near the glass transition temperature where the material behaves like brittle glass with significantly altered crystallinity.

There are many papers and phenomenological models on rubber thermoelasticity in the literature. Some classical, detailed studies are [59, 95-97]. Treloar's classic paper [95] in 1976 and Ogden's review [98] in 1986 give a good introduction to phenomenological models at the time of their publication and are still useful today. A thorough presentation can be found in a book published in 2000 by Holzapfel [99]. For reciprocating hydraulic seals specifically, a satisfactory constitutive model is the so-called Mooney-Rivlin model.

The Mooney-Rivlin model is based on the pioneering work of Mooney [100] and Rivlin [101] on finite elasticity in the 1940s. Its derivation is based on the elastic strain energy per unit volume, W, which is a function of the three principal stretches of deformation or extension ratios [98], that is, $W = W(\lambda_1, \lambda_2, \lambda_3)$, where $\lambda_i \equiv l_i / L_i$ (i = 1, 2, 3), l_i and L_i being deformed and reference length, respectively. It is assumed that the mechanical properties of rubber-like solids can be represented in terms of the energy function. Assuming isotropic solids and isothermal conditions, the energy must be independent of the coordinate system used (isotropy). Thus, it can be expressed in terms of the three strain invariants $I_1 = \lambda_1^2 + \lambda_2^2 + \lambda_3^2$, $I_2 = (\lambda_1 \lambda_2)^2 + (\lambda_2 \lambda_3)^2 + (\lambda_3 \lambda_1)^2$ and $I_3 = (\lambda_1 \lambda_2 \lambda_3)^2$, that is, $W = W(I_1, I_2, I_3)$. For an incompressible solid such as an elastomeric seal (which is nearly incompressible with typical Poisson's ratio of 0.499), the volume is constant, which means that $\lambda_1 \lambda_2 \lambda_3 = 1$, hence $I_3 = 1$. Many explicit forms of function W have been reported in the literature [98] but the Mooney-Rivlin function has been extensively used; it can be expressed as

$$W = \frac{c_1}{2}(I_1 - 3) + \frac{c_2}{2}(I_2 - 3) \tag{1}$$

where coefficients c_1 and c_2 are obtained experimentally at the temperature of interest. According to [95], this is the most general first-order expression in I_1 and I_2.

Given the strain energy function W, the Cauchy (true) principal stresses σ_i can be calculated from (see Eq. (6.69) in [99])

$$\sigma_i = \lambda_i \frac{\partial W}{\partial \lambda_i} - p_c \quad (i = 1, 2, 3) \tag{2}$$

where p_c is a hydrostatic pressure, which can be determined from equilibrium equations and boundary conditions. Using the Mooney-Rivlin function (Eq. (1)), Eq. (2) yields

$$\sigma_i = \lambda_i^2 \left[c_1 + c_2\left(\lambda_j^2 + \lambda_k^2\right)\right] - p_c \quad (i, j, k = 1, 2, 3 \text{ and } i \neq j \neq k \neq i) \tag{3}$$

Coefficients c_1 and c_2 in Eq. (3) are calculated from stress-strain data in uniaxial tension or compression of the rubber-like material such as in Figure 3. In uniaxial tension or compression, assuming that "1" is the load direction, $\sigma_2 = \sigma_3 = 0$ and $\lambda_2 = \lambda_3$. Setting $\sigma_1 \equiv \sigma$ and $\lambda_1 \equiv \lambda$, pressure p_c is calculated from Eq. (3) using either $\sigma_2 = 0$ or $\sigma_3 = 0$ and utilizing the incompressibility constraint $\lambda_1\lambda_2\lambda_3 = 1$. The result is $p_c = \left[c_1 + c_2\left(\lambda^2 + 1/\lambda\right)\right]/\lambda$. Then, Eq. (3) yields

$$\sigma = \lambda\left(c_1 + \frac{c_2}{\lambda}\right)\left(\lambda - \frac{1}{\lambda^2}\right) \tag{4}$$

The engineering (or nominal) stress, σ_{eng}, is equal to σ/λ. Thus, using Eq. (4),

$$\frac{\sigma_{eng}}{\lambda - \lambda^{-2}} = c_1 + \frac{1}{\lambda} c_2 \tag{5}$$

Equation (5) represents a straight line. Using engineering test data (σ_{eng}, λ) and plotting them as "reduced pressure" $\sigma_{eng}/\left(\lambda - \lambda^{-2}\right)$ versus the inverse of stretch, $1/\lambda$, coefficients c_1 and c_2 are easily calculated. If a stress-strain curve is not available and only the modulus of elasticity, E, is known, then, according to reference [102] (page 7-33), a reasonable approximation to use is $c_1 \cong 4c_2$ and $3(c_1 + c_2) \cong E$, resulting in $c_1 \cong 4E/15$ and $c_2 \cong E/15$.

According to [43], the Mooney-Rivlin model has been used for strains up to 200 per cent. This means that it is adequate for the problem of reciprocating hydraulic seals, where the maximum normal strain rarely exceeds 25 per cent and is more typically less than 15 per cent. Owing to its simplicity and satisfactory precision, the Mooney-Rivlin model has been adopted in many studies in the literature. Though it only requires calculating two constants, it can be further simplified to what is known as the neo-Hookean model. The strain energy function for the latter is $W = c(I_1 - 3)$ (c being a constant). The neo-Hookean model may be simpler and easier to apply, however, it is much less accurate than the Mooney-Rivlin model at higher strains. Therefore, if precision is a priority, the Mooney-Rivlin or a better model should be used.

Other phenomenological models have also been developed. According to Holzapfel [99], the (more complex) Ogden model [103] "*excellently replicates the finite strain behaviour of rubber-like materials*". However, it should be remembered that the differences between the various models become significant only at larger values of strain, which are normally not met in hydraulic reciprocating seals. Moreover, these models work under isothermal conditions. Some approximations in engineering calculations can be done when there is temperature variation, as demonstrated by Nikas and Sayles [5, 6] in the case of reciprocating rod seals for linear hydraulic actuators (Figure 2(a)). Models have also been developed for compressible materials [99], though they are more complex and probably not justified in the case of reciprocating seals because of the relatively low strains involved.

Further complications in the phenomenological modelling of rubber-like materials comes from inelastic effects such as hysteresis, frequency-dependent response, strain-stiffening at large stretch, scission of molecular cross-links at high temperature leading to time-dependent softening or permanent set, and stress-softening, the latter known as the Mullins effect. (The Mullins effect is observed in cyclic loading during the first and successive cycles at given strain when the stress drops, hence the term "stress-softening". This effect is important in engineering elastomeric parts. As it is related to fatigue, it clearly has implications in the life expectancy of elastomeric parts.) The aforementioned effects are more evident in rubbers hardened with fillers or particle-reinforced elastomers and there are several models developed in recent years to estimate elastomer behaviour in such cases – see for example [99, 104-108].

For the purposes of calculating the performance of hydraulic reciprocating seals, even the classic Hookean (linear elasticity) model with allowance of thermal effects may be adequate if the maximum normal strain does not exceed about 10 per cent [5, 6]. For strains larger than about 10 per cent and up to 15 per cent, Nikas and Sayles [5, 6] reported that a more advanced model should be used. Comparing the Hookean and the Mooney-Rivlin model predictions on rod seal leakage, hydrodynamic friction and extrusion predictions at temperatures of –54, +23 and +135 °C, and sealed pressure of up to 35 MPa, Nikas and Sayles [5, 6] reported a maximum difference in leakage of about 15 per cent at the highest temperature and high sealed pressure (of 25 MPa), although the differences are usually between zero and 5 per cent.

5.2 Contact mechanics of hydraulic reciprocating seals

The contact mechanics of hydraulic reciprocating seals involves the calculation of the contact pressure and tangential traction at a sealing contact. The latter is about calculating the shear stress or friction in the contact. The contact mechanics may also involve calculating the stress field in the body of a seal in order to locate zones of stress concentration and to link those with fatigue modelling or, simply, as a guide to make improvements in design in order to eliminate stress concentrations. Finally, the contact mechanics may also involve calculating the overall deformation and change of shape of the seal in dynamic conditions in order to establish potential performance issues such as extrusion, which may be of concern.

The aforementioned computational tasks are all complicated because the solid mechanics of polymeric seals is highly nonlinear and the typical boundary conditions in reciprocating sealing are complex and transient. Precise analytical solutions are not feasible. Analytical solutions can only be applied on the simplest of seal geometries such as rectangular and only approximately. Such analytical solutions on the static contact pressure assuming plane-strain conditions have been presented by Hooke et al. [29, 109] in 1966, Johannesson [20] in 1979, and Dragoni and Strozzi [110] in 1988 on O-ring rubber seals, Field and Nau [17] in 1975 on rectangular rubber seals, Strozzi [73] in 1986 on rectangular rounded seals, Johannesson and Kassfeldt [111] in 1989 on seals of arbitrary cross-section, and more recently by Nikas [1-6, 37-42] on rectangular rounded elastomeric seals and rotary vane seals, including composite (PTFE-elastomer) seals [2, 4, 42]. The contact pressure in most of those studies was calculated from the amount of surface overlap or interference and contact friction was usually neglected.

For complex seal geometries and boundary conditions, for example those pictured in figures 4 and 5, numerical solution is the only viable option. This means that FEA should be used whenever possible, though care should be taken to use finite elements formulated for incompressible materials to avoid meaningless results. FEA has been applied mainly since the 1970s in the study of hydraulic seals – see for example references [32, 64, 73, 75, 85, 112-121]. A bibliographic review of FEA of rubber-like materials, covering the period 1976-1997, can be found in reference [122]. Figure 12 shows an example of FEA analysis of a rubber O-ring, pressed between two frictionless plates.

5.3 Elastohydrodynamics and performance of hydraulic reciprocating seals

Hydraulic reciprocating seals typically operate with a thin lubricating film in their sealing contact (see for example Figures 9 and 11). The film separates the seal from its counterface, as for example in the simplified schematic in Figure 13, depicting a rod seal in a linear hydraulic actuator. The fluid film thickness varies from a few nanometres to a few micrometres. This form of lubrication is characterised as elastohydrodynamic (EHL). The fluid film is developed when lubricant enters the sealing contact by viscous shear and/or at high sealed pressure. The thickness of this film depends on many factors such as the stroking velocity, contact pressure distribution, surface roughness, lubricant viscosity and density at operating conditions (pressure, temperature), inlet geometry, degree of lubricant starvation, etc. As the main seal performance variables, namely the leakage rate and friction, both depend on the thickness of the sealing contact fluid film, the precise calculation of the film thickness is of major importance.

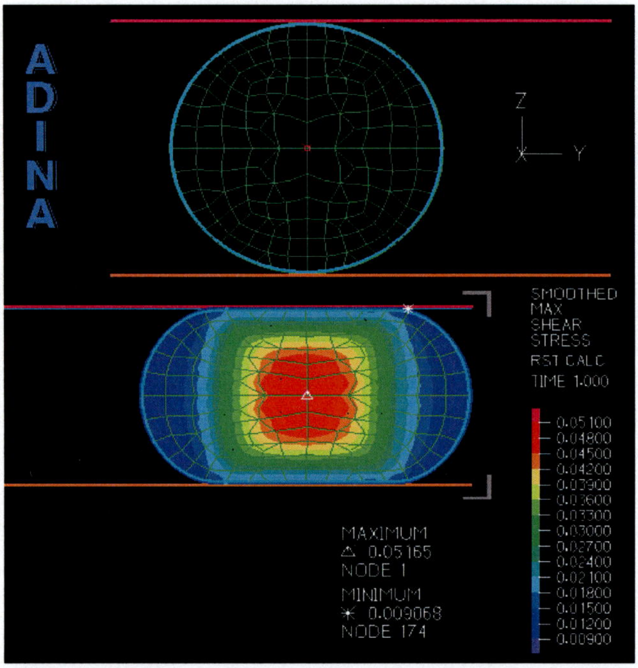

Figure 12. FEA example of a rubber O-ring pressed between two rigid, frictionless plates; original and deformed mesh shown with maximum shear stress distribution

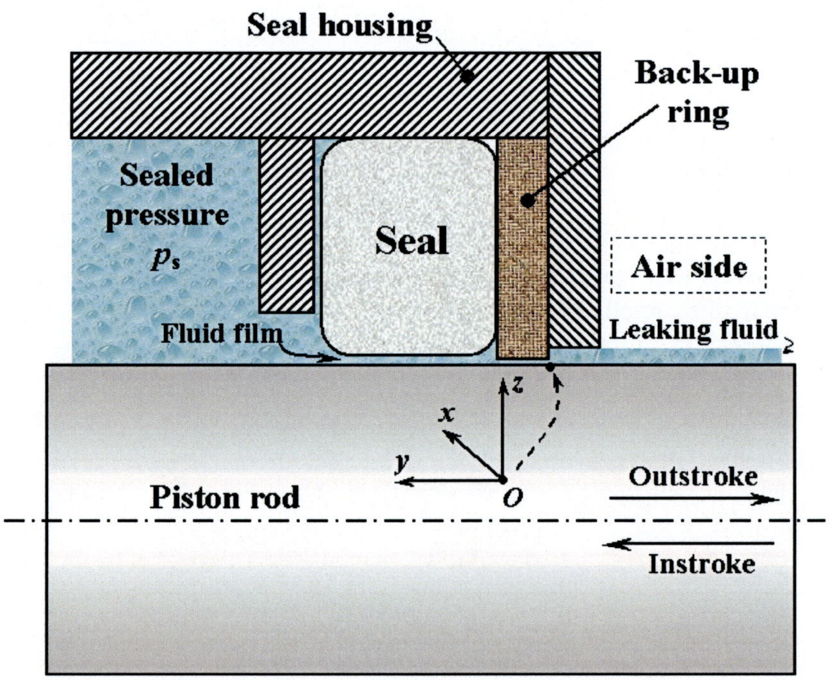

Figure 13. Schematic of rectangular elastomeric seal with back-up ring in a linear hydraulic actuator (only the upper half of the seal, ring and housing shown)

Calculating the film thickness distribution in the sealing contact is far from straightforward and simple. The calculation is based on a number of simplifying assumptions.

For example, fluid inertia is neglected because the film is very thin; frictional heating of the contact and fluid can be neglected when the stroking velocity is low, as for example in typical hydraulic actuators controlling aircraft landing gear; surface roughness effects can be neglected as a first approximation or because the low-stiffness seal roughness asperities are flattened under pressure. With such reasoning, several simplifications can be made prior to writing down a suitable equation to calculate the film thickness distribution. The aforementioned equation is derived from the Navier-Stokes equations of Fluid Mechanics after several simplifications and is the well-known Reynolds' equation [123] in a form that suits the particular application with assumptions accepted.

As fluid flow in the sealing contact under reciprocating motion is mainly one-dimensional, a suitable form of the Reynolds' equation is as follows [4]:

$$\frac{\partial}{\partial x}\left(\frac{\rho h^3}{\eta}\frac{\partial p}{\partial x}\right) = 6V\frac{\partial(\rho h)}{\partial x} + 12\underbrace{\frac{\partial(\rho h)}{\partial t}}_{\text{Transient term}} \tag{6}$$

where p is local contact pressure, h is local film thickness, V is the sum of the tangential velocities of the cooperating surfaces of the contact, t stands for time, and ρ and η are the mass density and dynamic viscosity of the lubricating fluid, respectively, which are functions of temperature, pressure and shear rate [42], although the latter is neglected because it is weak in the low-speed, low pressure applications met in reciprocating sealing. For low accelerations/decelerations or, generally, as an approximation, the last "transient term" in Eq. (6) is omitted and Eq. (6) then yields

$$\frac{\mathrm{d}}{\mathrm{d}x}\left(\frac{\rho h^3}{\eta}\frac{\mathrm{d}p}{\mathrm{d}x} - 6V\rho h\right) = 0 \tag{7}$$

which, obviously, holds for steady-state conditions, that is, operation with constant speed and sealed pressure. Almost all studies in the literature of reciprocating seals deal with the one-dimensional form of the Reynolds equation and the vast majority assume that the sealing contact is perfectly smooth. If surface roughness effects are accounted for, micro-EHL effects among neighbouring roughness asperities play some role in performance. Therefore, a more general, two-dimensional form of the Reynolds equation can be used, as was done in [1, 3, 5, 38-40] for rectangular rod seals.

The Reynolds equation may be accompanied by kinematical and boundary conditions. For example: (a) the no-slip boundary condition, dictating that the fluid velocity on a boundary wall is equal to the wall's velocity (zero slip); (b) the inlet condition: $\mathrm{d}p/\mathrm{d}x = 0$ "away" from the contact in the inlet zone, etc. For a detailed presentation of these topics, please refer to the literature, for example [1-6, 16-32, 38-42].

Once the mathematical problem is well-posed, the Reynolds equation can be solved numerically for the film thickness if the contact pressure distribution is known, or for the contact pressure if the film thickness distribution is known, or for both, in conjunction with contact mechanics equations, as explained in the previous section. Given the highly nonlinear nature of this problem, a successful solution is obtained only by using a convergence loop in an algorithm involving the lubrication and contact mechanics equations. Several solution methods have been published in the literature over a period of decades, varying in realism, sophistication and precision. Among the earliest studies are those of White and Denny [7], who assumed a tapered film profile and parabolic pressure distribution in deriving the film thickness, and that of Müller [8], who also used a tapered film profile, different between outstrokes and instrokes, and measured contact pressure distributions.

If the pressure distribution is somehow known (either measured or empirically assumed), Reynolds' Eq. (7) can be easily inverted to a cubic algebraic equation and solved for the film thickness. This is the so-called inverse solution of the Reynolds equation, which is usually attributed to Blok [124]. For details of this method in general, see reference [123]. In the study of reciprocating hydraulic seals the method has been applied by most researchers [16, 18, 19, 21, 22, 24-27, 64, 76, 77, 125, 126]. Despite its relative simplicity, the method is not without problems because it involves the tricky part of calculating the roots of a cubic equation and doing so for many points in a sealing contact. This is not an easy task from a numerical perspective and erroneous, imaginary roots can destabilise the solution process. The difficulties and pitfalls associated with this method in sealing have been detailed by Ruskell [23]; an interesting discussion can also be found in reference [26].

A modification of the method that avoids the cubic equation and solves a first-order, ordinary differential equation instead was postulated in reference [41] and further applied in references [2, 4, 42] on rectangular rounded reciprocating seals, including composite and rotary vane seals. According to the latter development, the equation to solve is [41]

$$\frac{\mathrm{d}H}{\mathrm{d}x} = \frac{H^3 \dfrac{\mathrm{d}^2 q}{\mathrm{d}x^2}}{6V - 3H^2 \dfrac{\mathrm{d}q}{\mathrm{d}x}} \tag{8}$$

where $H \equiv \rho h$ and $\mathrm{d}q/\mathrm{d}x = (dp/dx)/(\eta\rho^2)$. In order to solve Eq. (8) for H, a boundary condition is needed, that is, H must be known at any one point in the sealing contact. Such a suitable point is the extremum point (say $x = x_{\mathrm{m}}$) of q, where $\mathrm{d}q(x_{\mathrm{m}})/\mathrm{d}x = 0$ and, by the definition of q, $\mathrm{d}p(x_{\mathrm{m}})/\mathrm{d}x = 0$. With a known pressure distribution, the extremum point is easily located. Then, following the analysis of [41], the required boundary condition, which is the value of H at the extremum point of q, is calculated from

$$H_{\mathrm{m}} = \frac{2}{3}\rho_{\alpha} \left| \frac{2V\eta_{\alpha}}{\dfrac{dp_{\mathrm{in}}\left(x_{\mathrm{in}}^{(\alpha)}\right)}{dx_{\mathrm{in}}}} \right|^{1/2} \tag{9}$$

where index "α" refers to the inflexion point of q, p_{in} is the EHL inlet film pressure, and $x_{\mathrm{in}}^{(\alpha)}$ is the distance between the inflexion point and the nearest edge of the "dry" contact zone.

As is realised from Eq. (9), in order to calculate the film thickness with the inverse hydrodynamic theory, it is necessary to locate the inflexion point of a curve related to the contact pressure distribution. The correct location of the latter is crucial in the precise computation of the boundary condition to solve the film thickness equation and small errors lead to irrelevant results in terms of seal leakage and friction. Moreover, the role of the pressure gradient at the inflexion point (see denominator in Eq. (9)) on sealing performance is now revealed: the higher the gradient, the thinner the average film in the contact. This was quite early established in the literature and comprises the "secret" of minimising leakage, though, unfortunately, this increases hydrodynamic friction because thinner films result in more viscous drag [2, 4]. Thus, the inlet geometry of the seal and resulting local pressure distribution in that critical area is of paramount importance, a fact taken advantage of in modern seal designs with optimised profiles to suit even dynamic operating conditions.

The modified inverse hydrodynamic theory expressed via equations (8) and (9) has also been used with the more general transient Reynolds Eq. (6) as in reference [2]. The results produced by this and similar techniques are realistic but most published studies are based on static contact pressure distributions, which are realistic only at very low speed and for well supported seals such as rectangular seals supported by back-up rings on both sides. In cases where seals are allowed to move in their housing, even by a small amount (for example, O-rings), the inlet contact pressure has a dynamic variation and so does sealing performance. This is explained by the friction in the sealing contact, which deforms the seal, changing the inlet geometry dynamically. The problem is clearly coupled and solutions based on de-coupling it are more or less inaccurate when applied to dynamic sealing conditions. Unfortunately, to the best of the author's knowledge at the time of writing (2008), no published study has tackled the coupled problem. The reasons include computational complexity and time restrictions. Ideally, the problem must involve FEA for the contact mechanics of the seal combined with computational fluid dynamics or, equivalently, FEA with fluid-structure interactions. Such tasks require a high degree of sophistication and are time consuming, being rather unsuitable for parametric analyses involving hundreds of computer software executions.

Alternative methods have been applied by several researchers, though still not tackling the coupled problem in dynamic conditions. The said methods solve the Reynolds equation together with an elasticity equation for the seal, either simultaneously or serially in a number of iterations. Such "direct" methodologies were adopted in [1, 17, 76, 127]. In one of the earliest such studies, Field and Nau [17] in 1975 developed an elasticity equation from simple compression of a rectangular rod seal in smooth contact, including internal shear stresses in the material. After a lot of effort, they obtained results for outstrokes but failed to produce results for instrokes due to numerical instabilities. This is not surprising at all. The Reynolds equation is very sensitive to even very small errors in the film thickness (of sub-micrometre order of magnitude). When such errors are not eliminated in a numerical iteration scheme, they quickly destabilise the numerical procedure. This was reflected on the results [17] with wavy curves, which are indicative of numerical instability. Similar problems have been reported by Swales et al. [128].

Nikas [1, 3, 5, 38-40] also applied the direct solution method but extended this to the two-dimensional Reynolds equation [1] on rough contacts. He calculated a static contact pressure distribution for the rough contact with a columnar stress model but separated the pressure perturbations induced by roughness asperities in the numerical analysis, which are very weak in comparison with the bulk contact pressure, in order to achieve and accelerate convergence. Typical results from his theoretical analysis on film thickness are shown in Figure 14 for the rough contact of a rod seal and a piston rod. His approach was based on the solution of the Reynolds equation with a Successive Overrelaxation (SOR) method but that also encountered numerical problems.

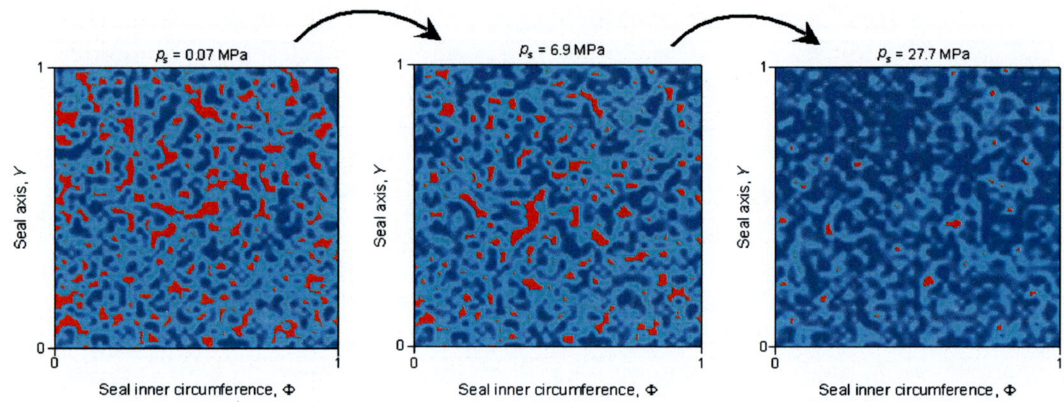

Figure 14. Film contour maps from the rough contact simulation of rod seal on piston rod. Lighter shaded areas indicate thinner film and red are areas of solid contact. Notice the partial film collapse at low sealed pressure p_s (left image)

A better yet much more complicated method was presented by Ruskell [32] in 1980. Ruskell's method overcame convergence problems by incorporating the Reynolds and seal elasticity equations into a single equation, which was solved numerically in a few iterations, usually up to six. His elasticity equation though was based on contact pressure distribution pre-calculated via FEA on a static, frictionless contact. This is of course not the real coupled EHL-mechanics problem but his method is still an efficient approach to a difficult problem. A similar approach was adopted by Prati and Strozzi [77]. A more sophisticated method, incorporating surface roughness and inter-asperity cavitation effects, was published in 2007 by Salant and co-workers [121, 129] for steady-state conditions of a reciprocating rod seal. The authors reported realistic results and emphasised the role of surface roughness in sealing performance.

Apart from the methods already described, methods developed for solving the EHL problem of soft solids in general have been published since the 1960s. The work of Hooke [130-132] is of particular interest to elastomeric reciprocating seals, as it deals with soft contacts under conditions of reciprocating motion. Though Hooke's approach is approximate in dealing with the EHL inlet and exit zones, it is mathematically sound and provides useful results on transient motion. Of great interest are his results on the issue of the reversal of motion, showing the expected film thinning at the edges of the contact, where most of seal wear takes place. Indeed, the precise numerical treatment of the reversal of motion in reciprocating seals is a complicated issue and remains challenging to date. A further analysis of this issue, though not confined to soft solids, was also presented by Hooke for line contacts using the one-dimensional Reynolds equation [133].

In order to avoid the numerical complexities in solving the transient Reynolds equation (Eq. (6)), Chang [134] proposed a simple and validated method, based on the solution of a first-order partial differential equation of one-dimensional wave propagation. This is essentially derived from Eq. (6) by setting the left-hand side term equal to zero and is sometimes referred to in the literature as the "reduced" Reynolds equation. The method provides a practical alternative to avoiding a full numerical solution, though it may be suitable only for well-supported seals, without significant flexing during the reversal of motion. Ikeuchi et al. [135] have also published a simplified method to solve the transient problem in the contact of compliant solids, including dynamic seals.

Further insight into the issue of transient lubrication of reciprocating seals was offered quite early by the theoretical simulations of Hirano and Kaneta [22, 24]. They studied the development and potential collapse of EHL films with pre-determined contact pressure distributions such as parabolic and Gaussian [22]. Their theoretical results confirmed their experimental results [13] on the criticality of the stroking-length-to-contact-width ratio in establishing a full elastohydrodynamic film, which must be at least 2. This is physically explained by considering the speed of wave propagation of the fluid from the contact inlet to the outlet, which is half the speed of the moving surface in the contact. Therefore, relatively short strokes result in partially collapsed films and increased wear of the seals. These results have been confirmed in several other studies. However, as Hirano and Kaneta pointed out [13], the film collapse (in the exact sense) is theoretical because it neglects the micro-EHL taking place between surface roughness asperities of the counterfaces in the contact, as for example the asperities of a rod seal and a piston rod.

Micro-EHL is indeed an efficient method of wear reduction and can be analysed on roughness scale, although its theoretical treatment and numerical analysis are both complex [136, 137]. In the author's experience (e.g. [41]), a proper treatment of roughness in elastomeric contacts should involve aspects such as inter-asperity cavitation, viscoelastic effects to explain stick-slip micro-scale phenomena such as Schallamach waves [45], and asperity adhesive forces [138, 139]. The latter, related to attractive molecular forces at the sealing interface among closely engaged roughness asperities, affects friction during start-up of motion, when the material is relaxed and the fluid film at the sealing interface is partially collapsed. Moreover, as friction is related to wear, proper roughness modelling will be beneficial in predicting seal wear more accurately. However, the topic of abrasive wear of elastomers rubbed on hard surfaces is vast and cannot be adequately discussed within the confines of this chapter. Interested readers are directed to references [49, 140, 141] for an introduction to the subject and, particularly to Zhang's book [45].

The real necessity of roughness modelling may be questionable considering that polymeric seals, when rubbed against hard, rough surfaces, can deposit a thin layer of polymer on the hard surface, filling up valleys between asperities of the hard surface [142, 143]. This is well known for PTFE for example and would effectively reduce the average roughness of the hard surface, giving some justification for a smooth-contact model. Furthermore, as already discussed in section 4.2, experimental results reported in the literature, for example [91, 92], show smoothening of the seal surface during running-in, even when sliding on very smooth counterface such as glass. Thus, it would appear that roughness modelling is useful as a theoretical improvement, with potentially more accurate friction predictions, but it is doubtful that it is absolutely critical to the seal designer. In well lubricated elastomeric contacts, the largest portion of friction comes from viscous shear and that can be realistically predicted with existing models of smooth EHL [41]. In contacts starved of lubricant, predictions are also realistic with a smooth-surface model [42], provided it is capable of calculating very small film thicknesses. Roughness modelling can thus be understood in terms of refining and fine-tuning an existing theoretical model.

5.4 Performance of hydraulic reciprocating seals and related issues

The main variables to consider when evaluating the performance of reciprocating seals are leakage, friction and extrusion. Leakage is a measure of how efficient sealing is. Friction is related to the resistance to motion and the associated power loss in the system. Finally, extrusion is related to the risk of damage to the seals from stress concentration and damage to their edges.

Leakage refers to the rate of mass of liquid passing through the sealing contact in the direction of reciprocation. As the fluid continuity equation (mass conservation) must be satisfied everywhere in the contact, leakage can be calculated at any one point. However, avoiding an arbitrary point selection, it is, generally, better to calculate the average leakage in the contact. Thus, in the case of the one-dimensional Reynolds equation (Eq. (6)), the mass leakage rate Q is [41]

$$Q = \frac{L}{W}\int_0^W \left(\frac{Vh}{2} - \frac{h^3}{12\eta}\frac{\mathrm{d}p}{\mathrm{d}x}\right)\rho \,\mathrm{d}x \qquad (10)$$

where W is the contact width in the direction of reciprocation and L is the contact size in the transverse direction. In a transient analysis, some variables in Eq. (10) are, naturally, functions of time, for example, speed V and local film thickness h. In the case of the two-dimensional Reynolds equation, see for example [1, 38] for the analysis of rod seals.

The frictional force on a reciprocating seal consists of a hydrodynamic component and other components related to adhesive or molecular forces such as van der Waals forces between closely engaged roughness asperities of the counterfaces in a sealing contact. In practise, the hydrodynamic force, which is greatest in thick-film lubricating conditions, is, almost always, the only one calculated, especially when the model refers to smooth instead of rough EHL. The hydrodynamic frictional force on the seal, F, is calculated by integrating the local viscous shear stress on the seal surface in the sealing contact. In the case of the one-dimensional Reynolds equation (Eq. (6)), the hydrodynamic frictional force is [41]

$$F = L\int_0^W \left(\frac{\eta V}{h} - \frac{h}{2}\frac{\mathrm{d}p}{\mathrm{d}x}\right)\mathrm{d}x \qquad (11)$$

In the case of the two-dimensional Reynolds equation, see for example [1, 38] for the analysis of rod seals. For other sources of friction, refer to [138, 139].

Apart from leakage and friction, reciprocating seals often suffer from extrusion damage. For example, in the case of rod seals in linear hydraulic actuators without back-up rings, the seal may get squeezed into the clearance between its housing and the piston rod as in Figure 15. The localised strain in the form of the extruded part is a zone of stress concentration and is readily affected by the sealed pressure and friction on the sealing contact. This localised deformation can lead to a cut or abrasion of the seal after a number of operating cycles, causing a premature end to the service life of the seal.

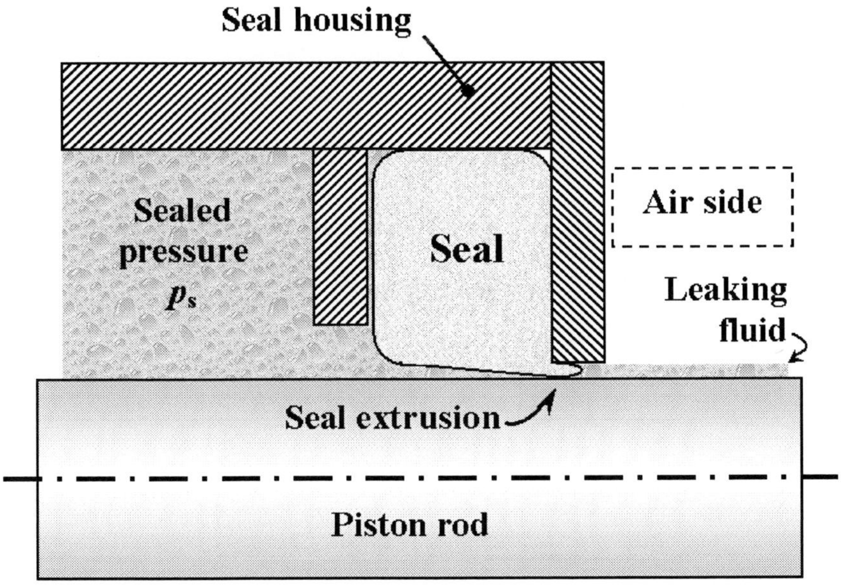

Figure 15. Extrusion of a rectangular, rounded, elastomeric, rod seal (only the upper half of the seal and housing shown)

The phenomenon was discussed quite early by White and Denny [7]. Nikas [37] analysed seal extrusion mathematically and produced algebraic equations giving the length, shape and contact pressure on the extruded part as a function of the operating conditions and corner geometry of the seal. It was thus demonstrated how to minimise extrusion. The results showed that extrusion is unavoidable for seals with sharp corners but it was not significantly influenced by the stroking velocity and the viscosity of the sealed fluid. Moreover, it was shown that extrusion rises with the sealed pressure and the rod-housing clearance; it is reduced with the seal width, the seal corner radius and the seal elastic modulus. As a result of those findings, it was suggested that the two most obvious solutions to minimise seal extrusion are (a) to use seals with rounded corners or chamfered ends and (b) to use anti-extrusion or back-up rings.

Back-up rings (Figure 13) are devices resembling washers. Their main task is to support elastomeric seals in high-pressure hydraulic systems, preventing seal gap extrusion. Thus, they are also known as anti-extrusion rings. They are also used to prevent seal roll deformation [144], which can cause total seal failure. As they normally affect seal deformation during operation, they should be accounted for when modelling seal performance. This is even more crucial if the rings are in contact with the reciprocating surface (e.g. a piston rod) and, thus, perform some sealing action. A mathematical analysis of back-up rings for rod seals was developed by Nikas [39], calculating their effects on leakage and friction of the seals for operating temperatures between –54 and +135 °C, and sealed pressures between 1 and 35 MPa. It was found that the contact pressure and average surface roughness of a back-up ring can be optimised to minimise the leakage-per-cycle of the system. It was also found that there exist a critical sealed pressure over which a back-up ring can become a more effective sealing element than the seal it supports. The effect of operating parameters on those results was also explored and several other conclusions were postulated in what was a computationally complex analysis. However, it should be emphasized that such results cannot be generalised because they depend on many factors such as the exact geometry of the solids involved (seal and rings), their dimensions and the operating conditions.

Seal extrusion can also be reduced by using composite seals. For example, consider the seal shown in Figure 16. It has a centrally placed elastomeric part bonded to two outer PTFE parts. The depicted seal is actually just the horizontal part of a goalpost-shaped vane seal developed for rotary vane actuators [2, 4] and operating in reciprocating motion. The PTFE (with glass fibres) gives the seal the needed rigidity at the edges to reduce extrusion and allows the corners to be sharp, minimising leakage. A theoretical analysis of this type of composite, rectangular, reciprocating seal was presented in [42] with an extensive parametric study of sealing performance and how it can be optimised in a broad range of temperatures. Results of leakage, friction and seal extrusion were presented on both starved and flooded contacts, and an optimum PTFE-to-seal volume ratio was calculated, based on given priorities, such that the composite seal outperforms the elastomeric seal of the same dimensions. This means that an elastomeric seal can be replaced by a composite seal of the same dimensions (thus fitting in existing housings without modifications), giving lower leakage, friction and extrusion.

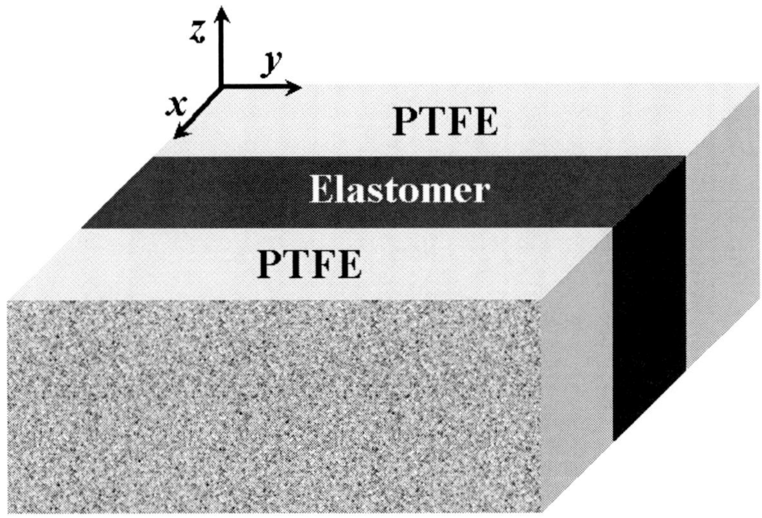

Figure 16. Example of composite seal [42]

Other means of improving sealing performance in a hydraulic system involve the fitting of two seals in a row. The seals in this arrangement are known as dual or tandem seals (Figure 17) and their primary goal is to minimise the leakage-per-cycle. The primary seal, which is the closest of the two to the high-pressure chamber, performs most of the sealing. The secondary seal merely wipes the fluid that has leaked from the primary seal. Various seal combinations can be had in such an arrangement, based on design priorities.

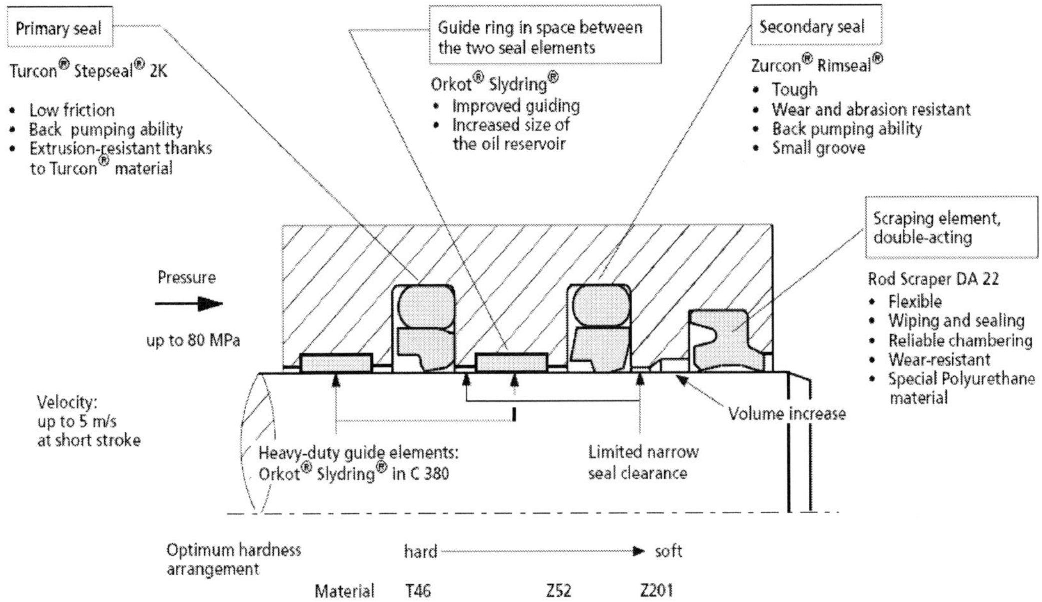

Figure 17. Tandem seal arrangement (courtesy of Trelleborg Sealing Solutions [55])

If properly designed, the tandem seal arrangement can reduce system leakage significantly. However, caution is advised to avoid designs promoting the development of an interseal pressure. The latter, mostly observed in identical-seal arrangements, is about the abrupt rise of the pressure at the interseal space to very high values, sometimes exceeding sealed pressure and causing damage to back-up rings and/or sealing failure. This phenomenon was quite early recognised and studied experimentally by Field and Nau in [145]. Nikas and Sayles [40] presented a mathematical analysis of identical tandem rod seals supported by back-up rings and studied their benefits in terms of leakage and friction, reporting on leakage reduction in the order of 50 to 70 per cent over a wide range of temperatures compared with a single-seal arrangement.

Furthermore, assuming that the interseal space is initially filled with air and using the van der Waals equation of state for that air, they analysed the evolution of the interseal pressure with the number of strokes. A typical result of their analysis is shown in Figure 18 where it is clear that the interseal pressure rises abruptly after about 1700 strokes (the analysis was stopped as the pressure exceeded system pressure after 1 more stroke). Nevertheless, the critical number of strokes to avoid damage can be predicted (about 1600 strokes in the case of Figure 18) and taken into account during operation to allow for servicing, for example, venting the interseal space. The critical number of strokes may be too large in some systems to pose any problem, as for example in hydraulic actuators controlling aircraft landing gear, where one thousand strokes take a rather long time.

It is also worth noticing in Figure 17 the use of a scraping element (right end of picture). This is primarily used to prevent solid contaminants entering the system and damaging the seals and gland bearings but it may also perform some fluid sealing, particularly in single-seal arrangements (unlike Figure 17). Therefore, its geometry plays a vital role in system performance [146].

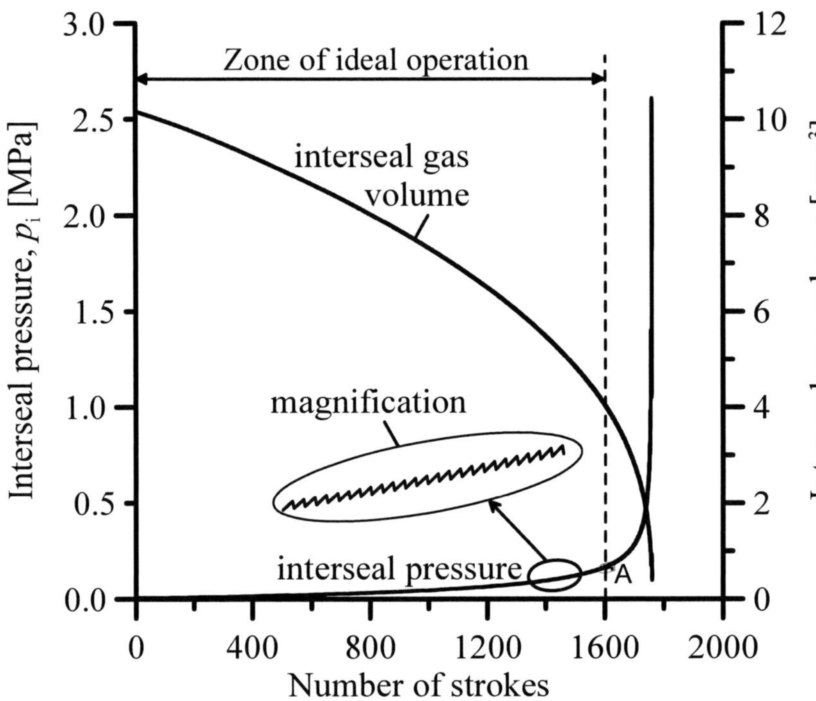

Figure 18. Variation of interseal pressure and gas volume with the number of strokes in a tandem seal arrangement of identical, rectangular, rounded elastomeric seals [40]

The presence of solid contaminants is detrimental not only to seal life in the long term, but also to system leakage in the short term. Debris particles can become embedded on a soft seal and scratch a hard piston rod. The scoring marks are, effectively, micro-channels that allow high-pressure liquid to escape. Unfortunately, it is not only relatively large debris that are capable of such damage. Remarkably, the wear of shafts and seals in tests performed by Tanoue et al. [147] in 1971 was significantly affected by sub-micrometre particles (< 0.25 μm) contained in used lubricating oils. Wear was proportional to particle concentration and it was reported that even for small particle concentrations, e.g. 0.2 per cent by weight, wear was significant.

6. Conclusion

Hydraulic reciprocating seals are critical elements of complex mechanical behaviour. Knowledge of their functioning is of paramount importance in hydraulic efficiency and safety. Evaluating the performance of reciprocating seals requires a combination of specialised mathematical tools from contact mechanics and tribology. This is a very difficult task and expert advice is required to achieve successful designs and reliable operation. As with other machine element applications, attention to detail is of great significance. This is often of little concern to the end-user who does not purchase the seals but the equipment fitted with the them. Nevertheless, basic knowledge of sealing mechanisms is advantageous or even sometimes necessary to avoid costly mistakes in high-risk applications and prevent catastrophes such as the destruction of space shuttle Challenger in 1986…

Acknowledgments

The author is grateful to Trelleborg Sealing Solutions (UK) and Smiths Group for their collaboration on sealing research at Imperial College London [3, 4] between 1999 and 2004. Special thanks are due to Mr Guy Burridge, Mr Robert Almond and Mr David Goddard of Smiths Group for sharing their vast technical experience with the author on reciprocating and vane seals. The author is also grateful to the Jacob Wallenberg Foundation (Sweden) for a research grant awarded for research in Materials Science in 2007 through the Royal Swedish Academy of Engineering Sciences, part of which was used for his financial support in writing this chapter.

References

[1] Nikas, G. K. (2003). *ASME J. Tribology*, 125, 60-69.

[2] Nikas, G. K., Burridge, G., Sayles, R. S. (2007). *Proc. IMechE, Part J: J. Engineering Tribology*, 221, 699-715.

[3] Nikas, G. K. (2001). *Determination of polymeric sealing principles for end user high reliability*. Technical Report DOW-08/01, Tribology Group, Department of Mechanical Engineering, Imperial College London, England.

[4] Nikas, G. K. (2004). *Research of fundamental sealing mechanisms needed for zero-leakage high-reliability rotary vane actuators*. Technical Report SMI-10/04, Tribology Group, Department of Mechanical Engineering, Imperial College London, England.

[5] Nikas, G. K., Sayles, R. S. (2004). *Tribology International*, 37, 651-660.

[6] Nikas, G. K., Sayles, R. S. (2005). *Sealing Technology*, 2005, 6-11.

[7] White, C. M., Denny, D. F. (1947). *The sealing mechanism of flexible packings*. Scientific and Technical Memorandum 3/47, UK Ministry of Supply.

[8] Müller, H. K. (1964). *Proc. 2nd Int. Conf. on Fluid Sealing*, paper B2, 13-28.

[9] Lawrie, J. M., O'Donoghue, J. P. (1964). *Proc. 2nd Int. Conf. on Fluid Sealing*, paper B6, 69-80.

[10] Field, G. J., Nau, B. S. (1972). *Proc. IMechE, Symposium on Elastohydrodynamic Lubrication (Leeds, England)*, paper C5, 29-36.

[11] Dowson, D., Swales, P. D. (1969). *Proc. 4th Int. Conf. on Fluid Sealing*, paper 1, 2-10.

[12] Kaneta, M., Todoroki, H., Nishikawa, H., Kanzaki, Y., Kawahara, Y. (2000). *ASME J. Tribol*, 122, 787-795.

[13] Hirano, F., Kaneta, M. (1971). *Proc. 5th Int. Conf. on Fluid Sealing*, paper G3, 33-48.

[14] Nau, B. S. (1971). *Proc. 5th Int. Conf. on Fluid Sealing*, paper G5, 81-96.

[15] Field, G. J., Nau, B. S. (1975). *Proc. 7th Int. Conf. on Fluid Sealing*, paper C1, 1-13.

[16] Dowson, D., Swales, P. D. (1967). *Proc. 3rd Int. Conf. on Fluid Sealing*, paper 1, 33-44.

[17] Field, G. J., Nau, B. S. (1975). *Transactions ASLE*, 18, 48-54.

[18] Hooke, C. J. (1967). *Proc. 3rd Int. Conf. on Fluid Sealing*, 45-56.

[19] Johannesson, H. L. (1983). *ASME J. Lubrication Technology*, 105, 288-296.

[20] Johannesson, H. (1978). *Proc. 5th Leeds-Lyon Symp. on Tribol*, 379-387.

[21] Karaszkiewicz, A. (1987). *Ind. Eng. Chem. Res*, 26, 2180-2185.

[22] Hirano, F., Kaneta, M. (1971). *Proc. 5th Int. Conf. on Fluid Sealing*, paper G2, 17-32.

[23] Ruskell, L. E. C. (1976). *Proc. Royal Society London A*, 349, 383-396.

[24] Hirano, F., Kaneta, M. (1969). *Proc. 4th Int. Conf. on Fluid Sealing*, paper 2, 11-20.

[25] Theyse, F. H. (1967). *Proc. 3rd Int. Conf. on Fluid Sealing*, paper F2, 17-32.

[26] Fazekas, G. A. (1976). *ASME J. Engineering for Industry*, 783-787.

[27] Karaszkiewicz A. (1985). *Ind. Eng. Chem. Prod. Res. Dev*, 24, 283-289.

[28] Medri, G., Prati, E., Strozzi, A. (1981). *Proc. 9th Int. Conf. on Fluid Sealing*, paper C1, 55-71.

[29] Hooke, C. J., Lines, D. J., O'Donoghue, J. P. (1966-67). *Proc. IMechE*, 181, 205-210.

[30] Johannesson, H. L. (1981). *Proc. 9th Int. Conf. on Fluid Sealing*, paper C3, 87-102.

[31] Chivers, T. C., Hunt, R. P. (1978). *Proc. 8th Int. Conf. on Fluid Sealing*, paper F3, 31-40.

[32] Ruskell, L. E. C. (1980). *IMechE J. Mechanical Engineering Science*, 22, 9-16.

[33] Nau, B. S. (1999). *Proc. IMechE, Part J: J. Engineering Tribology*, 213, 215-226.

[34] Nau, B. S. (1987). *Rubber Chemistry and Technology*, 60, 381-416.

[35] Field, G. J., Nau, B. S. (1973). *Proc. 6th Int. Conf. on Fluid Sealing*, paper C5, 45-56.

[36] Rana, A., Sayles, R. S., Nikas, G. K., Jalisi, I. (2001). An experimental technique for investigating the sealing principles of reciprocating elastomeric seals for use in linear hydraulic actuator assemblies. *Proc. 2nd World Tribology Congress*, Vienna, Austria, (on CD-ROM).

[37] Nikas, G. K. (2003). *Proc. IMechE, Part J: J. Engineering Tribology*, 217, 365-373.

[38] Nikas, G. K. (2003). *Proc. IMechE, Part J: J. Engineering Tribology*, 217, 461-473.

[39] Nikas, G. K. (2004). *Tribology International*, 37, 689-699.

[40] Nikas, G. K., Sayles, R. S. (2006). *Tribology International*, 39, 622-634.

[41] Nikas, G. K., Sayles, R. S. (2006). *Tribology International*, 39, 921-936.

[42] Nikas, G. K., Sayles, R. S. (2006). *Proc. IMechE, Part J, J. Engineering Tribology*, 220, 395-412.

[43] Müller, H. K., Nau, B. S. (1998). *Fluid Sealing Technology,* Mechanical Engineering Series 117, Marcel Dekker Inc. New York, NY, USA.

[44] Abouel-Kasem, A. (2006). *Sealing Technol*, 2006(11), 7-11.

[45] Zhang, S.W. (2004). *Tribology of Elastomers,* Tribology and Interface Engineering Series 47; Elsevier: London, England.

[46] Duhring, B., Iversen, G. (1999). *Proc. IMechE, Part J: J. Enginnering Tribology*, 213, 227-237.

[47] Bertolet, E. C. (1956). *Proc. 12th National Conference on Industrial Hydraulics* (USA), 10, 56-68.

[48] Atkins, L. E. (1975). *Proc. 7th Int. Conf. on Fluid Sealing*, paper E2, 11-24.

[49] Zhang, S. W. (1998). *Tribology International*, 31, 49-60.

[50] Stachowiak, G. W., Batchelor, A. W. (2005). *Engineering Tribology (third edition),* Elsevier Butterworth- Heinemann, London, UK.

[51] Li, W., Mays, S. (2003). *Proc. 17th Int. Conf. on Fluid Sealing*, 157-165.
[52] Lee, N.-J., Jang, J. (2000). *Compros. Sci. Technol*, 60, 209-217.
[53] Ganghoffer, J.-F., Brillard, A., De-Borst, R. (1998). *Math. Comp. Model*, 27, 23-49.
[54] Caddock, B. D., Evans, K. E. (1995). *Biomaterials*, 16, 1109-1115.
[55] *Hydraulic Seals – Linear, Rod seals.* Trelleborg Sealing Solutions catalogue, Edition April (2007). PDF file available from www.tss.trelleborg.com.
[56] *Hydraulic Seals – Linear, Piston seals.* Trelleborg Sealing Solutions catalogue, Edition April (2007). PDF file available from www.tss.trelleborg.com.
[57] Cnops, R. F. (1964). *Proc. 2nd Int. Conf. on Fluid Sealing*, paper B7, 81-91.
[58] Aston, M. W., Fletcher, W., Morrell, S. H. (1969). *Proc. 4th Int. Conf. on Fluid Sealing*, 64-75.
[59] Chadwick, P. (1974). *Philosophical Transactions Royal Society London A*, 276, 371-403.
[60] Grosch, K. A. (1963). *Proc. Royal Society London A*, 274, 21-39.
[61] Bhushan, B. (1999). *Principles and Applications of Tribology*, Wiley, New York, NY, USA.
[62] Flitney, R. K., Nau, B. S. (1988). *STLE Lubrication Engineering*, 44, 993-1000.
[63] Denny, D. F. (1959). *Wear*, 2, 264-272.
[64] Kanters, A. F. C. (1990). *On the calculation of leakage and friction of reciprocating elastomeric seals*. Ph.D. thesis, Eindhoven University of Technology, The Netherlands.
[65] Visscher, M., Kanters, A. F. C. (1990). *STLE Lubrication Engineering*, 46, 785-791.
[66] Visscher, M. (1992). *The measurement of the film thickness and roughness deformation of lubricated elastomers*. Ph.D. thesis, Eindhoven University of Technology, The Netherlands.
[67] Kambayashi, H., Ishiwata, H. (1964). *Proc. 2nd Int. Conf. on Fluid Sealing*, paper B3, 29-40.
[68] May, E. M. (1957). *App. Hyd*, 10, 110-114.
[69] Olssen, E. (1972). *Friction forces and oil leakage of O-rings on axially moving shafts*. Ph.D. thesis, Chalmers University of Technology, Sweden.
[70] Molari, P. G. (1973). *Proc. 6th Int. conf. on Fluid Sealing*, paper B2, 15-31.
[71] Austin, R. M., Flitney, R. K., Nau, B. S. (1978). *Proc. 8th Int. Conf. on Fluid Sealing*, paper J2, 11-20.
[72] Kawahara, Y., Ohtake, Y., Hirabayashi, H. (1981). *Proc. 9th Int. Conf. on Fluid Sealing*, paper C2, 73-85.
[73] Strozzi, A. (1986). *ASLE Transactions*, 29, 558-564.
[74] Johannesson, H. L., Kassfeldt, E. (1987). *Proc. 11th Int. Conf. on Fluid Sealing*, paper E4, 314-329.
[75] Lindgren, H. (1987). *Wear*, 115, 31-40.
[76] Field, G. J. (1973). *The elastohydrodynamic lubrication of rectangular section rubber seals under conditions of reciprocating motion.* Ph.D. thesis, City University London, England.
[77] Prati, E., Strozzi, A. (1984). *ASME J. Tribology*, 106, 505-512.
[78] Blok, H., Koens, H. J. (1965/66). *Proc. IMechE*, 180, 221-223.
[79] Krauter, A. I. (1982). *ASME J. Lubrication Technology*, 104, 455-459.

[80] Kassfeldt, E. (1987). *Analysis and design of hydraulic cylinder seals.* Ph.D. thesis, Luleå University of Technology, Sweden.
[81] Wernecke, P. W. (1987). *Proc. 11th Int. Conf. on Fluid Sealing*, paper E1, 249-277.
[82] Reddy, D., Nau, B. S. (1984). *Proc. 10th Int. Conf. on Fluid Sealing*, paper D3, 189-196.
[83] Iwanami, S., Tikamori, N. (1961). *Proc. 1st Int. Conf. on Fluid Sealing*, paper B2.
[84] Kaneta, M. (1986). *J. JSLE Int. Ed*, 7, 141-146.
[85] Lindgren, H. (1986). *Scraper ring properties and behaviour in hydraulic cylinders.* M.Sc. thesis, Chalmers University of Technology, Sweden.
[86] Karaszkiewicz, A. (1988). *Trib. Int*, 21, 361-367.
[87] Schrader, K. (1978). *Beitrage zur klaerung des abdichtvorganges gummielastischer abdichtungen axial verschiebbarer hydrostatischer bauteile*. Ph.D. thesis, Dresden University of Technology, Germany.
[88] Kanzaki, Y., Kawahara, Y., Kaneta, M. (1997). *Proc. 15th Int. Conf. on Fluid Sealing*, 79-95.
[89] Roberts, A. D., Tabor, D. (1968). *Wear*, 11, 163-166.
[90] Shouten, M. J. W., Dollevoet, R. P. B., de Laat, B. M. (1997). *Proc. 15th Int. Conf. on Fluid Sealing*, 111-131.
[91] Rana, A. S. (2005). *A tribological study of elastomeric reciprocating seals for hydraulic actuators.* Ph.D. thesis, Tribology Group, Department of Mechanical Engineering, Imperial College London, England.
[92] Rana, A. S., Sayles, R. S. (2004). An experimental study on the friction behaviour of aircraft hydraulic actuator elastomeric reciprocating seals. *Proc. 31st Leeds-Lyon Symposium on Tribology.*
[93] Reddyhoff, T., Dwyer-Joyce, R. (2006). *Sealing Technology*, 2006(7), 7-11.
[94] Meyer, K. H., von Susich, G., Valko, E. (1932). *Kolloidzeitschrift*, 59, 208-216.
[95] Treloar, L. R. G. (1976). *Proc. Royal Society London A*, 351, 301-330.
[96] Price, C. (1976). *Proc. Royal Society London A*, 351, 331-350.
[97] Chadwick, P., Creasy, C. F. M. (1984). *J. Mechanics and Physics of Solids*, 32, 337-357.
[98] Ogden, R. W. (1986). *Rubber Chemistry and Technology*, 59, 361-383.
[99] Holzapfel, G. A. (2000). *Nonlinear Solid Mechanics*, Wiley, New York, NY, USA.
[100] Mooney, M. (1940). *J. Applied Physics*, 11, 582-592.
[101] Rivlin, R. S. (1948). *Philosophical Transactions Royal Society London A*, 241, 379-397.
[102] Finite Element software MSc.Marc (version 2000) – Volume A. MSC software Corp, Munich, Germany. www.mscsoftware.com.
[103] Ogden, R. W. (1972). *Proc. Royal Society London A*, 326, 565-584.
[104] Horgan, C. O., Ogden, R. W., Saccomandi, G. (2004). *Proc. Royal Society London A*, 460, 1737-1754.
[105] Wineman, A., Shaw, J. (2004). *ASME J. Applied Mechanics*, 71, 769-773.
[106] Qi, H. J., Boyce, M. C. (2004). *J. Mechanics and Physics of Solids*, 52, 2187-2205.
[107] Dorfmann, A., Ogden, R. W. (2004). *International J. Solids and Structures*, 41, 1855-1878.
[108] Horgan, C. O., Schwartz, J. G. (2005). *J. Mechanics and Physics of Solids*, 53, 545-564.

[109] Hooke, C. J., Lines, D. J., O'Donoghue, J. P. (1967). *Proc. 3rd Int. Conf. on Fluid Sealing*, paper F4, 45-56.
[110] Dragoni, E., Strozzi, A. (1988). *ASME J. Tribology*, 110, 193-200.
[111] Johannesson, H. L., Kassfeldt, E. (1989). *Wear*, 130, 3-15.
[112] Medri, G., Molari, P. G., Strozzi, A. (1978). *Proc. 8th Int. Conf. on Fluid Sealing*, paper F2, 19-30.
[113] Medri, G., Strozzi, A. (1984). *Ind. Eng. Chem. Prod. Res*, 23, 596-600.
[114] Dragoni, E., Medri, G., Strozzi, A. (1987). *Proc. 11th Int. Conf. on Fluid Sealing*, paper B3, 160-174.
[115] George, A. F., Strozzi, A., Rich, J. F. (1987). *Proc. 11th Int. Conf. on Fluid Sealing*, paper B1, 117-137.
[116] Naderi, A., Albertson, K., Peng, S. (1994). *Proc. 46th National Conf. on Fluid Power, NFPA.* Chicago, TX, USA, 99-105.
[117] Peng, S., Sun, S., Albertson, K. (1996). *Proc. 47th National Conf. on Fluid Power, NFPA.* Chicago, TX, USA, 175-182.
[118] Claus, R. G. (2002). *Proc. 49th National Conf. on Fluid Power, NFPA.* Las Vegas, NV, USA, 383-389.
[119] Maser, N. B. (2006). *Numerical model of a reciprocating rod seal, including surface roughness and mixed lubrication.* M.Sc. thesis; Georgia Institute of Technology, School of Mechanical Engineering, Atlanta, GA, USA.
[120] Tang, J., Yang, W., Ding, Y.-M., Li, J., Zhang, Y., Lu, B.-T. (2007). *Lubrication Engineering, "Lubrication and Seal"(Chinese)*, 32, 36-39 (ISSN: 0254-0150).
[121] Salant, R. F., Maser, N., Yang, B. (2007). *ASME J. Tribology*, 129, 91-97.
[122] Mackerle, J. (1998). *Modelling and Simulation in Materials Science and Engineering*, 6, 171–198.
[123] Gohar, R. (2001). *Elastohydrodynamics (second edition)* Imperial College Press, London, UK.
[124] Blok, H. (1963). Inverse problems in hydrodynamic lubrication and design directives for lubricated flexible surfaces. *Proc. Int. Symposium on Lubrication and Wear.* Houston, TX, USA.
[125] Johannesson, H. L. (1989). *Wear*, 17-27.
[126] Kanters, A. F. C., Verest, J. F. M., Visscher, M. (1990). *STLE Tribology Transactions*, 33, 301-306.
[127] Yang, Y., Hughes, F. (1984). *ASLE Transactions*, 27, 197-202.
[128] Swales, P. D., Dowson, D., Latham, J. L. (1972). *Proc. IMechE, Symposium on Elastohydrodynamic Lubrication*, paper C4, 22-28.
[129] Salant, R. F. (2007). *Sealing Technology*, 2007(1), 7-11.
[130] Hooke, C. J., O'Donoghue, J. P. (1972). *J. Mechanical Engineering Science*, 14, 34-48.
[131] Hooke, C. J. (1986). *ASME J. Tribology*, 108, 545-550.
[132] Hooke, C. J. (1987). *Proc. 13th Leeds-Lyon Symposium on Tribology (1986).* Elsevier, Tribology Series 11, 299-305.
[133] Hooke, C. J. (1994). *Proc. IMechE, Part J, J. Engineering Tribology*, 208, 53-64.
[134] Chang, L. (2000). *STLE Tribology Transactions*, 43, 116-122.
[135] Ikeuchi, K., Fujita, S., Ohashi, M. (1998). *Tribology International*, 31, 613-618.

[136] Jin, Z. M., Dowson, D. (1997). *Proc. IMechE, Part C: J. Mechanical Engineering Science*, 211, 265-272.
[137] Kim, A. T., Seok, J., Tichy, J. A., Cale, T. S. (2003). *ASME J. Tribology*, 125, 448-451.
[138] Drutowski, R. C. (1969). *Proc. 4th Int. Conf. on Fluid Sealing*, 52-57.
[139] Wassink, D. B., Lenss, V. G., Levitt, J. A., Ludema, K. C. (2001). *ASME J. Tribology*, 123, 404-412.
[140] Fukahori, Y., Yamazaki, H. (1995). *Wear*, 188, 19-26.
[141] Zhang, S. W., Zhaochun, Y. (1997). *Tribology International*, 30, 839-843.
[142] Eiss Jr, N. S., Wood, K. C., Herold, J. A., Smyth, K. A. (1979). *ASME J. Lubrication Technology*, 101, 212-219.
[143] Eiss Jr, N. S., Smyth, K. A. (1981). *ASME J. Lubrication Technology*, 103, 266-273.
[144] Nwagboso, C. O. (1994). *STLE Lubrication Engineering*, 50, 66-72.
[145] Field, G. J., Nau, B. S. (1971). *Proc. 5th Int. Conf. on Fluid Sealing*, paper D2, 21-35.
[146] Peppiatt, N. (2003). *Sealing Technology*, 2003(12), 5-8.
[147] Tanoue, H., Ishiwata, H., Tada, H. (1971). *Proc. 5th Int. Conf. on Fluid Sealing*, paper D3, 37-48.

In: Tribology Research Trends
Editor: Taisho Hasegawa

ISBN: 978-1-60456-912-4

Chapter 2

THERMOTRIBOLOGY: FUNDAMENTALS AND CURRENT TRENDS

***P.N. Bogdanovich* [1] *and D.V. Tkachuk* [2]**

ABSTRACT

The chapter reviews briefly the history of heat problems in tribology from the founding father Prof. H. Blok to the present time. Blok pioneered the flash temperature concept and paved the way for further research in thermotribology. Basic models of frictional heating are outlined with particular attention to Blok's, Jaeger's and Archard's ones. Factors influencing the friction temperature and its distribution in the contact zone and its vicinity are considered. The effect of frictional heating on the tribological behavior of materials of different classes including the wear modes and regularities is discussed. Experimental techniques to measure the friction temperature and its distribution are described with special emphasis on the method of optical-electron scanning of the rubbing surfaces developed by the authors and up-to-date thermography techniques. The paper gives also the overview of the authors' research into the high-speed friction of metals, polymers and brittle inorganic materials. Special stress is set on the recent results of studying the wear of such brittle materials as glass and sapphire at abrasive machining. It is shown, in particular, that high thermal stresses resulted from frictional heating cause the brittle fracture (spalling) of glass even outside the contact area. The same phenomenon is observed at sapphire machining despite its much better thermal characteristics compared to glass. The experiments on sapphire and diamond cutting are described and the analysis of temperature fields in the cutting zone is presented. Modern trends in thermotribology are outlined.

1 Department of Materials Science, Belarus State University of Transport, Kirov St. 34, Gomel 246653, Belarus
2 2V.A. Belyi Metal-Polymer Research Institute, National Academy of Sciences of Belarus, Kirov St. 32A, Gomel 246050, Belarus

1. Introduction

The fact that the friction of solids was accompanied by their heating had been recognized even by the primitive man [1, 2]. Moreover, people used this side effect of friction to make a fire. Nowadays human beings encounter the phenomenon of frictional heating in their day-to-day life but for some of them the study of it has become their occupation. Among the scientists investigating heat problems in tribology the second Tribology Gold Medalist (1973) Prof. Harmen Blok (1910–2000) was the most famous; he has gained world recognition as the founding father of thermotribology. It was he who highlighted the flash temperature concept [3] and paved the way for further research in this field. This elegant concept has become the fundamental point of thermotribology and remains fruitful nowadays.

Research in thermotribology is promoted by numerous problems mechanical engineers face. Frictional heating and related thermal and thermomechanical phenomena may affect significantly the tribological behavior of sliding or rolling machine parts, first of all because high temperatures change the structure and chemical composition of the mated materials as well as the properties of the films covering their surfaces. For example, with temperature elevation such mechanical characteristics as the elastic modulus and hardness decrease that is especially critical for polymers and polymer-based materials [4]. The surface melting of one of the contacting materials may occur [5, 6], polymers may destruct [7], metals may oxidize [8, 9], boundary lubricating films may be disoriented or broken down [10] etc. Frictional heating may also be favorable to the failure of sliding members, e.g. the thermal cracking of working surfaces of brakes and face seals [11–14]; it may cause the sparking and severe wear of electrical brushes [15] and the seizure of gearings [3, 16]. The effect of the harmful consequences of thermal processes in friction contact should be minimized to increase the reliability and life of friction units and, finally, to improve the quality of machines. Much has been already done in this domain and these results are overviewed briefly in the present chapter.

2. Theoretical Studies of Frictional Heating

2.1. Distribution of Heat in Friction Contact

If two solids are in contact one of them becomes moving relatively to another when the external force applied to it does the work to overcome the friction force. Most mechanical energy spent by this tribosystem for overcoming friction transforms into heat. Hence, any tribosystem is dissipative from the physical viewpoint. This well-known experimental fact is illustrated by Figure 1. For example, Kostetskii studied the energy balance in dry friction and found that most friction work transformed into heat [17]. In engineering calculations of friction units it is usually assumed that all this work is converted into heat.

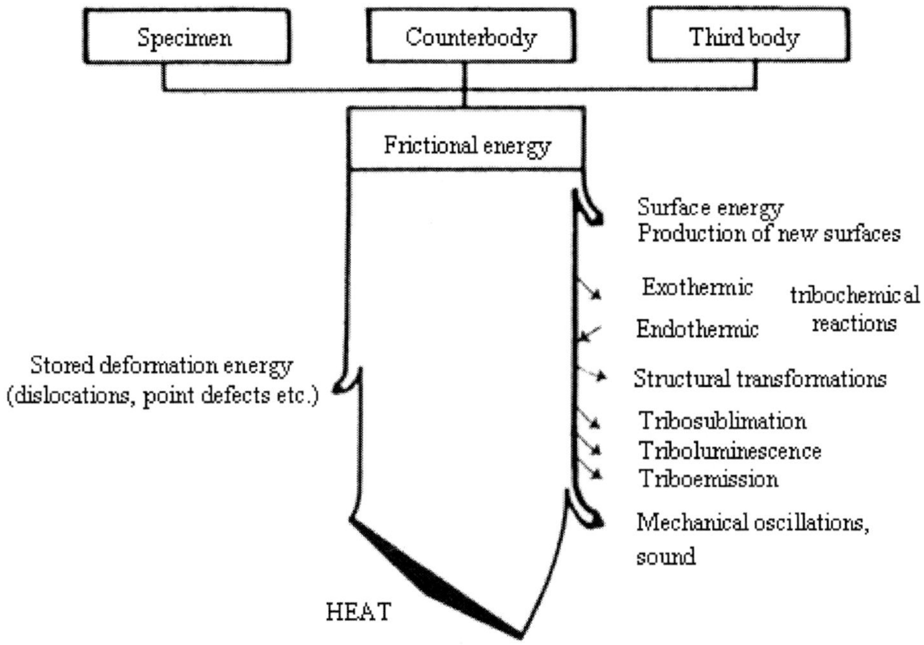

Figure 1. Different channels of friction energy dissipation (according to Uetz [18])

The heat generated in the friction contact is distributed between the mated solids in a certain proportion which is expressed by the heat partition factor α. It is the dimensionless fraction of the total heat flow directed into one of the solids. If the quantity of heat generated by friction over the unit contact area per unit time is q and no heat is removed to the environment then the heat flow directed into the upper solid (Figure 2) is $q_1 = \alpha q$ and the heat flow directed into the lower solid is $q_2 = (1 - \alpha)q$.

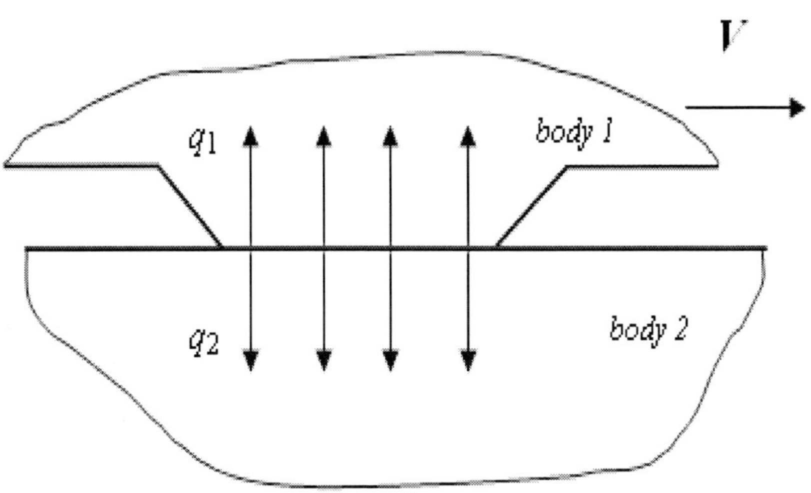

Figure 2. Distribution of heat flows between solids in contact

The term was introduced by Blok [3]. In the first measurements of the temperature on rubbing surfaces [19] Bowden and Ridler used the heat partition factor but did not obtain any formula for it.

Blok determined α from the condition of the equality of the maximal temperatures of the contacting surfaces within the limits of his assumptions which will be discussed below. For a single asperity sliding over a half-space at a low velocity ($v \leq 8a_1/25l$ or $Pe \leq 0.32$) he obtained the following expression:

$$\alpha = \frac{\lambda_1}{\lambda_1 + \lambda_2}$$

where a_1, λ_1 are the thermal diffusivity and thermal conductivity of the asperity material; λ_2 is the thermal conductivity of the half-space (counterbody) material; l is the size of the contact zone in the sliding direction; $\mathrm{Pe} = \frac{vl}{a}$ is the Peclet number which characterizes the proportion between the heat dissipated by convection and conduction; v is the sliding velocity.

At high sliding velocities ($v \geq 8a_1/l$ or $Pe \geq 8$ for rectangular and circle heat sources and $v \geq 40a_1/l$ or $Pe \geq 40$ for a strip source) α is expressed as follows:

$$\alpha = \frac{\lambda_1\sqrt{\pi}}{\lambda_1\sqrt{\pi} + \lambda_2\sqrt{\frac{16}{\mathrm{Pe}}}}.$$

Jaeger considered the heat sources shaped as a square, rectangle, and infinitely wide strip moving at a constant velocity [20]. Unlike Blok, he equated the average temperatures of the moving and stationary surfaces rather than their maximal temperatures. In case of an asperity having the rectangular cross-section and the heat-insulated lateral surface at $Pe > 20$ the heat partition factor is

$$\alpha = \frac{\lambda_1\sqrt{lv}}{1.25\lambda_2\sqrt{a_1} + \lambda_1\sqrt{lv}}.$$

It is seen that with increasing the sliding velocity the heat flow into the asperity proportional to $1 - \alpha$ decreases. It should be borne in mind that the asperity is heated frictionally and repeatedly enters contact with new cold areas of the half-space surface while the latter is heated by both the heat source (the two-dimensional contact area) and the asperity owing to heat conduction. This explains the fact that velocity increase favors heat removal to the half-space.

When convective heat transfer occurs on the lateral surface of the square asperity with a constant heat exchange coefficient σ the following Jaeger's formula is used:

$$\alpha = \frac{\lambda_1 \sqrt{v}}{\lambda_1 \sqrt{v} + 1.504\sqrt{a_1 \lambda_2 \sigma}}.$$

Levitskii calculated the heat partition factor for the contact geometry studied by Bowden when a stationary rod contacted the flat surface of a rotating disc. Replacing the disc with a heat-conductive half-space and assuming the rod cross-section to be rectangular and the heat flow to be linear as well as taking into account heat transfer from the free surface of the stationary body, he obtained the following expression [21]:

$$1 - \alpha = \frac{\sqrt{\pi\sigma}}{\sqrt{\pi\sigma} + \sqrt{\rho c v}}$$

where σ is the heat exchange coefficient; ρ and c are the density and specific heat of the stationary body material, respectively.

Like Blok and Jaeger, Charron proceeded from the condition of equality of the temperatures on the contacting surfaces. For the contact of two semi-infinite bodies or two thin rods with the heat-insulated lateral surfaces he derived the following formula for α [22]:

$$\alpha = \frac{\sqrt{\lambda_1 c_1 \rho_1}}{\sqrt{\lambda_1 c_1 \rho_1} + \sqrt{\lambda_2 c_2 \rho_2}}.$$

It is advisable to apply Charron's formula to friction units with the overlapping factor of about 1 (this term will be explained below) operating under stationary regimes [16, 23, 24]. Neglecting heat removal to the environment, Hasselgruber derived the following expression for non-stationary friction conditions [25]:

$$1 - \alpha = \frac{1}{1 + \frac{b_2 c_2}{b_1 c_1}\sqrt{\frac{a_2}{a_1}}}$$

where b_1 and b_2 are the dimensions of the contacting bodies in the direction normal to the friction surfaces.

Among the examples of non-stationary friction are short-time brakings. For this case Chichinadze et al. calculated α by the formula derived from the condition of equality of the average temperatures over the apparent contact area [14, 26]:

$$\alpha = \left[1 + \frac{h_1 \lambda_2}{h_2 \lambda_1}\left(\frac{\mathrm{Fo}_2 + \frac{1}{3}}{\mathrm{Fo}_1 + \frac{1}{3}}\right)\right]^{-1}$$

where h_1 and h_2 are the thicknesses of the contacting bodies; Fo_1 and Fo_2 are the Fourier numbers ($\mathrm{Fo} = \frac{at}{h^2}$ characterizes correlation between changes in the thermal conditions outside a body and the rate of the corresponding variations in the temperature field inside it; t is the time).

Paper [27] deals with the friction contact of a stationary rod of round cross-section and the cylindrical surface of a thin rotating disc. It is assumed that convective heat transfer to the environment occurs on the flat and cylindrical disc surfaces with a constant heat exchange coefficient. The quantities of heat directed to the rod and disc are determined from the hypothesis on equality of the average temperatures of the rod and disc within the contact area:

$$\alpha = \frac{\gamma}{1+\gamma};$$

$$\gamma = \frac{Q_d}{Q_r} = \frac{0.9\lambda_d \Delta \mathrm{th}\left(\sqrt{2\mathrm{Bi}_r}\,\frac{l}{R}\right)}{\lambda_r \sqrt{\mathrm{Bi}_r}\, T_{av}^*}$$

where Q_d, Q_r are the heat flows directed to the disc and rod, respectively; λ_d, λ_r are the thermal conductivities of the disc and rod, respectively; Δ is the ratio of the half-thickness of the disc to its radius; Bi_r is the Biot number for the rod ($\mathrm{Bi} = \frac{\sigma h}{\lambda}$ determines the accordance of the temperature distribution in a body to the temperature conditions in the environment; σ is the heat exchange coefficient; h is the body thickness); l is the rod length; R is the rod radius; T_{av}^* is the dimensionless averaged temperature on the disc lateral surface.

The authors of [28] studied the distribution of the frictional heat generated in the contact of two orthogonal cylinders or a cylinder and a rod of the finite length. They found that the quantity of heat spent for the heating of the stationary cylinder is about 30% of the quantity of heat entering the rotating cylinder.

It is assumed in the considered theoretical papers that the frictional heat is generated over the idealized interface of the bodies, i.e. the flat heat source model is involved. In this case the temperature maximum is located on the interface and the temperature inside the bodies decreases monotonously as the distance from it increases. This assumption is correct if dimensions of real contact spots (heat sources) are small comparatively to the spacings between them that is typical for the elastic contact of the bodies.

In closing of this paragraph it is pertinent to make two notes one of which concerns the idea to introduce the heat partition factor. As is mentioned in [4], this expedient is correct from the mathematical viewpoint since the frictional heat problem can be solved easily if the distribution of heat is known a priori. However, the expressions for α are obtained under the assumption that the maximal or average temperatures of the mated bodies are equal which is not quite justified physically. Thus, according to Blok's theory, at high sliding velocities the maxima of the temperature on the stationary and moving surfaces are located at different points of the contact area. Hence, it is wrongly to derive the formula for α from the condition of equality of the maximal temperatures in the contact center.

Another note relates to the role of heat transfer to the environment in the frictional heat distribution. A share of the frictional heat is dissipated through convection and this fact is taken into account in some expressions for α. The heat flow to the environment depends on the design features of a friction unit, particularly, on the ratio of the friction surface areas of the mated members which was called by Chichinadze the overlapping factor [29]:

$$K = \frac{A_{a_1}}{A_{a_2}}$$

where A_{a_1} and A_{a_2} are the apparent contact areas of the friction members.

Depending on the friction unit design K varies from 0 to 1. Two limiting cases (the end friction of two cylinder specimens and the friction of a thin rod over the flat surface of a disc) are shown in Figure 3.

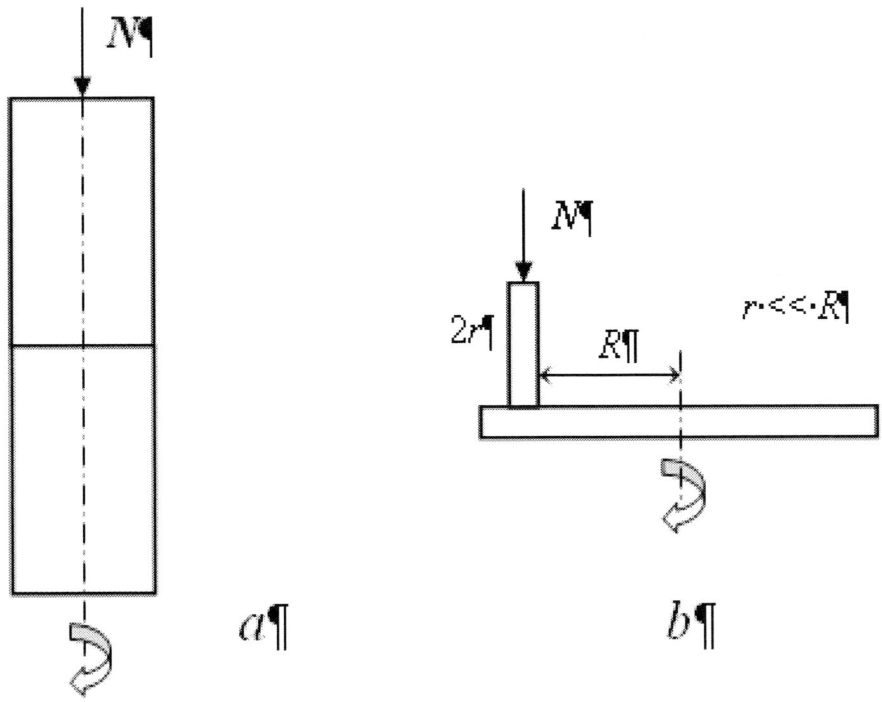

Figure 3. Contact of two bodies with different overlapping factors:
a – friction of cylinder ends; b – pin-on-disc contact

For the pin-on-disc contact (Figure 3, b) the overlapping factor is the ratio of the pin friction surface area to the area of the friction track on the disc surface:

$$K = r^2 / \left[(R + 2r)^2 - R^2 \right] = r/4(R + r).$$

2.2. Basic Models of Frictional Heating

The notions of the following three temperatures are used in thermotribology: the bulk temperature of a body T_b, the average surface temperature T_s, and the flash temperature T_f. The bulk temperature is the temperature averaged over the certain volume of one of the friction members, the average surface temperature is the temperature averaged over a thin surface layer of the body and the flash temperature is the local increment of the temperature in the contact of microasperities of two rubbing bodies.

The notion of the flash temperature reflects the fact that the friction contact is discrete. Since the real contact area is by three-four orders of magnitude less than the apparent contact area the heat generated in the friction zone is concentrated over small contact spots whose life is from a few milliseconds to hundredths of microseconds. Because of this the density of the heat flow over these spots can be very high resulting in a short-term local temperature rise reaching hundreds degrees.

Chichinadze proposed to consider the maximal temperature of the friction surface as the sum of the bulk, surface, and flash temperatures [14]:

$$T_{\max} = T_b + T_s + T_f .$$

In our opinion, this hypothesis is disputable. The matter is that the temperature is the measure of body heating, i.e. the response of the body to heat effect. All the temperatures in the sum are the results of the action of one and the same heat source, namely frictional heating. They correspond to different scale levels of the body structure at which frictional heating occurs. Concerning measurements of the friction temperature we note that the design and capabilities of the measuring instruments (e.g. the focal length of the optical system when using optical-electron methods) govern the area of localization and the concentration of heat. Which of the three mentioned temperatures is measured depends on the specific features of the experimental technique used.

Model representations are widely used when solving friction heat problems since it is difficult to obtain a solution of the heat conduction equation with the initial and boundary conditions corresponding to the real contact geometry, conditions of heat transfer to the environment, changes in the thermal and physical-mechanical characteristics of the mated materials etc. The use of assumptions is justified from the viewpoint of simplifying the procedure of solving the heat problem yet it poses some restrictions on the applicability of the models of frictional heating.

<u>Bowden's Model.</u> Bowden and Ridler were the first who determined the temperature in the friction zone involving the pin-on-disc geometry [19]. To solve the problem of finding the interface temperature they used the following assumptions:

1) both bodies had the smooth friction surfaces, i.e. contact occurred over the whole apparent area;
2) the pin was a semi-infinite rod with the one-dimensional temperature field within it;
3) according to Amonton's law, the friction coefficient did not depend on the load and sliding velocity;
4) the heat flow was stationary;

5) heat exchange with the environment occurred over the rod lateral surface obeying Newton's law with the known heat exchange coefficient σ.

For a round rod with the radius r Bowden obtained the following expression for the temperature rise on the friction surface:

$$T - T_0 = \frac{(1-\alpha)fNv}{\pi r\sqrt{2\sigma\lambda r}}$$

where T_0 = const is the ambient temperature; f is the friction coefficient; N is the normal load; v is the sliding velocity; λ is the thermal conductivity of the rod material.

Initially Bowden and Ridler carried out the experiment to measure the temperature of the friction surface by a natural thermocouple and then Bowden solved the heat problem formulated above. The linear dependence of the temperature on the sliding velocity and load agreed well with the experimental results for both dry and lubricated contacts. As for the dependence of the temperature on the thermal conductivity, it conformed poorer the experimental data. The reason was that Bowden assumed the rubbing surfaces to be perfectly smooth [30]. Moreover, in Bowden's formulation the heat problem was treated as a static one (the heat source was stationary) while real heat source moved. Therefore, this model is applicable to estimating the average surface temperature but unusable for determining local temperature increments or flash temperatures. However, the work of Bowden and Ridler was the first step in thermotribology and stimulated further research in this field.

Blok's Model. Blok proposed the first theory of the flash temperature which, unlike Bowden's model, was suitable for calculating the temperature of local heat sources within the real contact area and considered the motion of the sources. Despite its disadvantages, this theory has been used for several decades by tribologists who, in essence, have not overpassed its limits. Blok's fundamental papers [3, 31–34] have become classical.

Blok formulated the basic statements of its model as follows:

1) the friction surfaces are clean, i.e. free from any film separating them;
2) the size of a single contact spot is small compared to the dimensions of the mated bodies that allows one to consider them as infinitely large;
3) the heat generated in friction is absorbed completely by the bodies and no heat transfer to the environment occurs;
4) friction obeys Amonton's law and the friction coefficient is known. The distribution of the contact pressure is also known;
5) the specific heats of the bodies are unrestrictedly great.

The main results of Block's theory are the following. Initially he overlooked the heat source nature and considered sources of various shapes (strip, circular, and rectangular). He also assumed that the source intensity, i.e. the quantity of the heat generated by the unit area of the source per unit time, is distributed similarly to the contact pressure. The one-dimensional heat problem with the heat flow directed normally to the source was solved. The solutions obtained for both stationary and moving sources were extended to the case of frictional heating when the heat source was represented by a single thermally-insulated contact spot.

Blok's results were analyzed in detail and systematized by Schedrov [30], therefore, we give here the solutions of the heat problem only for the basic cases.

1. The asperity (body 2) contacting the half-space (body 1) does not move relatively to it. The contact spot is circular and the heat source intensity q is distributed uniformly over its surface. In practice this case corresponds to the rotation of a round vertical journal (cylinder end) contacting a plane. The flash temperature on the contact spot is calculated as follows:

$$T_f = \frac{qr}{\lambda_1 + \lambda_2}$$

where $q = fNv/\pi r^2$ is the density of the heat flow on the contact spot with the radius r.

2. Contact conditions are the same as in the above case but the source has the parabolic distribution of the intensity (the contact of a rotating sphere and a plane). The heat flow density is

$$q = q_0\left(1 - \frac{x^2}{r^2}\right)$$

where $q_0 = fp_0v$ is the intensity in the source center; p_0 is the pressure in the source center; x is the distance from the source center to the point at which the temperature is found.

The temperature in the center of the contact spot is maximal; it is determined as follows:

$$T_f = \frac{2}{3}\frac{q_0 r}{\lambda_1 + \lambda_2}.$$

3. The asperity contacting the half-space forms the circular contact spot and slides over the half-space at a velocity satisfying the condition $v \le \frac{4a_2}{25r}$ (low velocity). In this case the flash temperature can be calculated by the above formula for the stationary heat source.

4. The asperity contacting the half-space forms the circular contact spot and slides over the half-space at a velocity satisfying the condition $v \ge \frac{4a_2}{r}$ (high velocity). The temperature rise on the asperity surface in its center is

$$T_{f_2} = \frac{qr\psi}{\sqrt{2}\left(\lambda_1\psi + \sqrt{\pi}\lambda_2\right)}$$

and that on the half-space surface in the contact spot center is

$$T_{f_1} = \frac{qr\psi}{\lambda_1\psi + \sqrt{\pi}\lambda_2}$$

where $\psi = \sqrt{\frac{4a_2}{vr}}$ is the parameter depending on the sliding velocity and the size of the heat source.

According to Blok, the maximal contact temperature is the arithmetic mean of the temperatures T_{f_1} and T_{f_2}:

$$T_{\max} = \frac{1+\sqrt{2}}{2} \frac{qr\psi}{\lambda_1\psi + \sqrt{\pi}\lambda_2}.$$

5. The circular heat source slides over the surface of the half-space and the thermal conductivity of one of the bodies is negligible compared to that of the counterbody. If the source intensity is distributed uniformly and the sliding velocity is low the following formula is used:

$$T_f = \frac{2qr}{\sqrt{\pi}\lambda}.$$

At a high sliding velocity the maximal temperature is reached at the trailing edge of the contact spot:

$$T_f = \sqrt{2}\frac{qr\psi}{\sqrt{\pi}\lambda}.$$

In case of a rectangular heat source having the uniformly distributed intensity and sliding over the half-space surface the above formulas for the circular source are applicable with r replaced by the contact spot half-width which is half of the square side length.

The analysis of the expressions for the flash temperature obtained by Blok yields the temperature distribution over the section of the contact spot parallel to the sliding direction. In particular, under the assumption of the uniformly distributed source intensity q the temperature distributions for the stationary and moving sources are as shown in Figure 4.

If the heat source is stationary or moves with a low velocity the temperature distribution is symmetrical relatively to its center (Figure 4, curve 1). With increasing the sliding velocity the temperature maximum shifts in the direction opposite to the velocity. At high velocities the maximal temperature point is located at the trailing edge of the contact spot (curve 2).

Like other models of frictional heating, Blok's model idealizes the processes of heat generation and propagation in friction contact since it is based on the assumptions wide of reality (the juvenile state of the contacting surfaces, the absence of the mutual effect of single heat sources, the a priori knowledge of the heat source distribution, and the absence of convective heat transfer to the environment). In particular, the assumption of small sizes of heat sources is true, generally speaking, for the elastic contact of the bodies. Under heavy pressures saturated plastic contact occurs and the mutual effect of spots emitting heat energy becomes more significant, therefore, calculations by Blok's formulas yield undervalued flash temperatures.

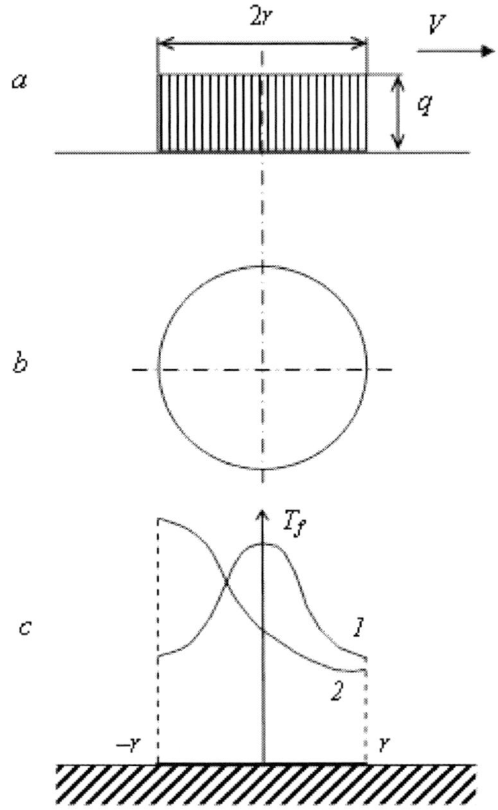

Figure 4. Temperature distribution over single contact spot along sliding direction: a, b – plane circular heat source (single contact spot); c – temperature distribution along source length (1 – distribution symmetrical relatively contact spot center for stationary source; 2 – distribution with maximum at trailing edge of contact spot for source moving with high velocity)

Blok postulated the law of the distribution of the intensity of a heat source over its area. However, in reality this distribution is unknown; it depends on the actual shape of contact spots and the law of the distribution of the real contact area [30].

Despite the artificiality of Blok's hypotheses and not very good agreement between the calculation results and experimental data these formulas provide the satisfactory qualitative estimate of the flash temperature.

Jaeger's Model. In essence, Jaeger's paper [20] does not change cardinally the basic postulates of Blok's theory and can be considered as its evolution. Like Blok, Jaeger assumed the contact surface to be flat but he treated the heat problem as a spatial one and found the temperature distribution for the rectangular and strip sources of a constant intensity both stationary and moving over the half-space surface at a constant velocity. This approach was superior to Blok's model since it allowed Jaeger to determine the temperature distribution not only at high and low but also at intermediate Peclet numbers with account for the mutual effect of single heat sources inside the contact area and even to find the temperature distribution under the half-space boundary in the vicinity of a source.

In addition, Jaeger solved the plane heat problem for a moving source with different distributions of its intensity (the intensity depended on the coordinates of a point inside the source). The maximal and average temperatures of the contact surface turned out to depend weakly on the mode of the source intensity distribution.

Using the obtained solutions Jaeger considered two cases of contact with heating important for tribology. In the first case an asperity on the surface of a hard half-space (body 2) slides over the surface of another hard half-space (body 1) forming the square contact area with the side 2l. The second case corresponds to the sliding of an infinite rod with the square cross-section (body 2) over the surface of a stationary half-space (body 1) when heat transfer to the environment with the coefficient of heat transfer σ occurs over the rod lateral surface.

The difference between Jaeger's and Blok's solutions is that Jaeger determined the heat partition factor α under the assumption that the average temperatures of the surfaces equaled rather than the maximal temperatures. The choice of this condition for finding α was explained by the fact that the average temperature was easier to calculate than the maximal temperature; in addition, it could be measured in the experiment.

For the first case the average contact temperature is

$$T_{av} = \begin{cases} 0.946\dfrac{ql}{\lambda_1 + \lambda_2} & \text{for low sliding velocities;} \\[2ex] \dfrac{1.064ql\sqrt{a_1}}{1.25\lambda_2\sqrt{a_1} + \lambda_1\sqrt{lv}} & \text{for high sliding velocities.} \end{cases}$$

For the second case the average contact temperature is

$$T_{av} = \begin{cases} \dfrac{0.946ql}{\lambda_1 + 1.338\sqrt{l\lambda_2\sigma}} & \text{for low sliding velocities;} \\[2ex] \dfrac{1.064q\sqrt{a_1 l}}{\lambda_1\sqrt{v} + 1.504\sqrt{a_1\lambda_2\sigma}} & \text{for high sliding velocities.} \end{cases}$$

The important advantage of Jaeger's model over Blok's one is that Jaeger solved the problem of temperature distribution in thickness of the surface layer of one of the bodies. This problem is of crucial significance in thermotribology. The dependence of the temperature on the depth under the friction surface is represented by the monotonously descending curve and the temperature decreases quite rapidly, i.e. the heat propagates over a shallow depth. The non-uniform heating of the thin surface layer of the body induces a high temperature gradient along the normal to the friction surface and because of this thermal stresses arising in the body can be very high. The thermal and thermal-stressed states of the surface layer can be controlled by varying its thermal characteristics using, for example, coatings with different conductivities.

Archard's Model. Archard proposed the model of the flash temperature simpler from the mathematical viewpoint [35]. He solved the friction heat problem practically in the same formulation as Blok and Jaeger, i.e. his assumptions were as follows: an asperity on the surface of one body slid over the counterface; the contact spot was treated as a flat circular heat source with the one-dimensional heat flows to the bodies; the heat partition factor was determined from the condition that the average temperatures of the mated surfaces equaled within the contact area; no heat was removed to the environment.

Archard considered the heat flow as the thermal current passing the area with a thermal resistance [36, 37] and used the electrical analogy. For the moving asperity the heat source is stationary and the average temperature within the source area is

$$T_{f_2} = \frac{Q_2}{4r\lambda_2}$$

where Q_2 is the heat flow directed to the asperity; r is the contact spot radius.

This formula is also true for the moving heat source (body 1) at low sliding velocities (Pe < 0.4) while for high velocities (Pe > 20) the flash temperature is determined as follows:

$$T_{f_1} = \frac{0.31Q_1}{\lambda_1 r}\sqrt{\frac{a_1}{vr}}$$

where Q_1 is the heat flow directed to the half-space.

The temperature at the interface is found from the condition

$$\frac{1}{T_{\max}} = \frac{1}{T_{f_1}} + \frac{1}{T_{f_2}}.$$

Archard also considered the contact area whose size depended on the normal load and proposed the approximate theory of temperatures it elastic and plastic contacts. He supposed that both bodies had the same mechanical and thermal properties and that at low sliding velocities all the frictional heat was evenly divided between them while at high velocities all the heat was absorbed by body 1 or half-space. Under these assumptions the contact spot temperature depends on the load, velocity, and properties of the materials in the following manner:

plastic deformation at low sliding velocities (Pe < 0.4)

$$T_f = \frac{f(\pi\sigma_y)^{\frac{1}{2}}}{8\lambda} N^{\frac{1}{2}} v$$

where f is the friction coefficient; σ_y is the material yield point;

plastic deformation at high sliding velocities (Pe > 20)

$$T_f = \frac{f}{3.25} \frac{(\pi \sigma_y)^{\frac{3}{4}}}{(\lambda \rho c)^{\frac{1}{2}}} N^{\frac{1}{4}} v^{\frac{1}{2}};$$

elastic deformation at low sliding velocities (Pe < 0.4)

$$T_f = \frac{f}{8.8\lambda} \left(\frac{E}{R} \right)^{\frac{1}{3}} N^{\frac{2}{3}} v$$

where E is the elastic modulus of the material; R is the curvature radius of the non-deformed spherical asperity (body 2);

elastic deformation at high sliding velocities (Pe > 20)

$$T_f = \frac{f}{3.8} \left(\frac{E}{\lambda \rho c R} \right)^{\frac{1}{2}} N^{\frac{1}{2}} v^{\frac{1}{2}}.$$

Model of Kuhlmann-Wilsdorf. The above models consider the heat sources of regular shapes. However, as the experimental studies have shown, real heat sources are mainly elliptical with the major axes parallel to the sliding direction [38–40]. Kuhlmann-Wilsdorf modified Blok's and Jaeger's models to calculate the flash temperature on elliptical contact spots [41–43]. The flash temperature for a plane elliptical source located on the interface is determined as follows:

$$T_f = \frac{qR'}{\dfrac{\lambda_1}{\Phi_1(v_1)\Phi_2(e, v_1)} + \dfrac{\lambda_2}{\Phi_1(v_2)\Phi_2(e, v_2)}}$$

where R′ is the characteristic contact spot size, $R' = \sqrt{\overline{A}/\pi}$; $\overline{A}$ is the average contact spot area; e is the ratio of the major axis of the contact spot parallel to the sliding velocity to the minor axis (ellipticity); Φ_1 is the function describing the velocity dependence of the temperature and varying from 0.1 to 2; Φ_2 is the function describing the dependence of the temperature on the spot ellipticity; v_1 and v_2 are the relative velocities of the first and second bodies, respectively, $v_{1,2} = \dfrac{\rho_{1,2} c_{1,2} v R}{\lambda_{1,2}}$; v is the sliding velocity.

At low velocities the flash temperature is maximal when the spot is slightly elongated in the velocity direction. If the sliding velocity is high (Pe varies from 1 to 10) the spots with the ellipticity e = 4…10 have the highest temperature. The temperature of the spots highly elongated in the sliding direction (e > 10) or the spots elongated in the perpendicular direction is lower than that of circular spots.

2.3. Other Theoretical Studies of Frictional Heating

It should be noted that Blok's flash temperature concept had been proposed when studying seizure in cylindrical involute gearings. However, this model considered thermal processes occurring on a single contact spot and difficulties arose in transition from the heat problem for a local spot to problems for full-sized friction units. Among them are the necessity to take into account the fact that the heat flow densities in real contact are variable and the proper selection of boundary conditions. The extension of the friction heat problem in Blok's formulation to real bodies is considered in detail in reviews [44–46]. In this connection we should note the results of Prof. A. Chichinadze and his disciples who studied thermal processes in clutches, brakes, and other friction units and developed frictional materials and methods of their testing [14, 47–50]. They obtained the fundamental dependences of the friction coefficient and wear rate on the temperature characteristics of contact.

The studies of Korovchinskii also merit notice here [51, 52]. In these papers he showed practically all basic achievements of research in the cutting and frictional heat carried out by the leading scientists in different countries. The specific examples of the solutions given by Korovchinskii embrace various shapes of moving contact spots (square, ellipse, etc.) and different plane and spatial (rectangle, oblate spheroid, etc.) surfaces of heat sources with constant and variable heat flow densities. His results formed a basis for solving the problems of heat distribution in specific full-sized friction units.

Papers [45, 46] summarize the results of the application of modified Blok's model to solving the heat problems for full-sized cylindrical involute gearings. The use of this model has allowed the authors to determine the location of the local wear zones or the "weakest" zones that was important for tribotechnology to perform the local heat treatment of machine parts, particularly, large gears in order to improve their wear resistance and life.

In essence, the studies of Blok and Jaeger were based on the heat source methods which were also applied by other tribologists. For example, Barber modified Jaeger's single-source solution to account for multiple heat sources [53]. Marscher reported his computational study of transient surface temperatures induced by multiple interacting heat sources and considered both one-dimensional and three-dimensional heat flows from the heat sources [54].

We emphasize again the significant restriction of the heat source methods, i.e. their applicability only to semi-infinite bodies, which makes it difficult to apply them to bodies of finite dimensions. This stimulated the development of integral transform techniques which involved numerical calculation methods because such complex problems could not be solved analytically. Most of these results were obtained by Ling and his coworkers and are reviewed in [55]. Particularly, he proposed the stochastic model in which a finite number of small contact spots were defined on the apparent contact area. The spots changed their positions in a random manner. The temperatures on the spots were found to exceed much the average surface temperature [56] and the surface temperature distributions were in good qualitative agreement with the experimental data. Another example of using the transform techniques are the studies of Floquet and his coworkers who extended their application to some three-dimensional geometries [57] and bearings [58].

Instead of using the heat partition factor most studies based on the integral transform methods involve the technique that guarantees the continuity of the temperature across the interface throughout the real contact area [59]. Another approach was proposed by Ryhming

[60] who supposed that all frictional heat was generated within the third body layer separating the two sliding bodies. The integral equations describing the heat flow within the third body were coupled with the equations for the sliding bodies by requiring the continuity of the temperature at the interfaces between the third body and the sliding bodies.

Progress in numerical methods has promoted surface temperature analysis, especially for sliding bodies of complex shape. Finite element methods have been used to model temperature distributions in both stationary and moving members of friction pairs in transient and quasi-steady state situations [61–63]. The technique is applicable to single and multiple contacts and interactions between the contacts can easily be taken into account. Moreover, it yields temperature fields in both sliding members in the vicinity of the contact area which can be used in the finite element analysis of stress distributions.

The finite element method was used to study the influence of subsurface heat generation on predicted temperature distributions [61]. It was shown that subsurface heating could produce temperature gradients different from those produced if all the heat was generated on the interface. Indeed, the member being stationary with respect to the heat source can experience a subsurface temperature exceeding the surface temperature near the contact leading edge [61]. Subsurface temperature peaks were shown to be feasible by Rozeanu and Pnueli [64, 65] and have been confirmed experimentally by Balakin [66]. An experimental evidence of subsurface heating will be also considered in Paragraph 3.2.

3. Experimental Studies in Thermotribology

3.1. Friction Temperature Measurements

The realization of temperature measurements at friction and machining of solids is a complex engineering task because of several factors. They include small dimensions of the areas on which thermal processes occur, the nonuniform temperature distribution in depth of the contacting solids and over their surfaces, a very short life (of the order of a few milliseconds) and the random in time and space appearance of heat sources [67]. The need for tribologists to overcome these difficulties has led to the development of a large number of methods for recording temperatures in the friction zone and corresponding instruments. A quite general but very arbitrary classification identifies two groups of the methods, i.e. the indirect and direct temperature measurement methods. Their capabilities and drawbacks are analyzed in detail in authors' paper [68]; here we briefly describe them.

Indirect Temperature Measurement Methods. The temperature in the friction zone can be determined, for example, by comparing the colors of heated portions of the specimen surface and a reference surface or by using melting temperature indicators [69, 70]. The fist method involves the substances sensitive to temperature changes which are introduced into one of the rubbing bodies as inserts or are applied to the friction surfaces. It is suitable only for the rough estimate of the temperature. The method of low-melting inserts is capable of providing information on the maximal temperature in the friction zone by using the substances with differing and a priori known melting temperatures.

The method based on the phenomenon of thermoelectron emission has a higher accuracy and a quite short response time [71, 72]. The number of the electrons with the energy that is sufficient for their escape from the metal increases as the temperature elevates. The electron stream from the surface is recorded by a photoelectron multiplier installed in one of the bodies at some distance under the friction surface. The surface temperature is determined from the electron stream density. The method is applicable only for the friction of metals and is capable of recording only the total energy from the apparent contact area. Its use is most appropriate for measuring the average surface temperature.

One of the experimental methods is based on the study of structural transformations in the surface layer material and serves to determine the maximal temperature in the friction zone [73]. Knowing the temperatures of polymorphic transformations of the metal and using X-ray diffraction to identify the type of structural changes in the surface layer caused by friction one can estimate the temperature reached on real contact spots. This method is incapable of determining the temperature which is below the lowest transformation temperature of the metal structure and is applicable only for a very narrow range of materials. The measurement accuracy depends on the interval between the neighboring transformation points.

The common drawback of the indirect methods is their low measurement accuracy.

Direct Temperature Measurement Methods. The direct methods of measuring friction temperatures are more widely applied in tribology. Among such instruments are thermocouples of various designs, i.e. artificial, natural, and semi-artificial. They can be used effectively to evaluate the average surface temperature [74, 75], however, the results must be considered to be approximate since the temperature may differ from the temperature on real contact spots by an order of magnitude. The reason is that the size of the thermocouple hot junction may exceed significantly the material volume within which the heat pulse energy is concentrated. A high temperature gradient along the normal to the friction surface makes it impossible to measure accurately the temperature on the surface if the thermocouple is installed in the subsurface layer.

The method of natural thermocouples is based on the fact that the bodies in contact (metals and alloys) may induce a thermo-emf when the contact zone is heated by friction; the bodies serve as the thermocouple electrodes. The temperature in the friction zone is determined from the recorded potential difference. The drawbacks of this technique are as follows: a low measurement accuracy, problems with calibration, the restricted field of application (metals and alloys only), the possibility of measuring only the average surface temperature, the dependence of measuring results on the real contact area and properties of the films covering the friction surfaces.

Artificial thermocouples are installed in the bulk of the rubbing bodies [76, 77]. The junction of two wires of different metals is located in the subsurface layer. Artificial thermocouples are also used to measure the cutting temperature. For example, paper [78] describes the application of chromel-alumel microthermocouples to measure the temperature in cutting of natural diamond crystals (Figure 5). Precalibrated microthermocouples 2 are held by epoxy glue layer 5 between the two halves of crystal 1 and then the entire system is installed in a diamond-cutting machine. The measurement error does not exceed 1.5%.

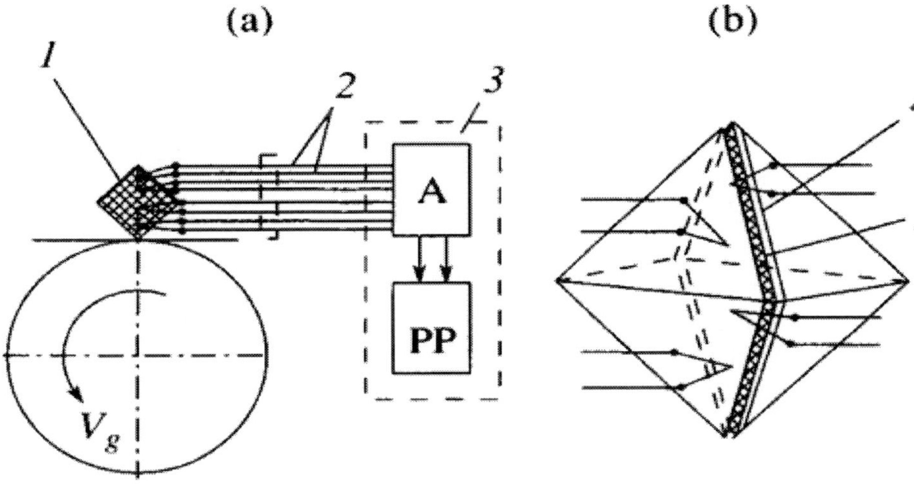

Figure 5. Schematic of measuring average temperature (a) in crystal cleavage zone (b): 1 – crystal; 2 – microthermocouples; 3 – recording device; 4 – cleavage plane; 5 – adhesive layer

Similar thermocouples were used to determine the temperature in the zones of fastening of corundum jewelry workpieces to holders and in the zones of their contact with the polishing disc [79]. To ensure reliable thermal contact of the thermocouple junction slots were cut to the same depth in the specimen. The slots were filled with the mixture of diamond micropowder and liquid glass (Figure 6). Figure 7 shows the schematic of measuring the temperature in grinding of silica ceramics by the artificial dual-electrode chromel-alumel thermocouple.

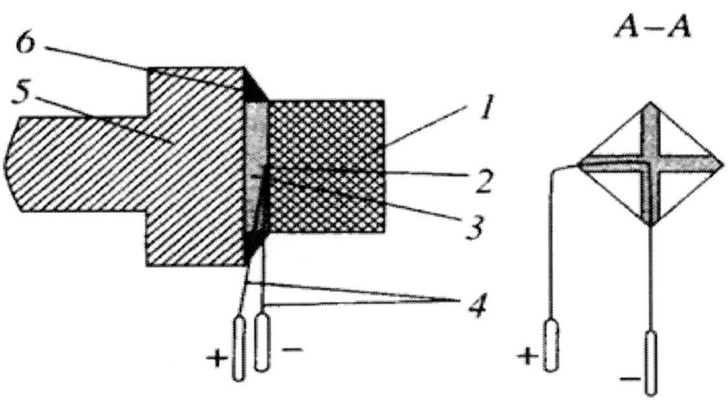

Figure 6. Schematic of thermocouple installation in corundum specimen: 1 – specimen; 2 – thermocouple junction; 3 – diamond powder with liquid glass; 4 – thermocouple; 5 – holder; 6 – mastic

In the sliding thermocouple method one electrode of the thermocouple is inserted into one of the rubbing bodies, the counterbody acts as the other electrode. Frictional heating induces the potential difference between the electrode and counterbody. The greater the potential difference the higher is the friction temperature. The method is applicable for metal–metal or metal–polymer pairs.

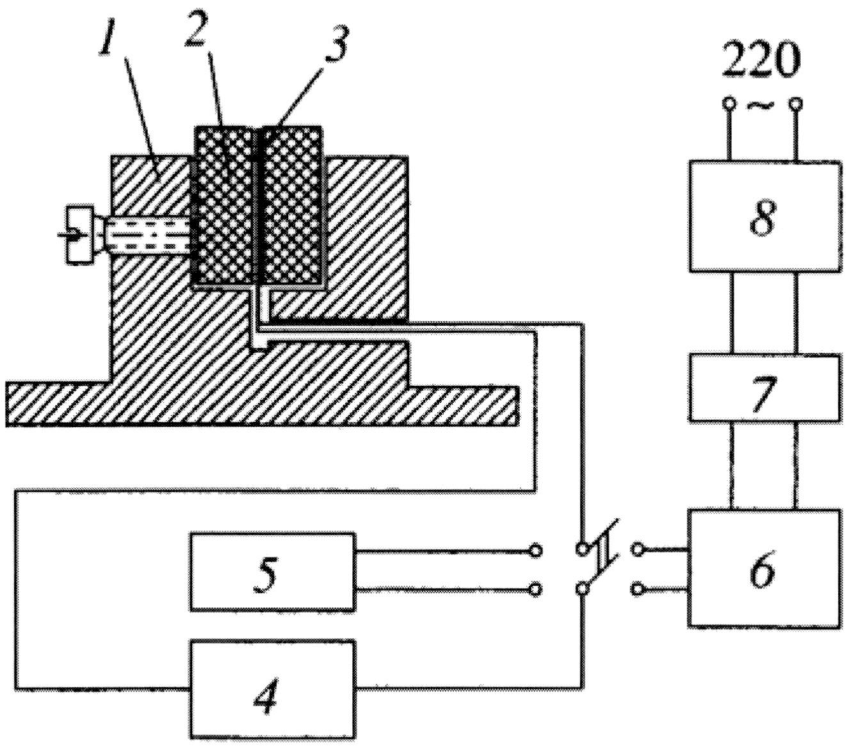

Figure 7. Schematic of measuring temperature in ceramics grinding: 1 – clamp; 2 – ceramic specimen; 3 – thermocouple; 4 – resistance box; 5 – ohmmeter; 6 – oscillograph; 7 – oscillograph power supply; 8 – voltage stabilizer

The drawbacks of the technique are the following: problems with calibration, the effect of friction-induced changes in the structure of the electrode and counterbody on the measurement results, the dependence of the results on the composition of films in the friction zone.

Semi-artificial thermocouples are also applied in which the current-conducting specimen serves as one element while the other element consists of two wires separated by a very thin insulating layer [80–82]. The main drawbacks of the method are the galling of the metal leading to shorting of the electrodes and the formation of a film of the non-conducting material on the counterface and the electrode ends.

The method of combined thermocouples is used to register fast-running thermal processes in the friction zone. Such thermocouples consist of two thermoelectrodes insulated one from another. The junction is formed in friction owing to the galling of the material of one electrode to the end of another electrode. The shorting of the electrodes is also possible by the metal counterface. This method has the same drawbacks as the method of sliding thermocouples.

Various designs of thermocouples and examples of their use to measure the temperature in brakes of railcars, subway trains, airplanes, and rocket launcher bogies are described in detail in review [83].

In the thermal resistance method the electrical resistance of a selected area of the friction surface is measured by applying the voltage to it and recording the resulting current. Changes in the current correspond to variations of the temperature in the area as the counterbody slides over it.

The above direct measurement methods yield averaged values of the temperature in the friction zone and are incapable of providing data on the geometry of local heat sources or the flash temperature, except for the artificial thermocouple method. In addition, they have a low measurement accuracy.

The pyrometric method is based on relationship between the temperature, the emissivity of the body, and the density of electric charges on the surface of the crystal receiving radiation [84]. Optical fibers are installed in one of the mated bodies to eliminate the influence of the mechanical load on the crystal but this introduces an error to the measurement results. The determination of the temperature under light friction regimes requires the use of additional measurement methods or the development of a detector for each specific case [85, 86]. Of interest is the study which has shown the possibility of measuring the distribution of the friction temperature in the bulk of the rubbing bodies (two cylinders whose ends were placed in contact) [87]. The temperature was estimated by the color on the display. In this study some important results were obtained, i.e. the nearly exponential variation of the temperature with increasing the depth under the friction surface, changes in the temperature distribution caused by the presence of a local heat source (wear particle) in the friction zone, and the dependence of the temperature distribution pattern on the load and velocity conditions. The advantage of the method is the possibility of observing fast-running thermal processes and measuring the temperature directly in the friction zone.

Various pyrometer types are used in tribological studies. For example, in vanishing-filament pyrometers a reference filament heated by electric current is placed in the visual field of the instrument which is focused on the study object. When the brightness temperatures of the filament and the object surface become the same the image of the filament disappears. The temperature of the filament is corrected with account for the body surface emissivity and the result is taken as the real temperature. The pyrometer can be used to measure the bulk temperature of the rubbing bodies while the rapidly varying surface temperature and the flash temperature on local contact spots cannot be recorded because of a long response time of the instrument.

The basic component of the pyrometer of another type, i.e. the pyrometric detector, is a photoelectric crystal plate. The voltage arising between the faces of a crystal as heat radiation from the contact zone falls on its surface is the measure of the friction temperature.

Pyrometers are generally used in tribology to study thermal processes in the contact of the bodies one of which is transparent for heat radiation. If the rubbing bodies are not transparent an optical fiber is installed in a hole made in one of them. Pyrometers record the total energy emitted by the friction surface in the instrument visual field. Therefore, the data on the number, size, and shape of local heat sources are necessary for the calculation of the flash temperature.

Despite the fact that many pyrometers have been developed and are produced commercially at the present time, the problem of creating the pyrometer detector has not yet been solved.

Among the instruments for direct friction temperature measurements are photocells and photomultipliers. Their operation is based on external photoeffect. The direct measurement of the temperature in the friction zone with the use of photocells was carried out by Bowden and Ridler [19]. The heat radiation generated in the zone of contact between a metal pin and a rotating glass disc passed through the disc and was recorded by a photocell installed from the back side of the disc on the pin axis. The error in this experiment resulted from the fact that the photocell recorded the energy of all temperature flashes which appeared simultaneously or nearly simultaneously on the emitting area. Therefore, for more accurate determination of the flash temperature it was important to register the heat radiation from the smallest possible contact area.

The capabilities for the direct measurement of the temperature in the contact of a metal with a transparent material expand significantly with the use of IR detectors. The primary advantage of these instruments is their short response time. They also have a high sensitivity and are capable of recording radiation from bodies with a low temperature. They contain a semiconductor detector whose conductivity varies depending on the intensity of the absorbed radiation. In doped semiconductors the heat radiation leads to the growth of the energy of electrons and their transition from the valence zone to the conductivity zone. The maximal sensitivity of the detector is reached when the radiation energy is equal to the width of the forbidden zone. In extrinsic IR detectors the radiation increases the concentration of charge carriers in the transition layer and the back current. The current strength in the circuit is the measure of the temperature.

The authors of [88] describe the application of an IR microscope and a sapphire lens acting as the counterbody to record the radiation from a spot of about 120 μm in diameter. However, the instrument did not provide the data on the geometry of heat sources. The authors used an optical scanning device which converted the heat radiation to an optical image to register the heat emitted by the area of $>$ 1.2 mm in diameter. The application of an IR microscope to measure the temperature in the contact of a magnetic tape and a recording head is described in [89]. The Barnes RM2A instrument is capable of recording heat radiation from small sources and has a short response time. It has limitations for the spot size and temperature resolution. The minimal spot size is 124 and 86 μm, respectively, for the objectives with the magnification 15 and 36×. When using the objective lens with the magnification 15× the temperature resolution is 0.5 °C for the DC mode and 9 °C for the AC mode. The greater the magnification of the lens the poorer is the temperature resolution. This technique yielded the linear dependence of the temperature on the contact load.

The authors of [90] developed an IR microscope and used it to study thermal processes in the contact of polymer composite and sapphire. It was integrated in a pin-on-disc tribometer (Figure 8). The heat radiation from the contact zone entered the instrument visual field and was compared with the ambient radiation level. The temperature was calculated from the radiation signal with account for the emissivity of the rubbing materials.

A multi-element IR sensor made of indium antimonide was used by the authors of [91] as the heat radiation detector whose signal was visualized by a quick-response thermal monitor. The radiation was transmitted from the source to the sensor by a fiber optic system installed in a hole in the ferrite specimen so that the diamond grain lost contact with the specimen at the instant it began to move beneath the hole. Thus, the measurement error was minimized since immediately after leaving contact the heated diamond grain entered the sensor visual field. Placing the sensors along the trajectory of the grain the authors obtained the dependence of its surface temperature on the time after contact with ferrite.

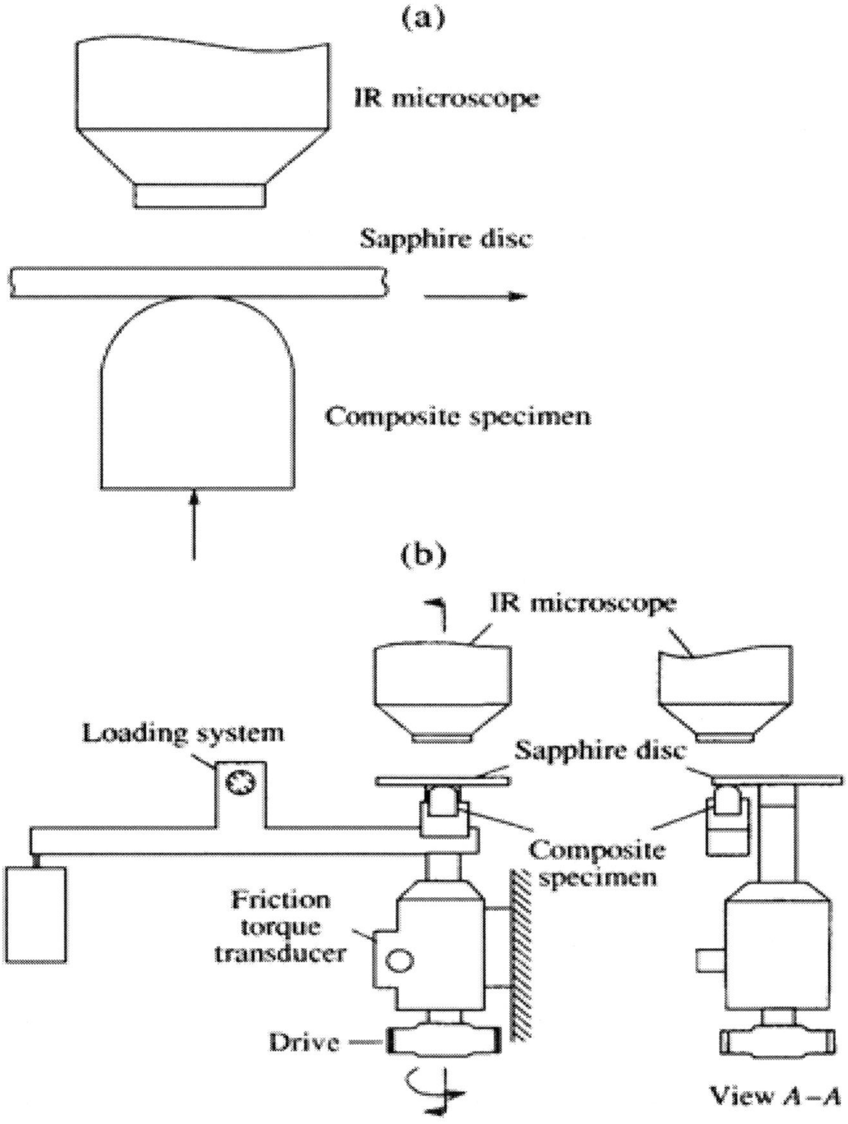

Figure 8. Contact geometry (a) and schematic of tribometer with IR microscope (b)

The dimensions of contact spots (heat sources) were evaluated when studying the friction of sapphire against steel using the photographic method. It was implemented on the basis of a photo camera with the exposure time from 4 ms to 1.5 s [38, 92]. The temperature of contact spots was determined by comparing their color with the color of the heated reference surface. The technique made it possible to obtain data on the shape and size of the contact spots, their distribution over the contact area, and the maximal temperature. However, the authors noted that the accuracy of temperature evaluation depended on the exposure time and the quality of the photos.

Short-response photoelectric pyrometers were used in studies [93, 94] to record temperature flashes lasting a few microseconds. This method is capable of evaluating the size of a contact spot or a group of spots located along the sliding velocity. The flash temperature is determined from the height of the oscillogram peak and the heat source size is found from the peak width. However, it is impossible to measure accurately the size and shape of spots along the entire contact zone and to determine the temperature distribution over a single contact spot. More complete information can be obtained by combining this technique with the above-mentioned photographic method.

The use of IR thermography is a promising trend in thermotribology. Fast progress in the methods and instruments for detecting and measuring heat radiation and growing interest of researchers in these techniques are caused by numerous advantages of their applications when studying the thermal state of rubbing solids in comparison with the other methods. Among them is the unlimited use of thermography since all solids with the temperature above the absolute zero emit electromagnetic energy a considerable share of which corresponds to the IR band. The other advantages include a high temperature and spatial resolution of the instruments, possibilities of remote measurements, and the broad range of temperatures which can be measured.

Thermal imagers transform the invisible heat radiation of an object to its visual image being the temperature distribution over the object surface. Different types of IR detectors are applied in these instruments, e.g. vidicons (receiving TV tubes with a photoresistive semiconductor layer), pyroelectric crystals, indium antimonide detectors etc. This method was used in [95] to record the temperature field in the pin-on-disc contact and to find its correlation with the mechanisms of steel wear. The effectiveness of using thermography to analyze the temperature distribution over the friction surface of polymer materials was shown in [96, 97]. The authors of [98] describe the application of this technique to determine the temperature field in the lubricated sliding contact. They present the temperature distribution and the time dependence of the temperature in the contact of a graphite rotor and fluorspar stator.

In study [99] an IR thermograph with the cooled detector was used to measure the temperature on the back surface of the chip in turning. The temperature measurement range of the instrument was from 50 to 2000 °C, the temperature resolution was 0.1 °C, and the measurement error was ±0.5%. The above-mentioned study of the temperature in the magnetic tape – recording head contact [89] employed, in addition to the IR microscope, an IR camera Thermovision 750 (AGEMA Co., Sweden) with the maximal scanning rate of 2500 lines per second and the temperature resolution of 0.5 °C. The minimal spot size was 1.5 mm. Another example of the application of this IR camera is described in the study of the role of internal friction in energy dissipation during the sliding of PTFE-based composite against steel [100]. A portable IR camera ThermoCAM PM300 (Inframetrics Co., USA) was used in [101] to determine the temperature field and temperature gradients in the contact of a railcar wheel and a brake block.

The analysis of the suitability of various temperature measurement methods for studying thermal processes at friction has allowed us to formulate the following requirements to measuring instruments: a short response time and a high linear resolution (capability of recording radiation from small contact spots) which are necessary to register fast-running thermal processes; a wide temperature measurement range (from the melting or destruction temperature of polymers to the melting temperature of metals) and a high sensitivity and resolution within it; the portability of the instruments intended for the diagnostics of real friction pairs; the convenience of the storage, presentation, and processing of experimental data. Some of the requirements are contradictory and it is difficult to satisfy them when using a universal measuring instrument. For example, the modern thermograph IR Snap Shot

(INFRARED SOLUTIONS Inc., USA) and similar instruments have the temperature measurement range from minus 30 to 1200 °C with the resolution no more than 0.3 °C and the measurement error ±2%, yet their scanning time exceeds 1.5 s. For this reason they are inapplicable to studying thermal processes on local contact spots whose life is by three orders of magnitude shorter than this scanning interval. Some instruments have a low measurement accuracy, a narrow temperature range or a poor spatial resolution. Therefore, with account for the noted advantages and disadvantages of the methods and instruments outlined above, the development of combined measuring systems is a promising trend in thermotribology. Such systems should include several instruments, e.g. a TV camera having a quick-response radiation detector and a special optical system with the set of objective lenses to vary the visual field, and a thermal imager or thermograph sensitive in the IR band. The application of digital measuring instruments provides a way for the storage of huge experimental data files, the transfer of them to a PC and the processing of them using specialized software (the plotting of temperature profiles and isotherms, the study of the kinetics of thermal processes etc.). Since these systems are not yet produced commercially researchers in thermotribology have a broad field of activity to create and use them. Below we present the example of such a system which has been developed at the V.A. Belyi Metal-Polymer Research Institute of the Belarus National Academy of Sciences and the Belarus State University of Transport.

3.2. Application of Optical-Electron Scanning Technique and IR Thermography to High-Speed Friction and Machining of Materials

Temperature Distribution over Contact Area and Contact Spots. We developed the friction apparatus comprising a high-speed friction machine and a system of temperature field registration (Figure 9). The friction machine allows the sliding velocity to be varied within the 0…100 m/s range [40, 102]. The friction coefficient is measured by strain gages. The ranges of the nominal pressure and sliding velocity are as follows: 0.1…0.8 MPa and 1…35 m/s for the sapphire – polymer pairs and 0.5…5.5 MPa and 1…80 m/s for the sapphire – metal and glass – metal pairs.

Thermal processes are studied using the registration system consisting of accessory lenses with different magnifications, an optical scanner, a monitoring device, a video tape-recorder, an amplifier, a device to form oscillograms of image brightness, and a digital oscillograph. An IR scanner “Thermovision-470” (AGEMA Co., Sweden) equipped with the additional optical system serves to investigate the temperature field in the friction zone of polymers. When studying the friction of metals against sapphire or when modeling the abrasive machining of glass and sapphire we used a TV camera. The scanner and camera are fastened to a rack which makes it possible to move them in the vertical and horizontal directions and to rotate them through 90 deg.

The heat radiation induced in the friction zone passes through the lens then the TV camera generates an electric signal. It is converted into a high frequency signal and transmitted to the monitoring device forming a TV image of the contact zone. The image is recorded by the video tape-recorder. The device to form oscillograms of image brightness connected to the video tape-recorder output produces the image brightness distribution along two sections (for example, along and across the sliding direction). The distribution is displayed by the digital oscillograph as the signal in millivolts.

The temperature distribution across a single hot spot is obtained by selecting a line on the monitor which corresponds to the required section (by the freeze-frame mode) forming a signal oscillogram of the line with respect to the zero level of the line sync pulse. The temperature distribution along the spot length is obtained by an oscillogram formed as a result of selecting instantaneous values of the signal of each line corresponding to a preset sync pulse duration.

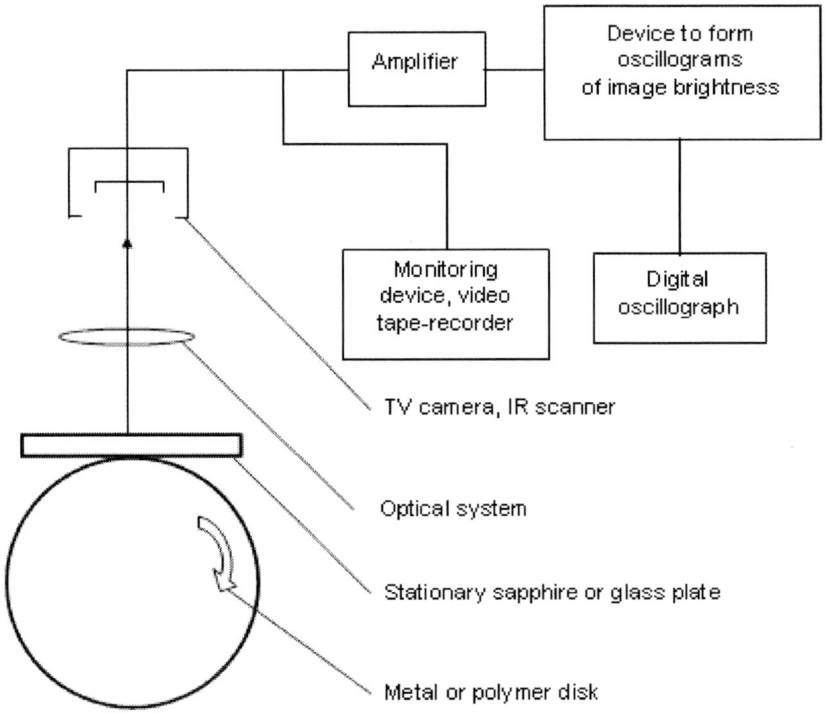

Figure 9. Contact geometry and schematic diagram of friction apparatus and temperature measuring system

The measuring system is calibrated using a furnace with an accuracy of temperature keeping of 0.1 °C. Discs made of the metals or polymers under study are placed into the furnace and the heating program is started. The IR scanner or the TV camera records the temperature field. The disk temperature is measured by high-precision thermocouples and the calibration curves are plotted. The radiation temperature is converted into the real temperature with account for the emissivity of the metals and polymers. Since it is difficult to consider temperature dependencies of the emissivities we use their values averaged over the studied temperature range. Sapphire absorption is taken into account by placing the sapphire plate between the camera or scanner lens and the furnace.

The temperature field recorded for the sapphire – steel pair (Figure 10) shows that the temperature of different hot spots can vary essentially and exceed the average value by almost two orders of magnitude [39]. The maximal temperature is found near the geometrical centre of the spot. The temperature distribution along the spot length (width) becomes asymmetrical if close neighbors appear. In some theoretical papers, e.g. in [3, 41], the maximal temperature position is shown to shift from the spot centre in the direction opposite to the sliding velocity if the spot is of regular shape. The shift grows with increasing velocity. The described temperature distribution is typical, for example, for three sites located in the middle of the 3D temperature plot (Figure 10). However, proceeding from the analysis of the temperature distribution over spots for all the pairs we can consider the maximum temperature point shift as accidental and depending mostly on the shape of asperities brought into contact. Distributions of the temperature along the spot length measured at different instants of the single spot life indicate that the maximal temperature point position is random but depends on the spot life.

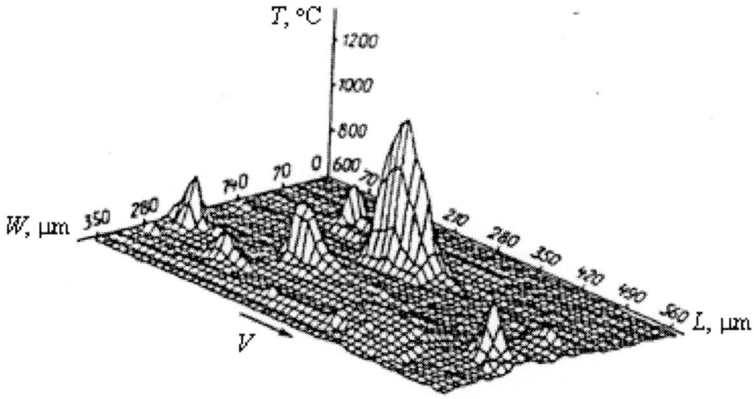

Figure 10. Temperature field in sapphire – steel contact under nominal pressure of 5.5 MPa and at sliding velocity of 22.5 m/s; (L and W are contact site length and width, respectively)

Using the system of temperature field registration described above we obtained the images of the contact zone with the visualized local heat sources (contact spots) an example of which is shown in Figure 11 [39, 102]. The images serve to determine the shape and dimensions of the heat sources which vary during the spot life; the variations depend on the properties of the mated materials and the load and velocity. The common regularity is that the spots are elongated in the sliding direction and that their shape is close to elliptical. For the sapphire – titanium, sapphire – steel, and sapphire – aluminum pairs the length of the spots reaches 250 μm while their width does not exceed 50 μm. In the sapphire – aluminum pair the heat sources are the most elongated and in the sapphire – titanium pair their shape is close to circular. The average size of the heated spots is 10…40 μm for all the pairs.

The images of the contact zone were also used to analyze the kinetics of the appearance and evolution of local heat sources. During the first loading cycles small spots appear (Figure 11, a, arrow 1). With time the contacting asperities are worn out leading to the expansion of the spots (Figure 11, b–d, arrow 1). Closely-spaced spots can merge into a new spot (Figure 11, b, c, arrow 1) whose brightness is, as a rule, higher than that of the parent spots. Large spots can break up into smaller ones (Figure 11, c, d, arrow 2) because the wear of the mated surfaces results in load redistribution in the contact. The life of the local heat sources is governed by the wear resistance of the pair materials, the load and velocity conditions, and the roughness of the surfaces. Depending on these factors it can vary by three or four orders of magnitude, i.e. from hundreds of milliseconds to tens of microseconds.

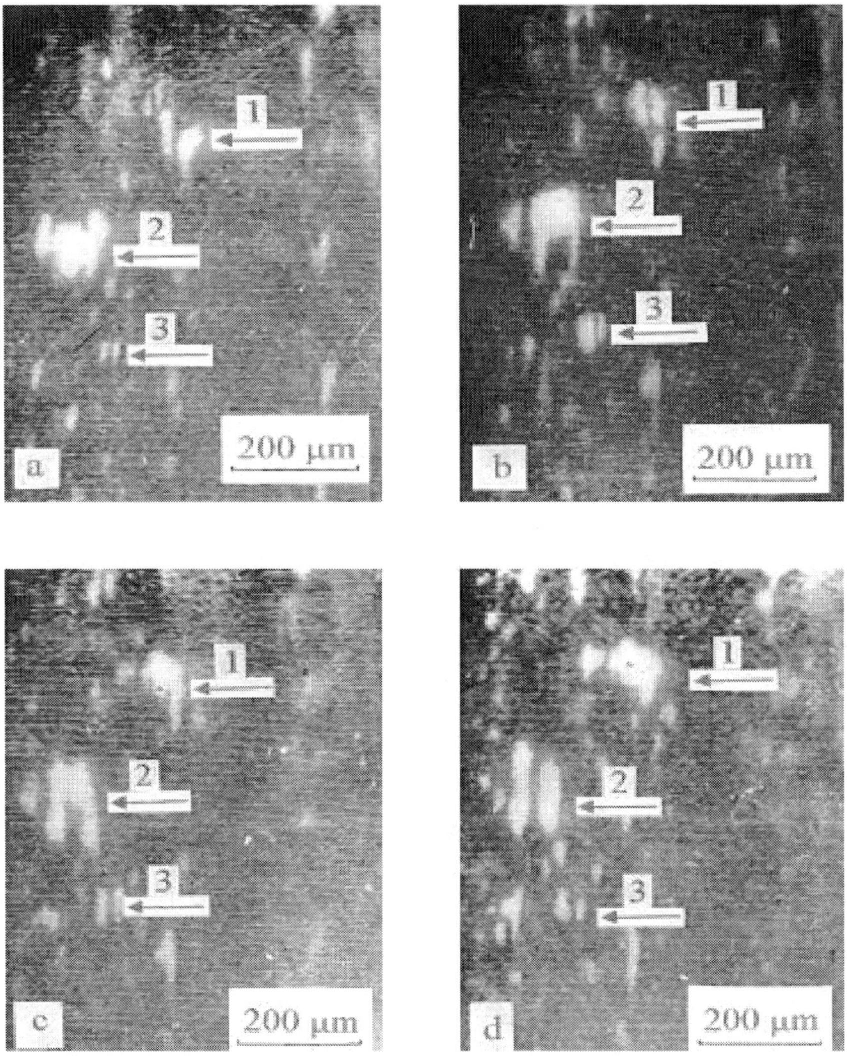

Figure 11. Kinetics of size and shape variation of hot spots in sapphire – titanium contact under load of 1.8 N and at sliding velocity of 47.5 m/s. Time interval between shots is 120 ms

The device to form oscillograms of image brightness produces the distribution of the spot brightness along the marker line perpendicular to the sliding direction. As the marker line is moved in the sliding direction, we can measure the temperature of any point within a single hot spot. We studied the maximal temperature of local heat sources and its dependence on the sliding conditions [40]. With the sapphire – titanium and sapphire – steel contacts no spot temperature equal to or exceeding the melting point of the metals is registered over the used load and velocity ranges. The surface topography of the steel discs worn out under different conditions is practically the same. Melting traces of asperity summits and their curvature radius variation are not detected. This fact proves the equality of the maximal spot temperature to the metal melting point. It is in complete agreement with a widely held view that the upper limit of the spot temperature should be the phase transition temperature or the melting (destruction) temperature of one of the mated materials [41, 91].

For the sapphire – copper contact the temperature of heavily loaded spots elevates with increasing the load and velocity. At 36 m/s and under > 0.5 N the disc material melts on contact spots. Under severer conditions the dimensions and the number of the spots with the temperature close to the copper melting point increase while the highest attainable temperature remains constant. The copper disc surface shows hollows of nearly hemispherical shape about 3 μm in diameter and 2…5 μm deep. They can probably result from copper melting on heavily loaded asperity summits or from metal transfer to sapphire. X-ray diffraction data show the presence of a copper film on the sapphire counterface. These results support but doubtfully that the spot temperature reaches the metal melting point in the sapphire – copper pair.

When sapphire rubs against aluminum the spot temperature exceeds significantly the metal melting point (660 °C) and can reach 1700 °C. The following mechanism of the phenomenon is proposed. Under the effect of the elevated temperature the disc material undergoes severe oxidation on real contact spots and an oxide film appears protecting the metal against direct contact with sapphire. So, sapphire rubs against aluminum oxide whose melting point is 2050 °C in contrast to other metal oxides under examination and exceeds much that of the parent metal. Thus, the upper spot temperature limit shifts towards higher temperatures. Besides, the oxidation of aluminum is accompanied by severe heat generation on contact spots being an additional source of the temperature rise.

Figure 12 shows the temperature distribution over the polyamide friction surface recorded by the IR scanner under various loads [40, 102]. The temperature distribution over a single contact spot and in its vicinity is presented by concentrically disposed closed ring-like strips under light normal loads and at low sliding velocities. Further on these strips are conventionally termed isotherms bearing in mind that the definite temperature range corresponds to each strip. The isotherms somewhat extend in the sliding direction. The temperature distribution over the spot area along this direction is presented by the curve almost symmetrical to the spot centre (Figure 12, d). The maximal temperature is registered in the spot centre; the temperature quickly decreases as the distance from the centre increases.

Since sapphire has high and polyamide low heat conductivities the friction surface area exposed to the heat flow from a local heat source is insignificant. It is about 3% compared with the apparent contact area. As the load increases (Figure 12, b), the region occupied by the heat wave expands and new contact spots appear leading to pressure redistribution in the friction zone. With increasing the pressure the number of the contact spots in the visual field of the IR scanner reaches three; their total area exceeds seven times the contact spot area in Figure 12, a. This increase in the heat emitting spot area may be due to the failure of asperities on the adjacent portions of the contact site outside the scanner visual field rather than to the growth of the normal load. It is accompanied by a higher load concentration on the contact site. The maximal temperature of the spot increases simultaneously with the load and the temperature distribution peak shifts towards the zone where the specimens come into contact (Figure 12, d).

Variation in the sliding velocity results in the significant alteration of the temperature field (Figure 12, c). As the velocity increases, the hot spots expand. It may be due to a greater share of the plastic deformation of the contacting asperities since the material is heated to a higher temperature. Though the duration of contact decreases, the heat wave propagates over the significantly vaster portion of the polyamide friction surface.

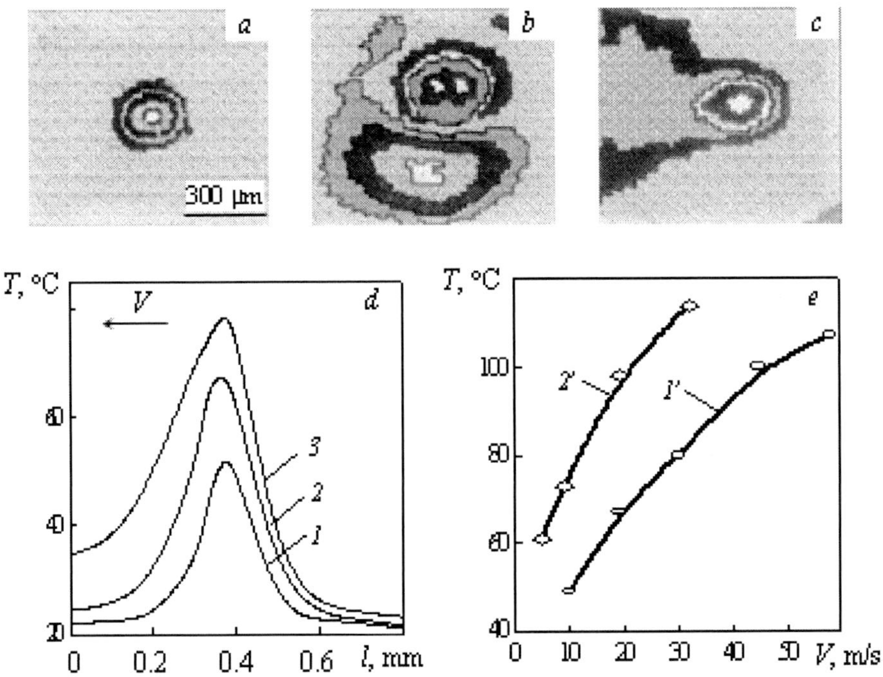

Figure 12. Temperature distribution in vicinity of contact spot (a–c) and along sliding direction (d), and maximal temperature (e) in sapphire – polyamide contact under nominal pressure of 0.29 MPa (1, 1′, 3) and 0.5 MPa (2, 2′) and at sliding velocity of 8 m/s (1, 2) and 30 m/s (3)

The isotherms close to the heat source extend in the sliding direction taking elliptical shape. The densest arrangement of the isotherms is observed ahead of the zone where the asperities come into contact. The temperature gradient along the spot length in this region reaches its maximum (550 °C/mm at 30 m/s). At a higher sliding velocity the peripheral isotherms open up and orient along the sliding direction while the quasi-stationary temperature region appears behind the zone where the asperities exit the contact. The temperature distribution along the sliding direction becomes skew with negative asymmetry (curve 3 in Figure 12, d).

When sapphire rubs against polystyrene and polyethylene the temperature distribution at the interface does not basically differ from the above described. The temperature fields on the friction surfaces of these materials differ only in the size of single contacts and regions of heat propagation due to their different mechanical and thermal properties.

In the contact of sapphire with the polymers load and sliding velocity growth is accompanied by the maximal spot temperature elevation. The effect of the velocity on the spot temperature in the sapphire – polyamide pair under various pressures is presented by the convex curves (Figure 12, e). For the sapphire – polystyrene pair the curves are more flat perhaps owing to the weaker effect of the bulk temperature of the contacting asperities and the velocity on the mechanical properties of polystyrene and the friction coefficient compared to polyamide and polyethylene. We note that similarly shaped velocity dependences of the maximal spot temperature have been obtained when studying thermal processes in the sapphire – metal contact.

A specific feature of polyethylene is that at the certain velocity depending on the load the temperature shows rapid augmentation. Fast temperature increase is apparently caused by the evolution of tribochemical reactions on the heaviest loaded contact spots accompanied by heat generation. Contact spots with the temperature as high as 150 °C have been registered on the polyethylene surface under the pressure of 0.8 MPa and at the velocity of 18 m/s. This fact can be explained as follows. When the spot temperature reaches the melting point of polyethylene its oxidation commences with heat generation and carbon oxide and dioxide formation. The oxidation of a thin surface layer of asperities at the moment of contact with the counterface elevates the spot temperature. Moreover, friction causes the simultaneous overall compression and pulse heating of the surface and subsurface layers near the contact spot centre. It is reported in [103] that the material in this region experiences structural modifications (the crystallinity rises and macromolecules become packed more densely) elevating the temperature required to melt the material in this volume.

Temperature Distribution in Depth of Rubbing Bodies. We also used the experimental apparatus described above to study the temperature under the friction surface of the specimens [39, 104]. In this case a glass plate contacts the flat surface of a rotating metal or glass disc (Figure 13). The temperature is measured along the marker line perpendicular to the sliding direction. The marker line can be shifted along the sliding line over the image of the contact area portion being examined. In reality the profiles of the mated surfaces are not seen since the linear resolution of the measuring system is 5 μm that is by an order or two larger than the profile arithmetic average roughness. Thus, we fail to determine the marker line location whether it is within the contact spot or between the spots. The section of the friction track corresponding to the marker position may contain several spots, hence, the temperature averaged over these spots is registered.

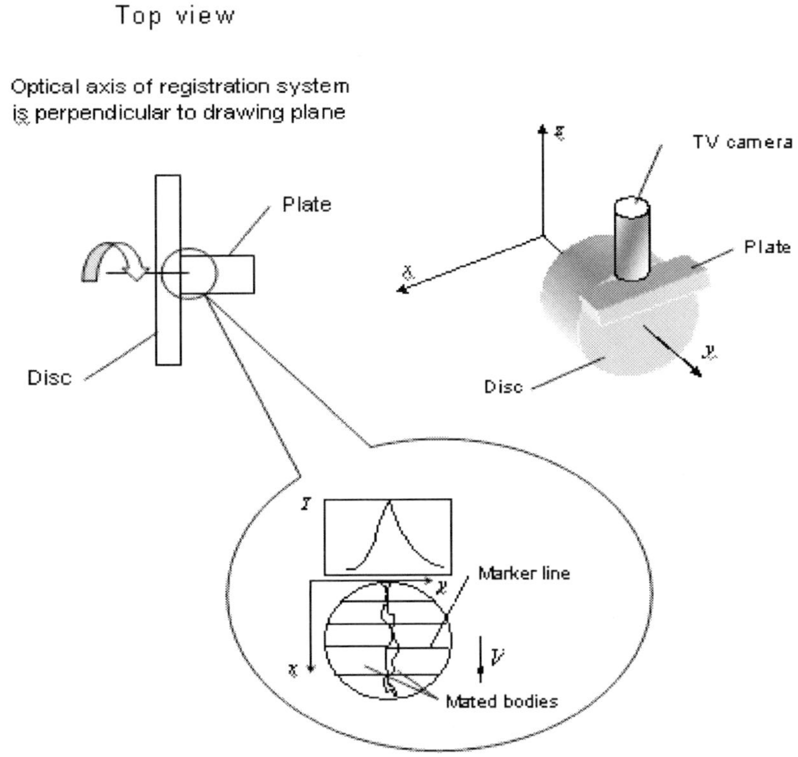

Figure 13. Schematic diagram explaining procedure of subsurface temperature measurement

As our studies have shown, the mode of the temperature distribution in depth of the surface layers and the position of the distribution curve maximum are governed by properties of the mated materials as well as by the load and velocity conditions. When the silica glass stationary specimen rubs against the titanium disc the temperature distribution is monotonous (Figure 14, curve 1). This proves that heat is generated within the glass surface layer a few microns thick.

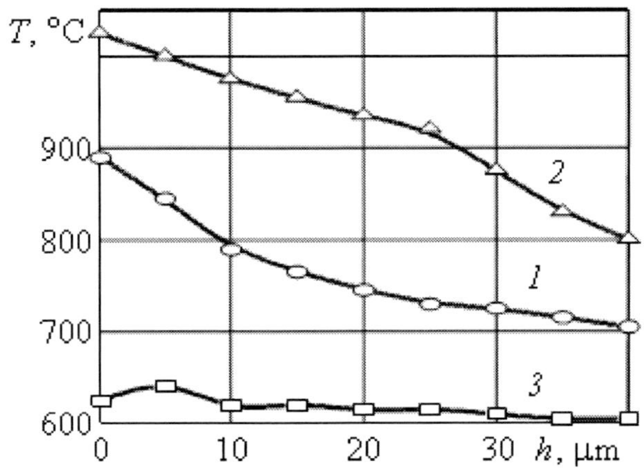

Figure 14. Temperature distribution in depth of silica glass stationary specimen in contact with titanium (1), aluminum (2), and steel disc (3) under nominal pressure of 0.1 MPa and at sliding velocity of 18 m/s

The silica glass – aluminum pair also shows the dependence without subsurface temperature maxima within the studied load and velocity ranges. At the velocity ≤ 45 m/s the temperature distribution can be conventionally divided into two portions (Figure 14, curve 2); temperature gradients for them can differ several times. The thickness of the severely heated and deformed surface layer can be roughly estimated by measuring the position of the conventional boundary between the steeper and flatter portions of the curve. It diminishes with decreasing the nominal pressure and increasing the sliding velocity. This can be attributed to a shallower penetration of counterface asperities into the glass surface layer and to a shorter life of friction junctions. At higher velocities the flat portion of the distribution curve is not registered because plastic deformation, hence, frictional heating are concentrated within a very thin surface layer of glass.

The temperature distribution with the subsurface maximum is typical for the silica glass – steel pair. The maximum is located ≈ 5 μm beneath the glass friction surface (Figure 14, curve 3). At the velocity ≤ 20 m/s the friction coefficient for this pair is below 0.25. Saverin's theoretical results [105] have shown that in this case the zone of the maximal tangential stresses lies at a depth of 10…12 μm under the surface in provision that the average contact spot diameter is 25…30 μm within the studied load and velocity ranges. Therefore, we can presume that most heat is generated under the glass surface in the zone where the tangential stresses are maximal.

The use of the developed optical-electron scanning technique has allowed us to obtain the temperature distribution in subsurface layers of both the stationary and rotating specimens made of silica glass. The distributions are presented in Figure 15 for different values of the pv factor. The curves are similar for both the disc and plate despite the fact that their cooling conditions are essentially different. As it has been shown by Rigney and Hirth [106], adhesive tractions resulted from the rupture of adhesion junctions between the contacting bodies may contribute to frictional heating through deformation they impart to the near-surface regions. Since like materials are in contact, both disc and plate undergo such plastic deformation. Therefore, heat is generated within a subsurface material volume in both specimens; the temperature distribution in depth consists of two portions with different temperature gradients (Figure 15, b). At a higher sliding velocity the deformed material volumes adjacent to adhesion junctions apparently contract so that heat is liberated within a thin surface layer and the flat portion of the distribution curve is not registered (Figure 15, a).

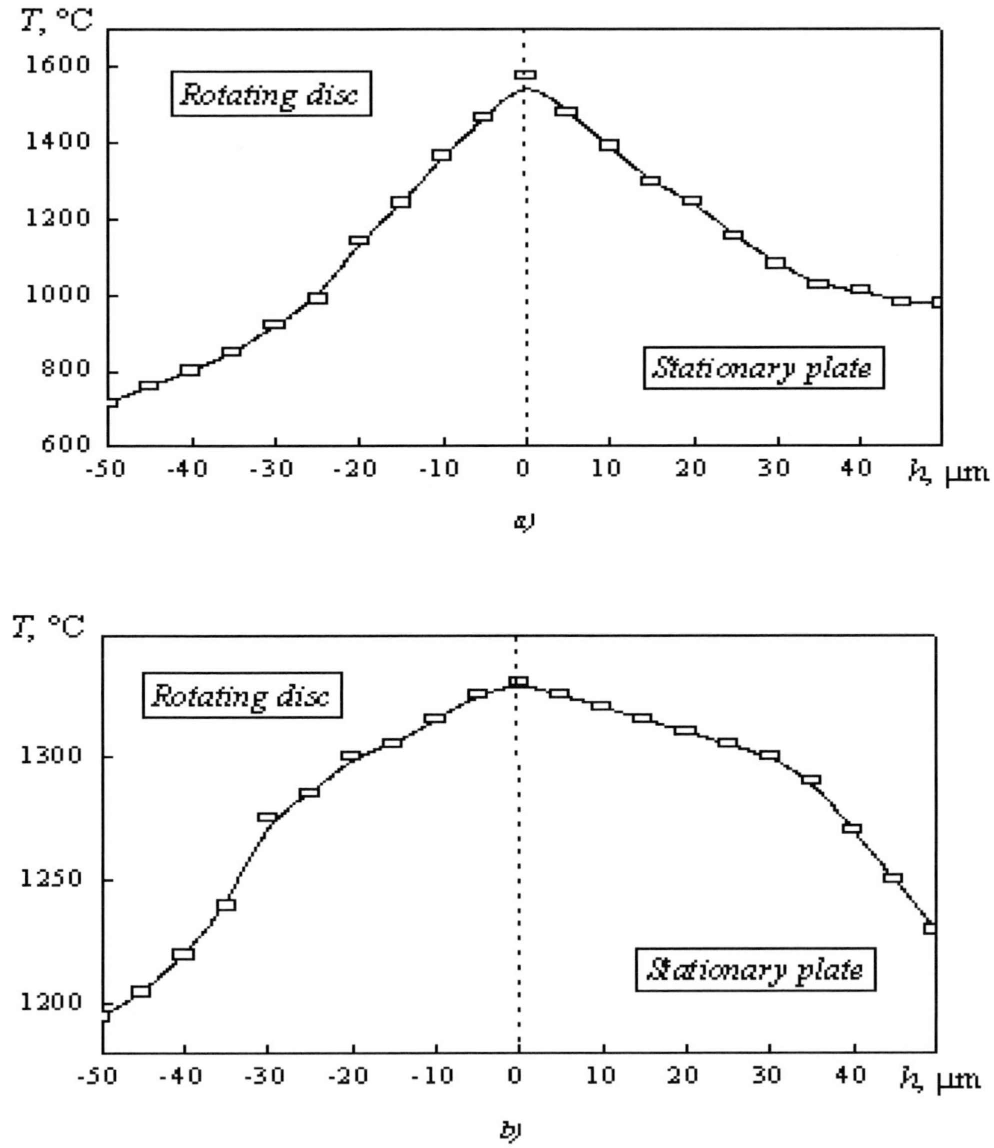

Figure 15. Temperature distribution in depth of members of silica glass – silica glass friction pair under nominal pressure of 0.1 MPa (a) and 0.3 MPa (b) and at velocity of 50 m/s (a) and 20 m/s (b)

Crystal Cutting. We also carried out the experiments on sapphire and diamond cutting to study regularities of thermal fields in the cutting zone and its vicinity including the temperature distribution and the kinetic dependence of the cutting temperature [107–110]. The contact geometry used at sapphire cutting is shown in Figure 16, a. Sapphire specimens were round plates (∅ 18 × 1 mm) contacting the edge of a rotating cutting disc. The discs 0.05…0.07 mm thick and 76 mm in diameter were made of tin bronze and charged every 2…3 min with diamond micropowder mixed with liquid binder containing 70…80% of paint oil and 20…30% of castor oil.

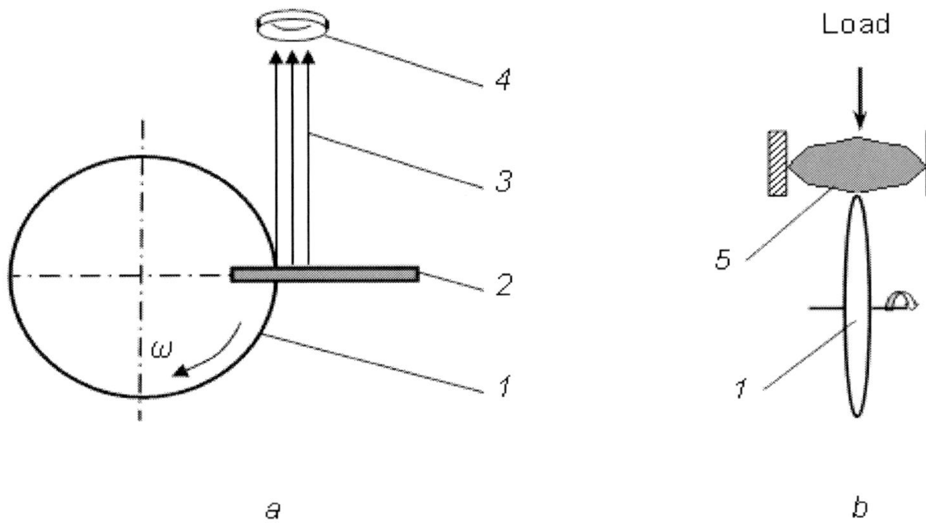

Figure 16. Schematic of specimen–cutting disc contact at sapphire cutting (side view) (a) and diamond cutting (top view) (b): 1 – cutting disc; 2 – sapphire plate; 3 – heat radiation; 4 – thermograph; 5 – diamond monocrystal

Experiments with diamond were carried out under real conditions of cutting of natural diamond monocrystals at brilliant manufacture of the "Kristall" factory (Gomel, Belarus). The standard production technology [111] was implemented in a diamond-cutting machine. The discs and abrasives were as described above. The rotational speed of the spindles was 12,000 rpm which corresponded to the linear velocity of 42.6 m/s. The load varied from 1.9 to 2.4 N. The schematic of the blank–cutting disc contact is shown in Figure 16, b.

The temperature field in the cutting zone was recorded with the thermograph IR Snap Shot model 525 (Infrared Solutions, Inc., USA). The signals were digitized and screened on the liquid-crystal display; it was also stored in the thermograph memory to process it further by a PC with the special software IR Snap View. One of the main advantages of the thermograph is its broad dynamical range allowing us to record regions whose temperatures differ significantly within one image. The basic technical characteristics of the thermograph are as follows: the sweep (the field angle) – 17.2 °C (horizontal) and 17.2 °C (vertical); the focus – from 260 mm to the infinite distance; the spectral range – 8…12 μm; the temperature measurement accuracy – 2 °C or 2% of the full range; the temperature sensitivity threshold at 30 °C – no more than 0.3 °C; the temperature measurement range – from –30 to 600 °C; the scanning time – < 1.5 s. The emissivity within the range 8…12 μm was taken 0.92 for diamond and 0.59 for sapphire in accordance with the reference data [112].

Figure 17, a shows a typical thermal image recorded in sapphire cutting at the velocity of 23.6 m/s and under the load of 0.65 N. It is seen that the maximal temperature region (the white area in the image) is located somewhat ahead of the zone of the cutting disc–sapphire contact rather than within it. This shift of the temperature maximum from the zone of contact between the sapphire plate (its contour is shown by dashed line 2) and abrasive grains can be caused by the following reason. The heat generated by cutting propagates in sapphire away from the contact zone but, owing to the high thermal conductivity of bronze cutting disc 1, the sapphire area immediately adjacent to the disc (the narrow strip) cools down. As a result, its temperature decreases. We note the non-round shape of the thermal image in the area to the left of the maximal temperature region. This can be explained by the fact that a great share of the cutting heat is absorbed by plate 3 of the sapphire specimen holder.

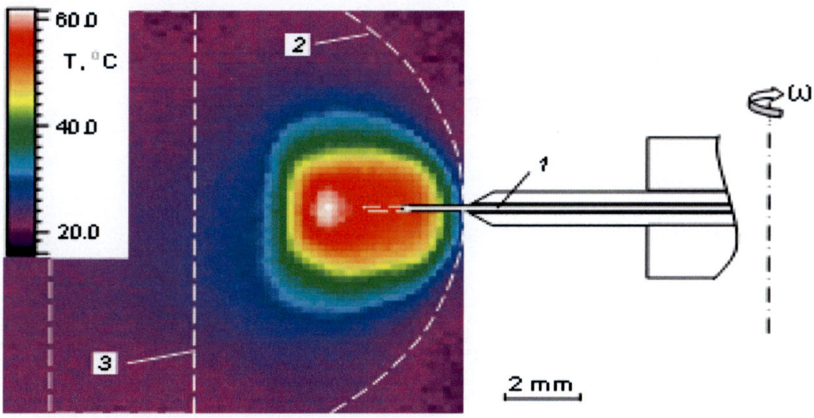

a

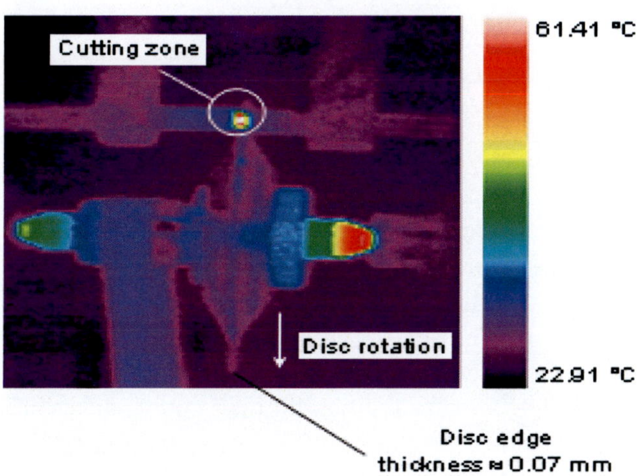

b

Figure 17. Thermal images recorded in sapphire cutting (a) and diamond cutting (b): 1 – cutting disc; 2 – sapphire plate contour; 3 – holder plate contour

To study the kinetics of thermal processes at sapphire cutting we recorded thermal images every 30 s starting from the beginning of cutting and used these data to plot the time dependences of the maximal surface temperature in the cutting zone T_m (Figure 18). With time the heating of the cutting zone becomes severer and T_m elevates. Under the load of 0.43 N it grows monotonously (curve 1) but after 180 s of cutting its rise becomes slower. This may be explained by the effect of two factors. The first factor is the accumulation of the cutting heat in sapphire and, possibly, in the working layer of the cutting disc; it dominates at $t < 180$ s and favors the surface temperature rise. The second factor, heat dissipation, is minor. The cutting heat is removed to the environment through both the sapphire plate and the rotating copper disc having the high thermal conductivity. With time the sapphire temperature increases thus intensifying heat removal and making the temperature growth rate lower.

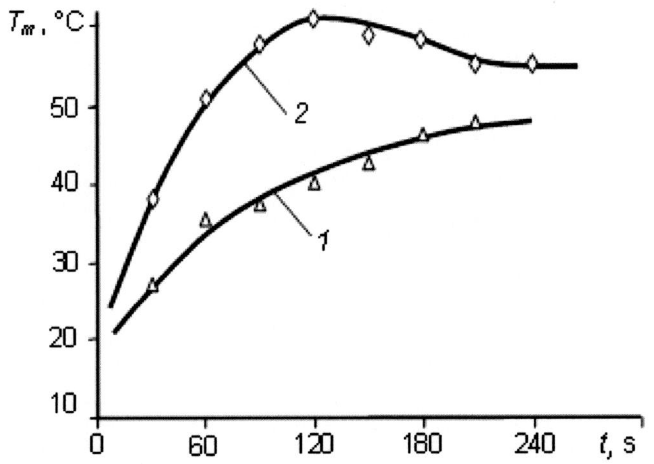

Figure 18. Kinetic curves of maximal surface temperature recorded in sapphire cutting at velocity of 24.6 m/s and under load of 0.43 N (1) and 0.65 N (2)

As the load increases to 0.65 N, the maximal surface temperature rises approximately 1.3 times. The dependence of T_m on the cutting duration is illustrated by the curve having the maximum at $t \approx 120$ s (Figure 18, curve 2). The left-hand branch of the curve results from increasing heat generation in the cutting zone with load growth. On reaching the maximum, the heat generated in cutting becomes equal to the heat removed to the environment. The slight temperature decrease at $t > 120$ s may be explained by the approach of the cutting disc to the holder plate favoring heat removal from the cutting zone.

The cutting rate of diamond is by an order of magnitude lower than that of sapphire since the extremely high hardness of the former not only makes the penetration depth of abrasive particles shallower, but also causes the wear and dulling of their cutting edges. The latter fact results in severe heat generation in the cutting zone increasing the diamond temperature. Under similar cutting regimes the T_m of diamond is 2…10 times higher than that of sapphire.

Figure 19 illustrates the typical time dependence of T_m during the period from the beginning to the end of the cutting of a diamond monocrystal. Initially the cutting depth is shallow and the diamond temperature exceeds a bit the ambient temperature. With time the cutting depth grows and the arc of contact becomes longer. Because of this, the duration of contact of abrasive grains with the crystal, hence, their cutting path increases favoring the severer heating of diamond. In addition, as the thickness of the non-cut part of the crystal decreases with time, the cutting heat tends to be accumulated in diamond rather than removed that also promotes cutting temperature rise. The temperature continues elevating as long as the arc of contact lengthens. Then the duration of abrasive grain–diamond contact shortens and the cutting temperature diminishes (the right-hand branch of the curve in Figure 19).

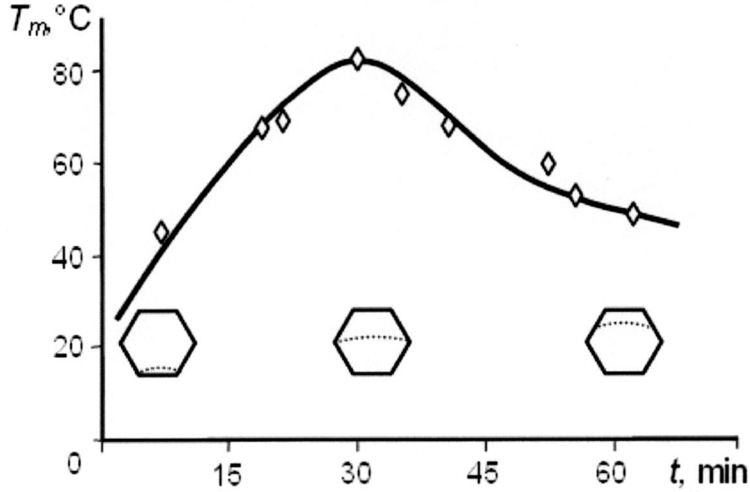

Figure 19. Typical kinetic curve of maximal surface temperature recorded in diamond cutting. Schematic below curve shows arc of contact between cutting disc and blank at different moments (dotted line)

The diamond temperature can reach high values. For example, Figure 20 represents the thermal image recorded 21 min after the beginning of cutting. It shows that the blank temperature exceeds 350 °C. In work [78] the average temperature in the zone of the cutting of natural diamond crystals was measured by chromel-alumel micro-thermocouples embedded into a model blank. The kinetic dependence of the temperature was similar to that shown in Figure 19. As for the temperature values, they did not exceed 200…220 °C even under the severer cutting regime (the disc linear velocity was 54.3 m/s and the load was 2.5 N) that was ≈ 1.5 times less than the values measured by the thermograph (Figure 20). This indicates that the temperature in the cutting zone is undervalued if it is measured by artificial thermocouples.

Assessing the applicability of the thermograph IR Snap Shot model 525 to the determination of the temperature field in the zone of diamond cutting we emphasize that such instruments measure the average temperature of the blank. Local temperatures on spots of contact between diamond powder grains and the crystal are apparently much higher. Under extreme cutting regimes thermal processes may contribute considerably to the wear of diamond not only inducing its graphitization but also accelerating the growth of fatigue cracks resulted from mechanical wear. It can be confirmed indirectly by the appearance of heated spots on the cutting disc edge which are registered in thermal images obtained under severe cutting regimes. In particular, for the crystal shown in Figure 20 this effect occurs within the cutting temperature range 262…355 °C and it is not observed under lighter loads, hence, at lower temperatures. The spot heated to about 90 °C corresponds apparently to a quite large wear particle of diamond.

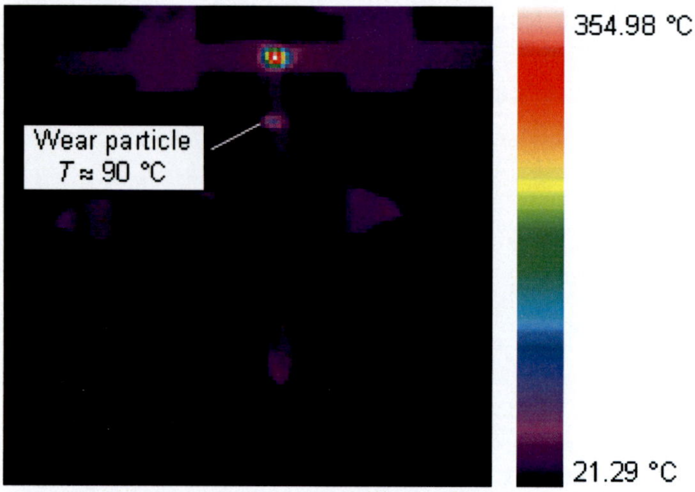

Figure 20. Thermal image recorded in cutting of stressed diamond under load of 2.4 N and showing large wear particle on cutting disc edge

4. Thermomechanical Effects Related to Sliding

Among the consequences of frictional heating are the so-called thermomechanical phenomena including thermal deformation around sliding contacts, changes in the contact geometry due to thermal deformation and thermoelastic instability, and thermal stress distribution around frictionally heated and thermally deformed contact spots. They influence considerably the thermal cracking, wear, and other failure modes of rubbing materials and deserve special consideration.

4.1. Thermal Deformations and Thermoelastic Instability

The stress-strain and thermal states of the friction zone are closely interrelated; this interrelation appears, particularly, as the so-called thermoelastic instability at which the nominally flat friction surface becomes consisting of a few “hot spots”. The pressure and temperature on these spots exceed considerably the apparent pressure and the average surface temperature intensifying the wear of the friction units with a great overlapping factor operating under heavy loads and at high sliding velocities.

The first direct evidence of the localization of frictional heating at macrolevel was obtained by Parker and Marshall [113] who used a low-temperature pyrometer to measure the temperature of the friction surface of a railcar wheel braked by a shoe. The measurements have shown the presence of heated areas whose size was between the spacing of roughness peaks and the apparent contact area. Similar results were later reported by Sibley and Allen [114] and by Santini and Kennedy [115] when testing the pads contacting rotating discs.

Barber confirmed experimentally the existence of macroscopic "hot spots" by measuring the temperature at different points of the railcar brake shoe surface by thermocouples [116, 117]. Each thermocouple yielded a fluctuating signal that indicated an instability caused by the cyclic loading of some surface areas. In Barber's model thermoelastic instability is believed to result from the simultaneous occurrence of the thermal expansion and wear of the material on the contact area. The stability of contact depends on elastic and thermal characteristics of the stationary member material and on the friction and wear coefficients of the pair.

Paper [118] deals with the sliding of the face of a rectangular plate over a rigid half-space. In contrast to the above model, according to the authors of this study, an unstable thermoelastic state may occur in the absence of wear. In this case the thermal expansion of the material is compensated by heat conduction to the half-space. Instability is shown to occur only at the critical sliding velocity governed by the thermal conductivities, thermal diffusivities, thermal expansion coefficients, and elastic moduli of the mated materials as well as by the size of the moving member. This model was expanded to the case of sliding with wear and the critical velocity was shown to exceed that for contact without wear. The experimental verification of the model carried out by Dow and Stockwell [119] has shown a good agreement between the theoretical and experimental results.

The model of Dow and Burton was improved in papers [120, 121] describing the effect of the thermal conductivities of the contacting materials on the stability of systems like face seals. If the thermal conductivities are the same instability occurs only at a very high critical velocity exceeding much the operating conditions of real seals. In contrary, the pair of an insulator and a conducting material is always unstable since a reasonable critical velocity exists at any friction coefficient.

It is noted in review of face seals [122] that most of them operate under hydrodynamic lubrication. The fluid film separating the contacting surfaces causes their nonuniform heating and thermoelastic deformation. The results reported in [123] show good agreement between the experimental values of the critical velocities and those calculated for a constant film thickness. The improved model presented in [124, 125] takes into account changes in the film profile and mean thickness with varying the velocity and load.

The real three-dimensional contact geometry of face seals was considered by Lebeck [126, 127]. It has been shown that the critical velocity characterizing the system stability exceeds that of the two-dimensional system because of heat removal from the moving member compensating its thermal expansion.

According to the data reported in [128–130], thermoelastic instability occurs in operation of friction units with high overlapping factors under heavy pressures and at high sliding velocities. Typical examples are face seals and brakes showing thermal cracking on the surface and in subsurface layers of the mated materials.

Using the above-mentioned IR scanner "Thermovision-470" we carried out the model experiment to study thermoelastic instability in the contact of a flat surface with a surface having regular waviness [131]. The latter was produced by glass spheres glued on a sapphire plate and contacted the cylindrical surface of a steel disc. Figure 21 presents the temperature field in the contact at different instants. There are four temperature peaks at the apparent contact area. Their location corresponds to that of the contact spots on the friction surface. It is seen that the mechanical and thermal loads on contact spots vary cyclically. The duration of the cycle equals 80 ms.

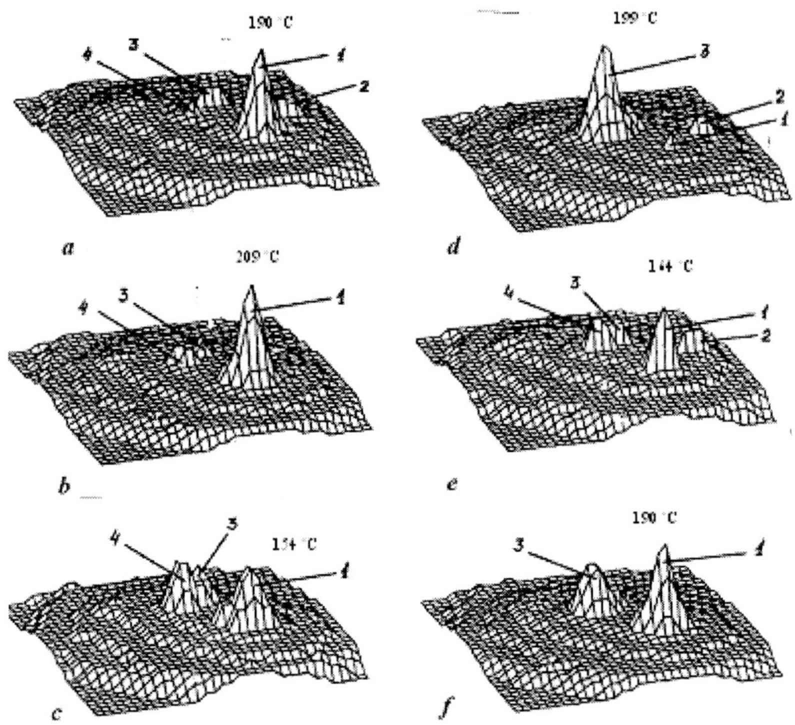

Figure 21. Temperature field in contact of flat and wavy surfaces under pressure of 0.5 MPa and at velocity of 30 m/s

Regularities of variation in the "hot spot" temperature correlate with the basic conclusions of Barber's model. So, the severe frictional heating of spot 1 leads to the thermal expansion of the spot and its vicinity. Moreover, the plastic flow and shear of the asperity material take place. But the real pressure is that the wear rate does not exceed the expansion rate. This offers the further heating and bulging of the material (Figure 21, a, b), and the spot area shrinks. As a result, the real pressure, hence, the wear rate increases until the expansion is compensated by the wear of the expanded material. The mechanical load acts elsewhere and the above process can be observed on spots 3 and 4.

It should be emphasized that temperature rise is longer than the period when the temperature decreases (Figure 21, b–e for spot 3 and d–f and a–d for spot 1). It is believed that the heating of the asperity is the discontinuous process caused by the sequence of heat pulses of a short duration. The pulses result from interactions with counterface asperities. The asperity under examination undergoes cooling simultaneously with heating and between acts of loading. According to paper [132], about 10,000 of loading cycles are required for the attainment of a significant temperature rise on the contact spot when a periodic heat source acts. At high stresses and temperatures the shear and plastic deformation of asperity summits can occur after a few loading cycles. Then the asperities are completely or partially unloaded for a short time. That is the way the temperature of the contact spot decreases with a higher rate than it does during frictional heating.

4.2. Thermal Stresses Near Friction Contact Zone and Thermally-Induced Failure Modes

Frictional heating is localized on real contact spots and temperature gradients may induce high thermal stresses in their vicinity. They are added to mechanical stresses and the total stress may exceed the material strength. The first study to determine thermal stresses near a frictionally heated contact was work of Mow and Cheng [133]. They applied the integral transform technique to a fast-moving band heat source on an elastic half-space. The same problem was also considered by Yang who used various integral transform techniques [134] and by Mercier and his coworkers who involved numerical methods [135]. It was shown that very high compressive thermal stresses act in the vicinity of the heat source and have the maximum on the interface.

For tribological practice the problem of thermal stresses in friction contact is important from the viewpoint of their influence on the failure of sliding members, primarily thermal cracking (or heat checking) occurring in brakes and face seals. One of the studies in this field is paper [13] whose authors measured the dimensions of macroscopic hot spots on the failed seal surface and used finite element analysis to calculate the temperature and thermoelastic stresses near the spots. The thermal stresses were shown to be very high and exceed mechanical stresses. Similar results were obtained by Tseng and Burton for a uniform heat source moving over the surface of a two-dimensional body [136]. It is also reported in paper [137] that in the two-dimensional case thermomechanical stresses are mainly compressive on the interface but beneath it a principal tensile stress occurs.

Of interest are the studies of Evtushenko and his co-workers reported in papers [138–142]. The authors investigated the relation between the temperature field in the contact zone and thermal stresses induced by the non-stationary heating of this zone. It is shown that with increasing the sliding velocity the temperature field in the half-space is localized in a thin surface layer. The tensile thermal stresses exceed compressive isothermal stresses. In later studies [143, 144] Evtushenko investigated the effect of the local frictional heating of the half-space surface on the stress intensity factor in the vicinity of an internal and edge crack and a periodical system of such cracks. His recent paper [145] deals with the thermal cracking of materials induced by frictional heating. High compressive transversal stresses are shown to arise in the material subsurface layer; with time they decrease and change their sign, i.e. become tensile. When the stresses exceed the strength of the material its thermal cracking occurs. In [146] Evtushenko related thermal stresses in friction contact to the wear of ductile and brittle materials, described the way to construct wear maps for certain material combinations and proposed a criterion of thermomechanical wear. This criterion allows one to identify the region of friction parameters where thermomechanical wear occurs; outside it other wear modes may run.

Here we should mention again the problem of thermal cracking which intensifies the failure of various friction units, e.g. face seals and brakes [11–14]. Thermal cracks are usually perpendicular to the sliding direction and approximately equally spaced. One of the mechanisms of their formation is as follows. Elevated temperatures on the interface produce very high compressive stresses here promoting the plastic flow of the material which may result in tensile stresses. These tensile stresses may induce thermal cracking by the fracture of brittle inclusions in the material or by low-cycle fatigue caused by the repeated action of thermal load on one and the same surface area.

In some situations several modes of material failure may occur simultaneously in friction. One example is our experiments on the abrasion of non-metal inorganic materials which are widely applied in electronics and optical engineering [108, 109, 147, 148]. The required surface quality of components made of them is reached by using different procedures of abrasive machining accompanied by severe frictional heating. For such brittle materials the problem is especially vital since, depending on the severity of pulse thermal effects, they show either ductility or brittleness because the threshold of brittle fracture is strongly governed by the temperature and its gradient (see, e.g. [146]).

The experiments were carried out using the friction apparatus described in Paragraph 3.2. It was adapted for the modeling of the cutting and abrasion of hard and superhard materials (the cutting and grinding of diamond crystals, the grinding and polishing of optical components etc.) at a velocity of the sliding of abrasive grains over the surface machined up to 100 m/s. The study objects were rectangular plates of silicate glass (75 × 26 × 1.5 mm) and round plates of sapphire (∅ 18 × 1 mm) contacting the cylindrical surface of a rotating disc. The contact geometry corresponded to that shown in Figure 9. We used a steel 45 disc 127 mm in diameter and 5 mm thick with a layer of abrasive particles applied onto its working surface prior to each experiment. The abrasive was silicon carbide with the particle size of 400 μm mixed with castor oil. The damage of the machined surfaces was examined by using a metallographic microscope.

The photos presented in Figure 22 illustrate the middle part of the friction track proving that sliding velocity increase changes the microrelief of the glass worn surface and causes transition from one dominating wear mode to another. This is because not only the mechanical properties of glass vary under the increasing severity of frictional heating but also the role of thermal and contact stresses in glass failure also changes. At $V = 26$ m/s the contact temperature does not reach any high values, so far the specimen material undergoes basically two wear modes: abrasive wear and brittle fatigue fracture (Figure 22, a), both resulted from the cyclic effect of thermal and contact stresses. The occurrence of abrasive wear is confirmed by scars and deep grooves parallel to the sliding direction of abrasive grains.

The fatigue wear mode can be judged from the hollows scattered over the friction surface which result from the fatigue spalling of the material between the crossing microcracks. The debris appeared are shaped as polyhedrons; some of them contain microcracks. The microcracks in glass perpendicular to the sliding direction of grains also speak in favour of fatigue wear (a microcrack is shown by arrow 1 in Figure 22, a).

Along with the failure modes described above the brittle spalling of the material occurs which occupies large enough areas of the friction surface (Figure 22, a, arrows 2).

As the sliding velocity increases, the effect of the temperature on glass wear becomes more pronounced; frictional heating causes not only higher thermal stresses but also the softening of the glass surface layer reducing the elastic modulus and hardness of glass. The sites on which this occurs show smooth lengthwise grooves whose bottom contains a comparatively little amount of brittle fracture traces (Figure 22, b, in the arrow direction). This is the result of abrasive wear arising from the low-cycle fatigue of glass.

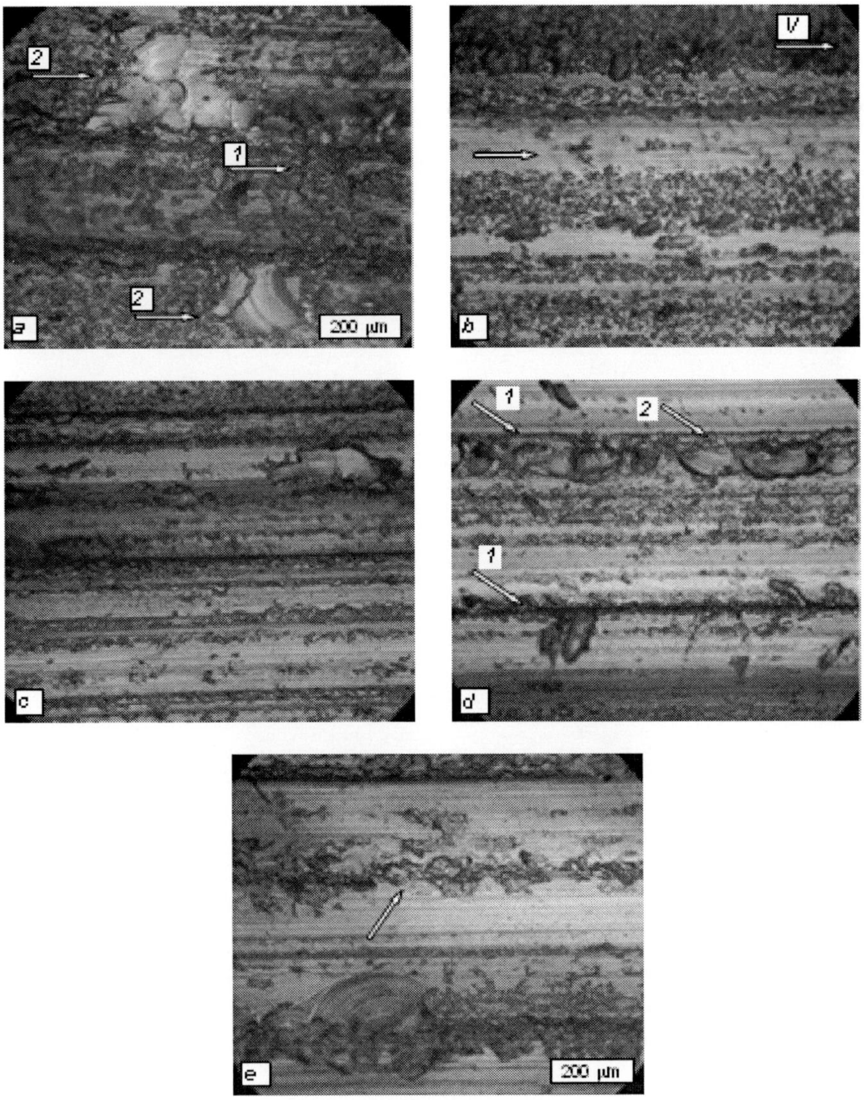

Figure 22. Optical images of friction track (middle part) on glass surface after abrasive wear at different sliding velocities: a – 26 m/s; b – 36 m/s; c – 55 m/s; d – 80 m/s; e – 105 m/s

With increasing the sliding velocity up to 55 m/s the contact spot temperature elevates making heat-induced changes in the glass mechanical properties more essential. The contribution of low-cycle fatigue into the total failure mechanism of the material increments leading to a greater number of the longitudinal smooth strips on the friction track and a smaller area of brittle fracture sites (Figure 22, c).

At sliding velocities > 80 m/s almost the whole surface of the friction track acquires a smoothed relief. However, the material undergoes brittle fracture along the deep grooves (Figure 22, d, arrows 1), brittle spalling on the groove edges being the proof (Figure 22, d, arrow 2). These grooves may appear after a single passage of an abrasive particle (possibly an agglomerate of the particles) whose penetration depth is so great that the sum of the contact and thermal stresses may exceed the material strength. As a result, the material undergoes brittle fracture with chips appearing ahead of the embedded abrasive particle and spalling over the groove edges rather than is deformed plastically. A similar failure pattern is observed at velocities > 100 m/s (Figure 22, e).

The pattern of the material failure on the periphery of the friction track differs much from the above-described one. This is connected, most probably, with different conditions of the thermal loading of the middle and peripheral track parts. At the sliding velocity of 26 m/s the zone where abrasive particles enter contact with glass (Figure 23, a) experiences a weaker heat effect in contrast to other track parts since the abrasive particles that have left the contact zone and have performed almost a full revolution manage to dissipate the heat accumulated at the previous contact with glass and enter in contact again having a lower temperature. In the process of travelling over the friction zone a particle gets heated and its temperature elevates as it approaches the contact exit.

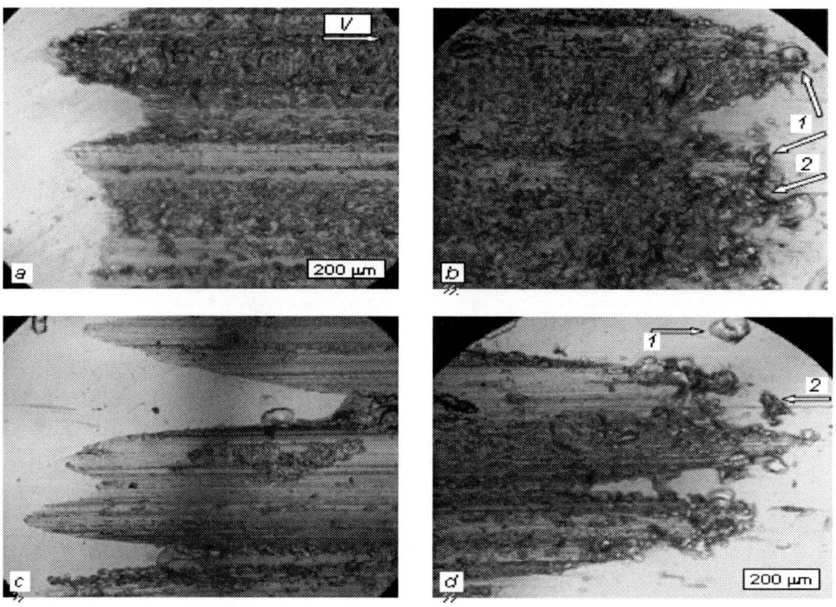

Figure 23. Optical images of friction track on glass surface after abrasive wear at different sliding velocities: a, b – 26 m/s; c, d – 55 m/s; a, c – contact entry; b, d – contact exit

Thus, the position of the maximal frictional temperature shifts towards the contact exit causing the severer fracture of glass in this zone (Figure 23, b). A thicker network of fatigue cracks is observed here and the contact site boundary shows spherical damage traces (arrows 1). The microcracks perpendicular to the velocity vector are seen here (arrow 2). Similar differences in glass fracture pattern at the opposite sites of the friction track are found at higher sliding velocities (55 m/s) (Figure 23, c, d). A distinctive feature is only that in this case the thermal stresses are so high that they are capable of inducing glass thermal cracking outside the contact area. For example, we have detected a spherical-like spot of glass thermal cracking in the region adjacent to the contact exit where friction did not occur (arrow 1 in Figure 23, d). There are also the damaged areas of irregular shape (arrow 2).

With increasing the velocity of abrasive grains up to 80 m/s the pattern of glass fracture in the zone of entering contact does not change essentially (Figure 24, a) while the opposite part of the friction track undergoes much severer fracture (Figure 24, b). The zone adjacent to the contact site boundary not subjected to friction contains the fragment of the thermally damaged material shaped as a chain of spalled spots (shown by the arrow). The local areas of the glass surface remote from the contact boundary also experience thermal cracking (Figure 24, d, arrow 1). This proves the effect of high thermal stresses at the contact exit and in adjoining regions.

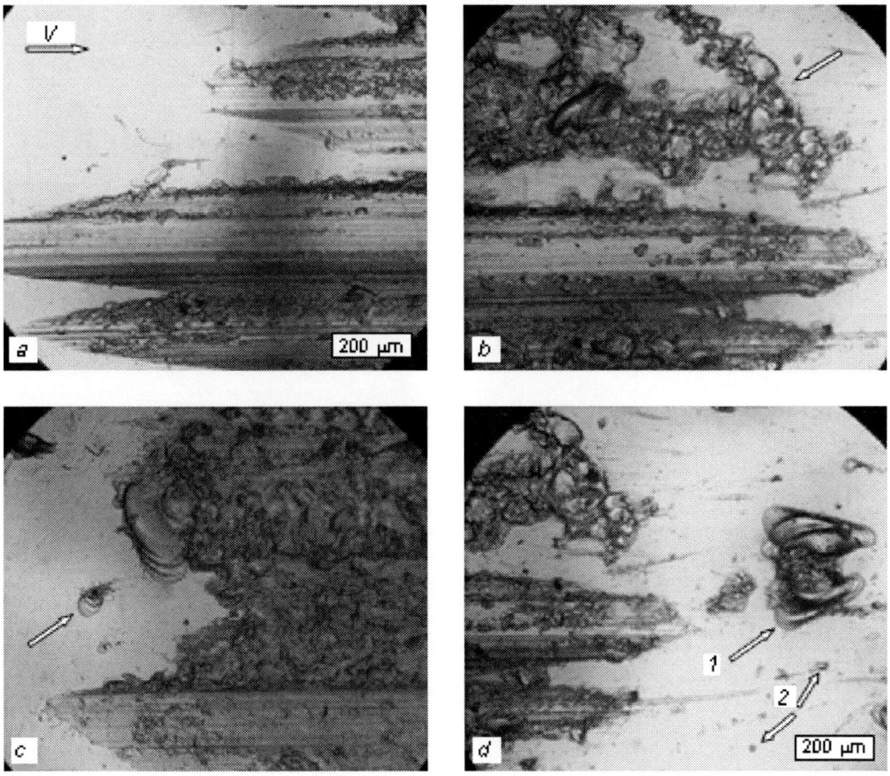

Figure 24. Optical images of friction track on glass surface after abrasive wear at different sliding velocities: a, b, d – 80 m/s; c – 105 m/s; a, c – contact entry; b, d – contact exit

Elevated temperatures in the abrasive particle – glass contact may lead to local glass melting. In some cases microvolumes of molten glass are removed from the friction track and deposit on the specimen surface as droplets (arrows 2 in Figure 24, d).

At sliding velocities > 100 m/s the pattern of the abrasive wear of glass does not essentially differ from the described above. The only specific feature is the much severer fatigue fracture of the material in the zone of entering contact as compared to fracture at V = 80 m/s. In addition, small spots of local glass damage are seen in the regions adjoining the contact entry (Figure 24, c). The reason is that with increasing the disc rotation speed (sliding velocity) the abrasive particles heated in the friction zone and entering contact repeatedly do not manage to cool down that raises thermal stresses near the contact site boundary.

The difference of the wear of sapphire from that of silicate glass is that the former does not experience local melting. The dominating wear mode of sapphire is its brittle spalling in the regions bounded by fatigue cracks (Figure 25, a, b). The fatigue fracture of sapphire is observed even outside the contact site (the arrow Figure 25, c).

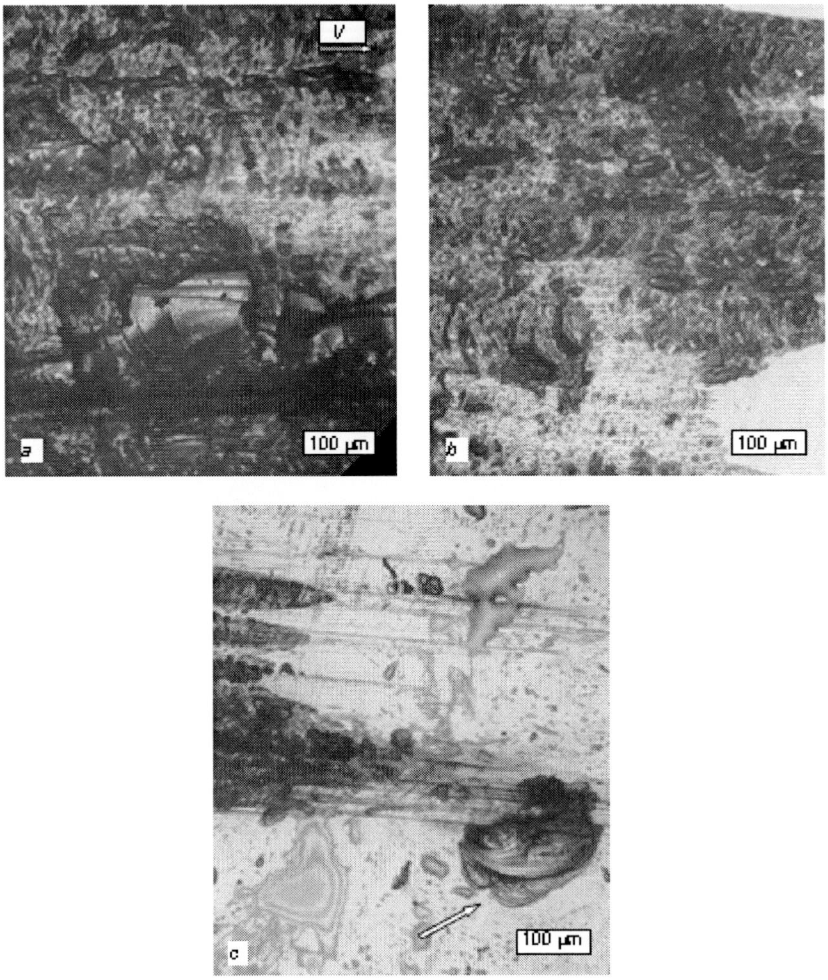

Figure 25. Optical images of friction track on sapphire surface after abrasive wear under load of 1.3 N and at different sliding velocities: a, b – 21 m/s; c – 33.5 m/s; a – middle part; b, c – contact exit

5. Conclusion

We have presented only a brief overview of the achievements in thermotribology. Studies in this field are extensive and specialized, so that this chapter can not encompass all the results obtained. The interested readers are referred to numerous publications in the related journals – "Tribology International", "Wear", "International Journal of Heat and Mass Transfer", "Transactions of ASME. Journal of Tribology", "Journal of Friction and Wear" (the English translation of "Trenie i Iznos"), to name just a few. Many relevant collections of papers and conference proceedings are also published yearly. Here we only conclude that thermotribology has become one of the fast-developing directions of the science of friction, wear, and lubrication having a promising future. New achievements in this field will be promoted, in our opinion, by the aspiration of researchers to get a deeper insight into the physical mechanisms of frictional heating. This can be made feasible on the basis of up-to-date thermography techniques and rapidly developing numerical computational methods which are a powerful instrument for the adequate understanding of thermal processes in friction.

REFERENCES

[1] Kragelskii, I. V., Schedrov, V. S. (1956). *Development of the Science of Friction,* USSR Academy of Sciences Publishing: Moscow, pp 5-6.

[2] Myshkin, N. K., Sviridenok, A. I., Tkachuk, D. V. (2007). *Proc. Int. Conf. "POLYCOMTRIB–2007".* 7-8.

[3] Blok, H. (1937). *Proc. Inst. Mech. Eng.* 2, 222-235.

[4] Myshkin, N. K., Petrokovets, M. I. (2007). *Friction, Wear, Lubrication. Physical Principles and Engineering Applications of Tribology,* Fizmatlit: Moscow, pp 95-96.

[5] Montgomery, R. S. (1976). *Wear.* 36, 275-298.

[6] Carignan, F. J., Rabinowicz, E. (1979). *ASLE Trans*, 24, 451-459.

[7] Pogosyan, A. K. (1977). *Friction and Wear of Filled Polymers;* Nauka: Moscow, pp 8-9.

[8] Quinn, T. F. J. (1983). *Tribol. Int*, 16, 257-271.

[9] Quinn, T. F. J. (1983). *Tribol. Int*, 16, 305-315.

[10] Matveevskii, R. M. (1971). *Temperature Stability of Boundary Lubricating Films and Solid Lubricant Coatings in Friction of Metals and Alloys;* Nauka: Moscow, pp 13-14.

[11] Dow, T. A. (1980). *Wear*, 59, 213-221.

[12] Bill, R. C., Ludwig, L. P. (1980). *Wear*, 59, 165-189.

[13] Kennedy, F. E., Karpe, S. A. (1982). *Wear*, 79, 21-36.

[14] Chichinadze, A. V. (1967). *Calculation and Study of External Friction at Braking;* Nauka: Moscow, pp 11-13.

[15] Dow, T. A., Kannel, J. W. (1982). *Wear*, 79, 93-105.

[16] Drozdov, Yu. N., Pavlov, V. G., Puchkov, V. N. (1986). *Friction and Wear under Extreme Conditions;* Mashinostroenie: Moscow, pp 34-36.

[17] Kostetskii, B. I., Linnik, Yu. I. (1968). *Proc. USSR Acad. Sci*, 183, 1052-1055.

[18] Uetz, H., Föhl, J. (1978). *Wear*, 49, 253-264.

[19] Bowden, F. P., Ridler, K. E. W. (1936). *Proc. Roy. Soc. Ser. A*, 154, 640-647.

[20] Jaeger, J. C. (1942). *Proc. Roy. Soc. NSW*, 56, 203-224.

[21] Levitskii, M. P. (1949). *J. Techn. Phys*, 19, 1010-1014.

[22] Charron, F. (1943). *Publ. scient. et techn. Ministere air*, 182, 34-41.

[23] Pyzhevich, L. M. (1964). *Calculation of Frictional Brakes;* Mashinostroenie: Moscow, pp 88-90.

[24] Balakin, V. A. (1982). *Friction and Wear at High Sliding Velocities,* Mashinostroenie: Moscow, pp 45-50.

[25] Hasselgruber, H. (1963). *Konstruktion*, 2, 136-141.

[26] Ginzburg, A. G. (1974). *Calculation and Testing of Friction Pairs* Mashinostroenie: Moscow, pp 23-28.

[27] Evtushenko, A. A., Ukhanskaya, O. M., Chapovskaya, R. B. (1997). *J. Eng. Phys. Thermophys*, 70, 495-500.

[28] Evtushenko, A. A., Ukhanskaya, O. M. (1993). *Phys.-Chem. Mech. Mater* , 2, 89-96.

[29] Kragelskii, I. V., Chichinadze, A. V. (1954). *Industrial Lab*, 53, 607-610.
[30] Schedrov, V. S. (1955). *Frict. Wear Mach*, 10, 155-296.
[31] Blok, H. (1937). *Proc. Inst. Mech. Eng*, 2, 18-22.
[32] Blok, H. (1955). *Appl. Scient. Res. Sect. A*, 2-3, 151-181.
[33] Blok, H. (1987). *Proc. Int. Conf. Tribology "Frict., Lubr., Wear"*, 1, 1-8.
[34] Blok, H. (1984). *Proc. 3rd Workshop on Thermomechanical Effects in Sliding Systems*, 8-12.
[35] Archard, J. F. (1958). *Wear*, 2, 438-455.
[36] Bowden, F. P., Tabor, D. (1964). *The Friction and Lubrication of Solids,* Clarendon: Oxford, pp 223-224.
[37] Holm, R. (1948). *J. Appl. Phys*, 19, 361-366.
[38] Quinn, T. F. J., Winer, W. O. (1987). *Trans. ASME. J. Tribol*, 110, 290-295.
[39] Tkachuk, D. V., Bogdanovich, P. N., Belov, V. M. (2004). *Tribotest J*, 11, 125-135.
[40] Bogdanovich, P. N., Tkachuk, D. V. (2006). *Tribol. Int*, 39, 1355-1360.
[41] Kuhlmann-Wilsdorf, D. (1985). *Wear*, 105, 187-198.
[42] Kuhlmann-Wilsdorf, D. (1986). *Wear*, 107, 71-90.
[43] Kuhlmann-Wilsdorf, D. (1987). *Materials Science and Engineering*, 93, 119-133.
[44] Gurskii, B. E., Chichinadze, A. V. (2007). *J. Frict. Wear*, 28, 316-329.
[45] Gurskii, B. E., Chichinadze, A. V. (2007). *J. Frict. Wear*, 28, 395-400.
[46] Gurskii, B. E., Chichinadze, A. V. (2007). *J. Frict. Wear*, 28, 452-456.
[47] Chichinadze, A. V., Troyanovskaya, G. I. (1957). *Temperature Field, Friction Coefficient, and Wear Coefficient of Frictional Pairs;* Nauka: Moscow, pp 39-40.
[48] Kragelskii, I. V., Chupilko, G. G., Chichinadze, A. V. (1955). *Friction Processes in Aircraft Brakes. Selection of Frictional Pairs* USSR Academy of Sciences Publishing: Moscow, pp 122-128.
[49] Polyakov, N. V., Chichinadze, A. V. (1983). *J. Frict. Wear*, 4, 17-22.
[50] Chichinadze, A. V., Buyanovskii, I. A., Gurskii, B. E. (2002). *J. Frict. Wear*, 3, 90-97.
[51] Korovchinskii, M. V. (1966). *New in Theory of Friction*, 98-145.
[52] Korovchinskii, M. V. (1968). *New in Theory of Friction*, 5-72.
[53] Barber, J. R. (1970). *Int. J. Heat Mass Transfer*, 13, 857-869.
[54] Marscher, W. D. (1982). *ASLE Trans*, 25, 157-184.
[55] Ling, F. F. (1973). *Surface Mechanics;* Wiley: New York, pp 337-354.
[56] Ling, F. F. (1969). *J. Lubr. Technol*, 91, 397-405.
[57] Floquet, A.,Play, D. (1981). *J. Lubr. Technol*, 103, 243-252.
[58] Floquet, A., Play, D., Godet, M. (1977). *J. Lubr. Technol*, 99, 277-283.
[59] Ling, F. F. (1959). *Z. Angew. Math. Phys*, 10, 461-474.
[60] Ryhming, I. L. (1979). *Acta Mech*, 32, 261-274.
[61] Kennedy, F. E. (1982). *J. Lubr. Technol*, 104, 582-588.
[62] Kennedy, F. E., Ling, F. F. (1974). *J. Lubr. Technol*, 96, 497-508.
[63] Kennedy, F. E. (1981). *J. Lubr. Technol*, 103, 97-103.
[64] Rozeanu, L., Pnueli, D. (1978). *J. Lubr. Technol*, 100, 479-485.
[65] Rozeanu, L., Pnueli, D. (1980). *J. Lubr. Technol*, 102, 545-551.
[66] Balakin, V. A. (1981). *J. Eng. Phys*, 40, 660-665.
[67] Bogdanovich, P. N., Prushak, V. Ya. (1999). *Friction and Wear in Machines;* Vysheishaya Shkola: Minsk, pp 176-185.
[68] Bogdanovich, P. N., Tkachuk, D. V., Belov, V. M. (2006). *J. Frict. Wear*, 4, 74-83.

[69] Sladkov, A. Z. (1972). *Industrial Lab*, 2, 25-29.
[70] Ainbinder, S. B., Loginova, A. Ya. (1982). *J. Frict. Wear, 1,* 19-24.
[71] Zinenko, S. A., Karapetyan, S. S., Silin, A. A. (1982). *J. Frict. Wear*, 3, 102-108.
[72] Zinenko, S. A., Karapetyan, S. S., Silin, A. A. (1982). *J. Frict. Wear*, 4, 57-60.
[73] Starchenko, Yu. P., Movchun, V. M., Miletskii, A. V. (1989). *J. Frict. Wear*, 2, 53-59.
[74] Fitzgerald, M. R., Neal, P. B. (1992). *Trans. ASME. J. Tribol*, 114, 122-130.
[75] Benabdallah, S. M. H., Gauvin, R. (1989). *Proc. 5th Int. Congr. Tribol*, 3, 391-397.
[76] Gordon, A. N. (1976). *Accuracy of Contact Temperature Measurement Methods;* Mashinostroenie: Moscow, pp 22-23.
[77] Danielyan, A. M. (1964). *Tool Heating and Wear in Metal Cutting;* Mashinostroenie: Moscow, pp 58-60.
[78] Epifanov, V. I., Kononenko, V. I., Soltan, A. V. (1980). *J. Superhard Mater*, 2, 3-4.
[79] Rogov, V. V., Filatov, Yu. D., Karapuzov, V. R. (1978). *Synthetic Diamonds*, 3, 49-53.
[80] Nosko, A. L., Romashko, A. M., Kozhemyakina, V. D. (1982). *J. Frict. Wear*, 6, 96-101.
[81] Guskov, V. I. (1974). *Mach.-Build. Bull*, 4, 40-43.
[82] Keglin, B. G., Tikhomirov, V. P. (1990). *J. Frict. Wear*, 1, 60-63.
[83] Lysenok, Yu. V., Balakin, V. A. (2003). *J. Frict. Wear*, 2, 57-63.
[84] Goryunov, V. M., Maksimov, M. M., Piskunov, Yu. M. (1984). *J. Frict. Wear*, 1, 118-120.
[85] Rovinskii, D. Ya., Panaioti, I. I. (1975). *Powder Metallurgy*, 6, 93-98.
[86] Rovinskii, D. Ya., Fedorchenko, I. M., Kremenchugskii, L. S. (1984). *J. Frict. Wear*, 4, 36-40.
[87] Konyukhov, E. S., Lirman, M. V., Shpinyak, M. N. (1967). *Industrial Lab*, 5, 835-836.
[88] Chandrasekar, S., Farris, T. N., Bhushan, B. (1990). *Trans. ASME. J. Tribol*, 112, 535-540.
[89] Bhushan, B. (1990). *Tribology and Mechanics of Magnetic Storage Devices;* Springer-Verlag: New York, pp 399-403.
[90] Tripathy, B. S., Furey, M. J. (1993). *Wear*, 162-164, 385-396.
[91] Gulino, R., Bair, S., Winer, W. O., Bhushan, B. (1986). *Trans. ASME. J. Tribol*, 108, 29-34.
[92] Winer, W. O. (1984). *Proc. NASA Conf*, 2, 535-548.
[93] Chichinadze, A. V., Goryunov, V. M., Ginzburg, A. G. et al. (1996). *Probl. Mach.-Build. Autom*, 6, 40-50.
[94] Chichinadze, A. V., Matveevskii, R. M., Braun, E. D. (1986). *Materials in Tribology of Non-Stationary Processes;* Mashinostroenie: Moscow, pp 123-124.
[95] Wang, Y., Lei, T., Yan, M., Gao, C. (1992). *J. Appl. Phys*, 25, 165-169.
[96] Erhard, G., Weis, Ch. (1987). *Konstruktion*, 11, 423-430.
[97] Masahiro, S., Terumi, I., Yoshizo, O. (1994). *Proc. 3rd Asian Symp. Visual*, 563-567.
[98] Reungoat, D., Tournerie, B. (1991). *Mec., mater., elec*, 438, 31-32.
[99] Young, H. T. (1996). *Wear*, 201, 117-120.
[100] Wieleba, W. (2005). *Wear*, 258, 870-876.
[101] Barzdanis, Yu. V. (2002). *Method for Determining Bulk Temperature Fields and Their Gradients in "Wheel–Brake Block" Pair. PhD Thesis;* RGUPS: Rostov-on-Don, pp 25-28.

[102] Bogdanovich, P. N., Belov, V. M., Tkachuk, D. V. (2006). *Int. J. Appl. Mech. Eng*, 1, 5-13.
[103] Ainbinder, S. B., Tyunina, E. L., Tsirule, K. I. (1981). *Properties of Polymers in Various Stress States;* Khimiya: Moscow, pp 176-177.
[104] Tkachuk, D. V., Bogdanovich, P. N. (2004). *Int. J. Appl. Mech. Eng*, 9, 177-182.
[105] Saverin, M. M. (1946). *Contact Strength of Material under Simultaneous Action of Normal and Tangential Loads;* Mashgiz: Moscow, pp 46-51.
[106] Rigney, D. A., Hirth, J. P. (1979). *Wear*, 53, 345-370.
[107] Bogdanovich, P. N., Belov, V. M., Tkachuk, D. V. (2007). *J. Frict. Wear*, 28, 44-50.
[108] Bogdanovich, P., Tkachuk, D. (2007). *Acta Mech. Automat*, 1, 15-18.
[109] Tkachuk, D. V., Bogdanovich, P. N., Bliznets, D. A. (2007). *J. Frict. Wear*, 28, 532-540.
[110] Bogdanovich, P. N., Tkachuk, D. V., Bliznets, D. A. *Int. J. Heat Mass Transfer*. In press.
[111] Epifanov, V. I., Pesina, A. Ya., Zykov, L. V. (1982). *Technology of Machining of Diamonds into Brilliants;* Vysshaya Shkola: Moscow, pp 145-156.
[112] Lieneweg, F. (1980). *Measurement of Temperatures in Engineering;* Metallurgiya: Moscow, pp 456-457.
[113] Parker, R. C., Marshall, P. R. (1948). *Proc. Inst. Mech. Eng*, 158, 209-229.
[114] Sibley, L. B., Allen, C. M. (1961). *ASME Paper 61-LUBS-15.*
[115] Santini, J. J., Kennedy, F. E. (1975). *Lubr. Eng*, 31, 402-417.
[116] Barber, J. R. (1967). *Wear*, 10, 155-159.
[117] Barber, J. R. (1969). *Proc. Roy. Soc. Ser. A*, 312, 381-394.
[118] Dow, T. A., Burton, R. A. (1972). *Wear*, 19, 315-328.
[119] Dow, T. A., Stockwell, R. D. (1977). *J. Lubr. Techno*, 95, 359-364.
[120] Burton, R. A., Nerlikar, V., Kilaparti, S. R. (1973). *Wear*, 24, 177-188.
[121] Burton, R. A. (1973). *Wear*, 24, 189-198.
[122] Nau, B. S. (1967). *Proc. 3rd Int. Conf. Fluid Sealing*, 34-48.
[123] Banerjee, B. N., Burton, R. A. (1976). *Nature*, 261, 399-400.
[124] Banerjee, B. N., Burton, R. A. (1978). *J. Mech. Eng. Sci*, 20, 239-242.
[125] Banerjee, B. N. (1980). *Wear*, 59, 89-110.
[126] Lebeck, A. O. (1976). *J. Lubr. Technol*, 98, 277-285.
[127] Lebeck, A. O. (1980). *Wear*, 59, 121-133.
[128] Dow, T. A. (1980). *Wear*, 59, 31-52.
[129] Netzel, J. P. (1980). *Wear*, 59, 135-148.
[130] Kennedy, F. E. (1980). *Wear*, 59, 149-164.
[131] Bogdanovich, P. N., Tkachuk, D. V. (1997). *Proc. Belarus Ac. Sci*, 3, 119-122.
[132] Petrokovets, M. I. (1994). *J. Frict. Wear*, 4, 114-117.
[133] Mow, V. C., Cheng, H. S. (1967). *Z. Angew. Math. Phys*, 18, 500-507.
[134] Yang, C. C. (1971). *Int. J. Eng. Sci*, 9, 507-520.
[135] Mercier, R. J., Malkin, S., Mollendorf, J. C. (1978). *J. Eng. Ind*, 100, 43-48.
[136] Tseng, M. C., Burton, R. A. (1982). *Wear*, 79, 1-9.
[137] Ju, F. D., Huang, J. H. (1982). *Wear*, 79, 107-1189.
[138] Evtushenko, A. A., Chapovska, R. B. (1994). *J. Frict. Wear*, 4, 23-28.
[139] Yevtushenko, A. A., Chapovska, R. B. (1996). *Int. J. Mech. Sci*, 38, 1103-1116.
[140] Evtushenko, A. A., Ivanik, E. G. (1996). *J. Eng. Phys. Thermophys*, 69, 61-66.

[141] Evtushenko, A. A., Ukhanskaya, O. M. (1994). *J. Eng. Phys. Thermophys*, 66, 563-570.

[142] Yevtushenko, A. A., Ivanyk, E. G. (1996). *Wear*, 197, 160-168.

[143] Evtushenko, A., Zelenyak, V. (1998). *J. Eng. Phys. Thermophys, 71,* 1141-1146.

[144] Konechny, S., Evtushenko, A., Zelenyak, V. (2001). *J. Frict. Wear*, 1, 31-37.

[145] Evtushenko, A., Kutsei, M. (2006). *J. Frict. Wear*, 2, 9-16.

[146] Yevtushenko, A. A. (2005). *Proc. 3rd Symp. Failure Mater. Struct*, 447-450.

[147] Tkachuk, D. V., Bogdanovich, P. N., Bliznets, D. A. (2007). *Abstr. Int. Sci. Conf. "Polycomtrib-2007"*. 192-193.

[148] Bogdanovich, P. N., Tkachuk, D. V., Bliznets, D. A. *Wear.* In press.

In: Tribology Research Trends
Editor: Taisho Hasegawa
ISBN: 978-1-60456-912-4

Chapter 3

TRIBOLOGY AND BIOTRIBOCORROSION OF ARTIFICIAL JOINT PROSTHESES

Yu Yan*
School of Mechanicl Engineering, University of Leeds, Leeds, UK

ABSTRACT

Since the introduction of medical implants into human bodies, corrosion and wear have been regarded as key issues for their long-term durability. There has been a recent renewed interest in the use of large diameter metal-on-metal (MoM) hips, primarily because of the reduced volumetric wear compared with the well-established polyethylene-on-metal joints. Long term durability of MoM joints relies on control of both their *corrosion resistance* (relating to ion release) and *wear behaviour* (relating to creation of nanometre-scale wear debris). Concerns about the potential risk of released metal ions to the biological environment (patient) are of great importance. In this respect tribocorrosion is a serious consideration in joint performance.

One of the key metal ion release processes for metallic hip replacements – tribocorrosion, has not been investigated in any systematic manner. In this present study an electrochemical cell integrated into a reciprocating tribo-meter was designed and employed to enable evaluation of the corrosion and tribocorrosion behaviour in simulated synovial fluids. A range of electrochemical methods were used in the assessment of materials under biotribocorrosion systems and results were supported by surface analysis SEM (Scanning Electron Microscopy) and XPS X-ray Photoelectron Microscopy) and bulk solution analysis techniques ICP-MS (Inductively Coupled Plasma Mass Spectroscopy).

The material degradation rate is strongly dependent upon the charge transfer (*corrosion*), the mechanical damage (*tribology*) and also their interactions (*tribocorrosion*) in these simulated biological environments. Corrosion/tribocorrosion plays a very important role in the degradation processes. 20%-30% material damage is attributed to corrosion-related processes in the steady state after a 35%-45% material loss

* University of Leeds, Leeds, UK. LS2 9JT.

due to corrosion in the running-in state. The development of the tribofilm (oxides/hydroxides/organometallic complexes) is responsible for the lower wear rate and lower friction in the steady state. Material properties (hardness, microstructure and wettability) all influence biotribocorrosion behaviour. This chapter will discuss the known factors and challenges in this quickly expending area.

1. Introduction

The phenomena of tribology have been realized for thousands of years. In 1966, the well-known 'Jost report' was released to British government. Since then, the word 'tribology' has been widely used and research on this area has been greatly explored. Tribology has become an interdisciplinary area. It is linked with materials, chemistry, physics and even biology. It was my privilege to be invited to attend the meeting '50++ Tribology the Next 50 Years' in the headquarter of the Institute of Mechanical Engineers in London. Dr. Peter Jost, Duncan Dowson and others reviewed the past, present and future of Tribology and identified possible topics for the next a few decades in Tribology.

- Biotribology: Tribological studies of systems in the body alongside those on synovial joints. These include the skin, hair, teeth, eyes, blood flow, hear valves and artificial hearts. These will be influential in the engineering, health care and cosmetic industries. Biomimetitcs is also an emerging area, using natural world biological inspiration to develop novel engineering solutions.
- Micro- and Nano-tribology: This includes assemblies of atoms and molecules having at least one nano-scale dimension. Four areas at present relating to nano-tribology include probes, structure, processes and simulation of tribological systems.
- Lubricant: Increased emphasis has been on energy efficiency and carbon dioxide emissions. New additives will be greener and more complicated (nanoparticles).
- Coatings: It is clear that coatings have huge potential to reduce wear and friction in a large range of applications and can provide solutions where lubricants can not be used, such as in space.
- Modelling: General property modelling where basic mechanical properties are linked to wear behaviour (yield strength, work hardening, ductility, fracture toughness etc.). This is another approach to understanding tribological behaviour.

In this chapter, a small branch of tribology – biotribocorrosion is discussed.

1.1. Background of Arthroplasty Implant

Arthritis causes long term health problems for more than one in seven adults and is the second most common cause of absence from work in both men and women [1]. 10% of the population did and will receive one (or more) joint replacement(s) during their lifetime in the UK [2]. Several million total hip replacements with a market value of about 2 billion pounds are implanted annually worldwide. Figure 1.1. summarizes the implant market by the

Swedish National Hip Arthroplasty Register in 2004 [1]. 25,000 hip replacements (91% primary and 9% revision) and 22,000 knee replacements (94% primary and 6% revision) had been made in England in 2003 (between April and December) [2]. The major group receiving joint implants in the UK is the population of age range from 45-70 (Figure 1.2.). However, more and more young and active patients demand to have their joint (hip/knee) replaced due to arthritis and accidental damage. It requires the replacement to last longer and 'safer'. Therefore, new generation Metal-on-Metal (MoM) joint replacements have been considered as an alternative to the commonly used Metal-on-Polyethylene (MoP) implants [3-5].

A survey by NJR (National Joints Registry of the UK) shows the most popular femoral head material is metal, which was implanted in 76.3% primary hip replacements [2]. Some concerns remain regarding the levels of metallic ions released *in vivo* from these MoM prostheses. Even though the implant is expected to last 10-20 years or longer, 10% of the implants need revision within 5 years due to various reasons [1-5]. However, wear is one of the major reasons attributed to the failure or revision of implants. Corrosion is regarded as the source of released ions. Therefore, to improve the wear resistance and biocompatibility of implant materials is the focus of most implant industries and manufacturers.

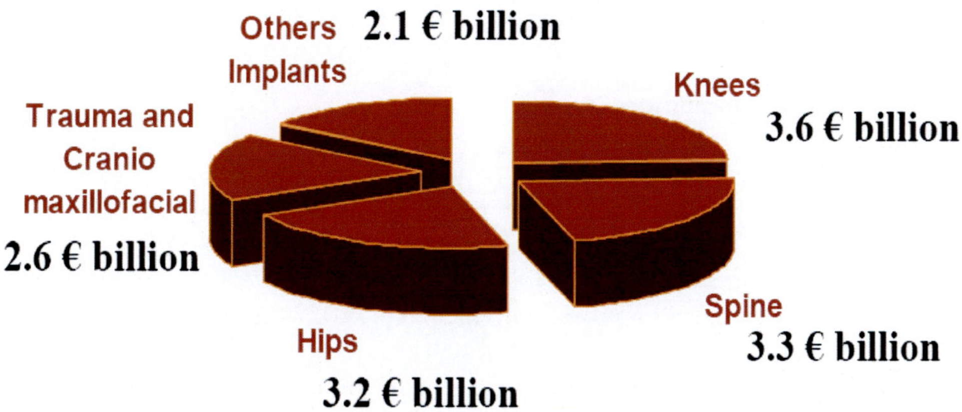

Figure 1.1. Summary of the implant market worldwide.

Biomaterials are defined by Williams [6] as non-viable materials used in medical devices, intended to interact with biological systems. Biomaterials cover all classes of materials for various needs (e.g. dental, orthopaedic and cardiovascular). Implantable devices intended for major load-bearing applications (primarily in orthopaedic and dentistry) are made mainly from metals, ceramics or polymers [7-9]. Metals and alloys have a wide range of applications in biomedical devices. Devices from metals and alloys can be used for fracture fixation, partial and total joint replacement, external splints and heart valves [7]. Their high modulus, yield point and ductility make them suitable for such applications. Although pure metals are sometimes used, alloys frequently provide improvement in material properties. The disadvantage of metals is that they are susceptible to chemical and electrochemical degradation. However, polymers and ceramics can also be subjected to corrosion attack [8]. Three groups dominate biomedical metals: iron-based stainless steel, cobalt-based alloys and Titanium and Titanium alloys.

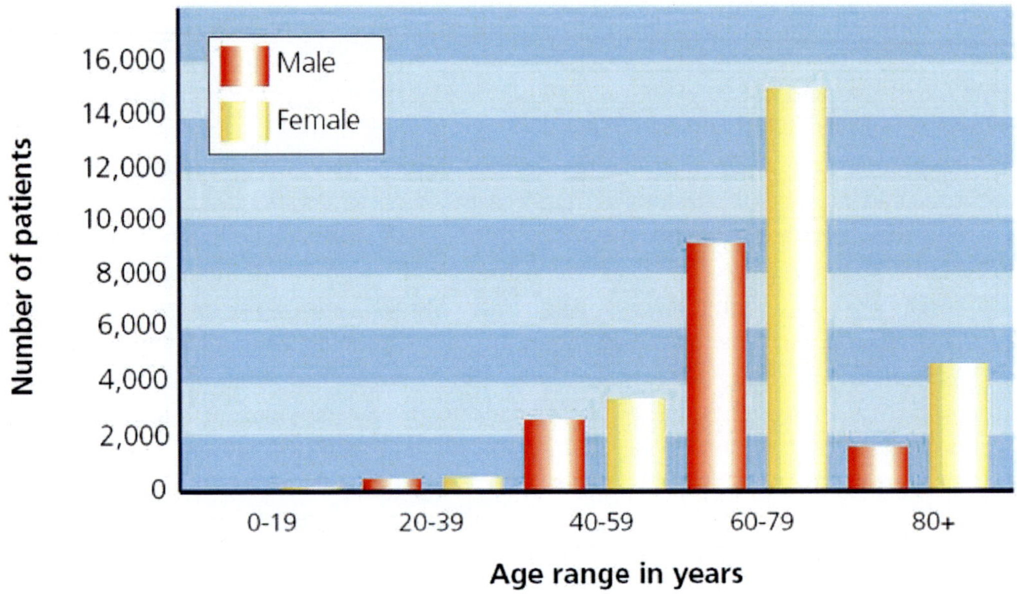

Figure 1.2. Age range of hip replacement recipients in the UK in 2003.

1.2. Total Joint Replacement

Total Joint Replacement (TJR), or joint arthroplasty, is a surgical procedure in which the entire joint is removed and replaced with a prosthetic joint. Some typical TJR implants are shown in Figure 1.3. Due to the increasing osteoarthrosis and similar disabling conditions, total artificial replacement of human joints has become a widely used treatment. It is performed to release pain and improve joint function [10]. In TJR, the most popular types are Total Hip Replacement (THR) (Figure 1.3(a) and (b)) and Total Knee Replacement (TKR) (Figure 1.3(d)). TJR as normally only performed on patients who were over 60 years old in the early periods. However, nowadays, more and more young and active patients are requiring TJR surgeries. Improvement and development of safer, longer lasting and better functioning implants are expected for such applications [10, 11].

Because TJR are under the fluctuating and cyclically repeated forces caused by gravity and muscular action, mechanical characteristics such as strength elasticity, toughness and ductility are relevant factors [12]. Figure 1.4. shows the force change during a cycle of walk. A peak load of more than 4 times bodyweight was obtained when the action of heel strike was made. Two phases in the human hip joint are classified during a normal walking cycle. In the stance phase, the hip joint carries very high load and the relative movement in the joint (between femoral head and socket) is little. In the swing phase, even though the load on the joint is lower than in the stance phase, the movement of the joint is greater. Some authors believe that the material degradation progresses (wear) severely in this phase [13]. The correct counterface, material combinations, surface finish, diameter of femoral head and clearance are important aspects in minimizing friction, wear and corrosion [12]. Low friction, low wear and good biocompatibility are desirable characteristics for TJR prostheses.

From a historical study, the first recorded attempt to replace the hip joint was made by Gluck from Berlin in 1880. The prosthesis was made of ivory. Then the designs intended to just replace femoral head by different materials. In this period, acrylic materials and metals were involved [13]. The TJR have developed largely over approximately the last 60 years. In the early period of TJR, a number of configurations of Metal-on-Metal (MoM) hip prostheses were invented and used.

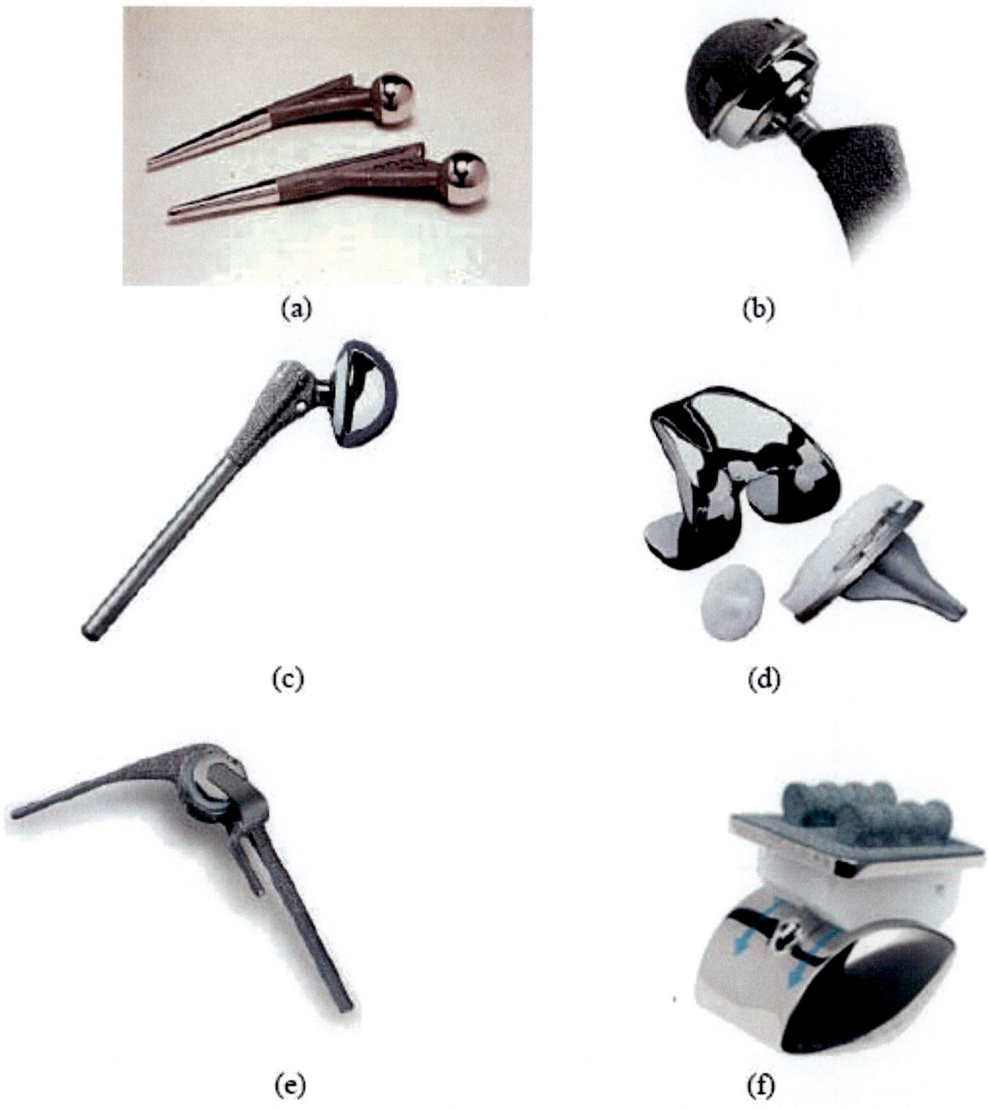

Figure 1.3. Typical TJR implants (a) (b) total hip replacement (c) shoulder replacement (d) knee replacement (e) elbow replacement (f) ankle replacement.

They are normally referred to as *the first generation MoM bearings*, which were manufactured from as-cast cobalt chromium alloys. Among them, the McKee-Farrar prosthesis was the most widely used design [14]. While some of the early McKee implants experienced short term failures, others survived for service periods of 20 or 30 years [15]. In the late 1950s and early 1960s, Charnley tried polyethylene (PTFE and UHMWPE) as an alternative acetabular material, a design which remains to this day. It achieved low friction and it involved a stainless steel (later cobalt chromium alloy) femoral component [16].

However, problems of loosening and osteolysis are associated with this Metal-on-Polyethylene (MoP) design. The drive to improve the long-term service life of TJR has resulted in interest in using hard-on-hard bearing couples. Developed and modified Ceramic-on-Ceramic (CoC), MoM and Ceramic-on-Metal (CoM) bearing systems were introduced and have attracted many investigations [17].

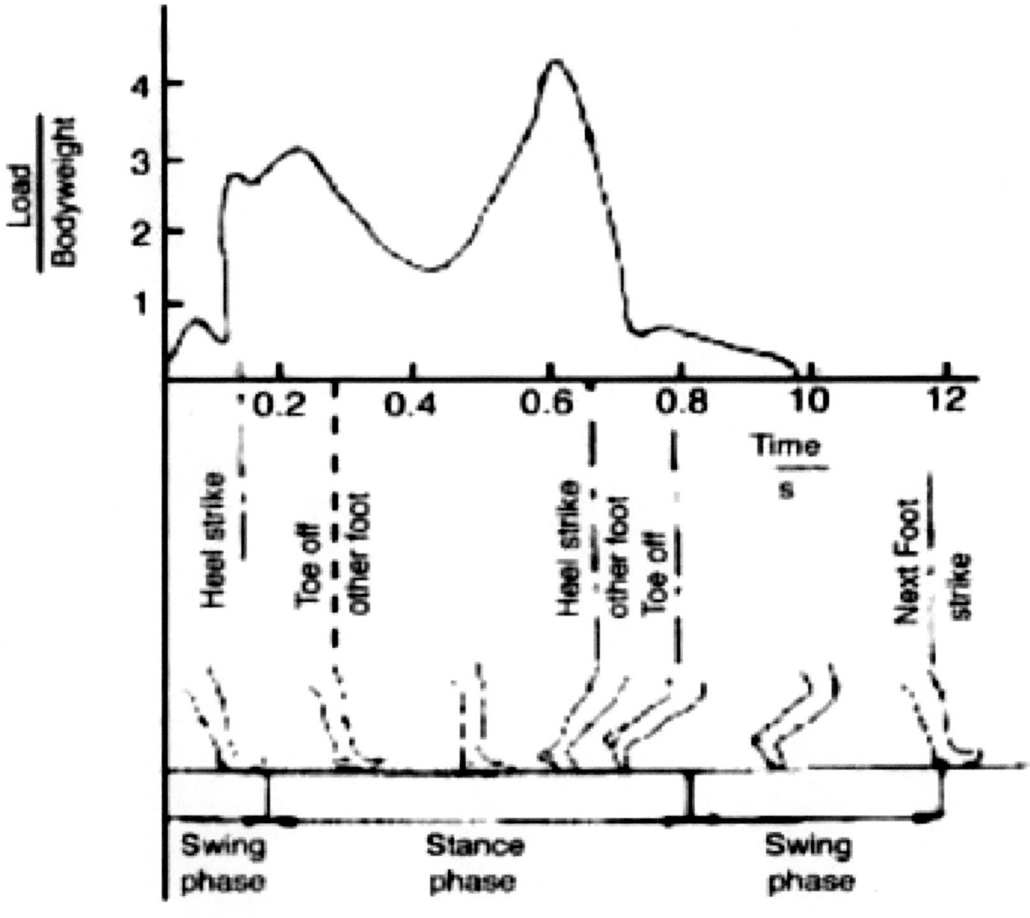

Figure 1.4. Hip joint force during one cycle of walk [10].

Alumina and zirconia ceramics were introduced into orthopaedic surgeries in the early 1970s. From clinical and laboratory results, for CoC types of TJR, especially THR, low wear rates were noticed. CoC benefits from higher scratch resistance properties and better wettalility and therefore enhanced lubrication properties [18]. However, complications due to the ceramic brittle fracture, acetabular loosening and ceramic degradation have been reported. In addition, it is very difficult to remove all of the ceramic fragments from the surrounding tissues if CoC fails and revision surgery is required [17]. Modification of surfaces of CoC components has been attempted and results are promising [18]. Many authors believe that CoC prostheses are an effective option for younger and more active patients. The history of hip replacement is shown in Figure 1.5.

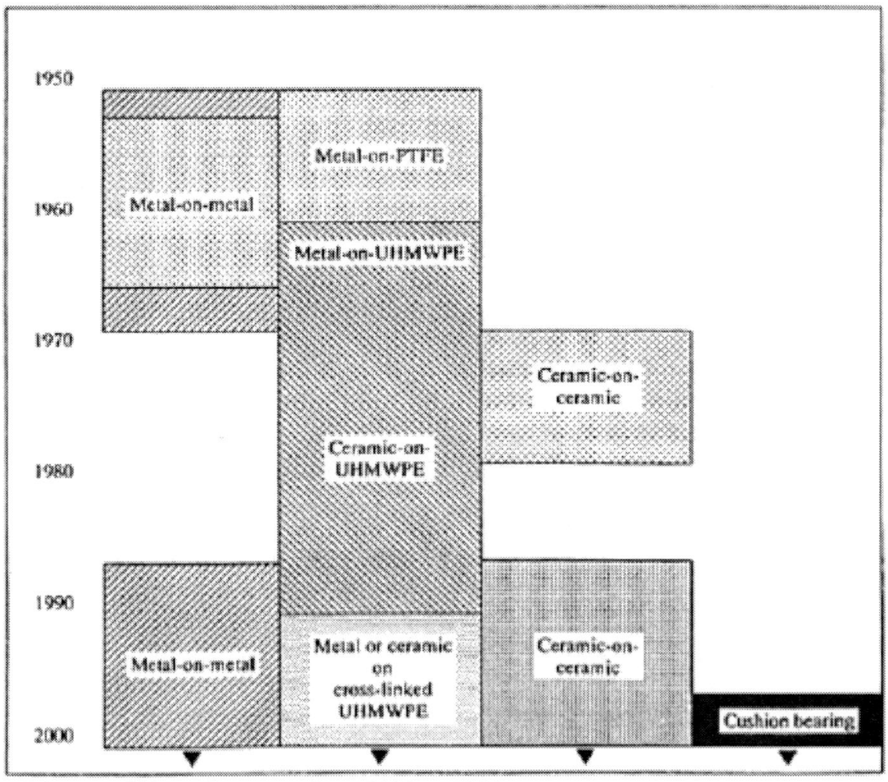

Figure 1.5. History of hip replacement from 1950 to 2000.

1.3 Tribology Behaviour of MoM TJR

The wear rates for MoM, particularly for CoCrMo MoM implants were very individual depended. The results are gathered from both *in vivo* and *in vitro* simulator tests. MoM implant material degradation is influenced by many factors such as macrogeometry (diameter and clearance), deviation from sphericity and alloy composition. In this section, consideration of MoM implant configuration design, the role of materials and different testing methods are reviewed. The discussion of their effect on MoM tribology behaviour is made.

For most tests, wear means the total material degradation and is commonly accepted due to mechanical processes. Wear rate then refers to the volumetric or gravimetric material loss during certain periods of time. However, *wear* (mechanical process) and *corrosion* (chemical and electrochemical processes) can not be separated in real MoM implants therefore in this section, the term 'wear' refers to the damage caused due to both mechanical and electrochemical processes. Details of studies and attempts to separate these effects will be reviewed in the later section.

Before they go to the simulator testing stage, candidate materials should be assessed by more simplified methods to screen or rank materials tribological behaviour or combinations. From a tribological point of view, in MoM artificial joints, sliding wear is the dominant wear mechanism. Pin-on-disk or ball-on-disk tests have been extensively carried out worldwide [19-22]. Materials which have high wear in pin-on-disk experiments are expected to have high wear rates when they used in joint replacement.

Dowson *et al.* [23] found that the predicted film thickness increased steadily as the implant diameter increased and by the time a film thickness of about 14 μm was established (femoral head diameter 36 mm), the wear rate is exceeding low (<0.08 mm^3/million cycles). Increase in the head diameter caused the wear rate to decline as the lubrication regimes developed from boundary to mixed. Another advantage of a large diameter head is that it can prevent dislocation of the femoral ball when the operation is performed [24].

The amount of ball-socket clearance in MoM hip joint implant also can influence the wear rate of a metal-on-metal bearing. Jin *et al.* [25] suggested that a small clearance can achieve full fluid film lubrication between ball and socket, therefore increases the contact area and reduces friction and wear. The increase of the volumetric wear as a function of the decrease of clearance was confirmed by Scholes *et al.* [26]. However, the smallest clearance limit of 20 μm and optimal clearance of 100-150 μm are recommended [27].

2. TRIBOCORROSION - BASIC THEORY

Tribocorrosion is defined as the chemical, electrochemical and mechanical processes leading to a degradation of materials under tribological contacts in a corrosive environment [28, 29]. It is an irreversible transformation.

Material degradation due to the combination of mechanical and electrochemical may occur under a variety of conditions. A sliding movement between two surfaces under two bodies or three bodies contact is a common cause of tribocorrosion. With micromotion involved, fretting-corrosion is a special type of tribocorrosion. Tribocorrosion is also observed in ball bearing under rolling contact. Particle impact or impingement attack can also result in a combined mechanical, chemical, electrochemical attack of material.

As discussed earlier, a stable passive film on metallic materials surface can protect materials from corrosion attack. Under mechanical relative movement, oxide films on passive metals can be locally or even completely destroyed. An accelerated corrosion has been found by different studies. It is generally believed that wear will enhance corrosion rate.

The electrochemical approach offers the opportunities to control the surface chemistry of the material under tribological contact. Mischler *et al.* [28] monitored the current change by holding a certain potential. The oxides of materials were disrupted by mechanical movement. The fundamental work suggested that the passive film on mechanical degradation played an important role.

The relationship between friction and electrochemical processes attracted a number of authors. Okazaki [30] noticed that friction can change biomaterials corrosion behaviour. 316L, CoCrMo and Ti6Al4V showed current fluctuation under tribological contact. Ponthiaux *et al.* [31] believed that a decreased friction was due to a third-body effect by the released corrosion products particles. Higher friction was found where a large tendency to passivation prevails. One important conclusion from this work is that between unworn and worn area, a galvanic couple may take place.

Garcia *et al.* [32] promoted an equation to model the current flow between the two areas. An anodic potential assumed to be provided and it is expressed as:

$$I = fA_a \int_0^{\frac{1}{f}} i_a(t)dt + f(A - A_a) \int_0^{\frac{1}{f}} i_p(t)dt \qquad \text{(Eq.1.1)}$$

where I is the current flowing through an electrode of total area A. A_a it the worn are. A-A_a then presents the passive area without tribological contact. The repassivation current density is $i_a(t)$. The $i_p(t)$ corresponds to the anodic current density for passive material polarized at that given anodic potential. The contact frequency is f.

Goldberg *et al.* [33] suggested an equation by which the current transient and repassivation current by single scratch test could be modelled and predicted. Oxide film fractured due to a mechanical sliding.

$$I(t) = I_{peak} \exp\left[\frac{-(t - t_0)}{\tau}\right] + I_\infty \qquad \text{(Eq.1.2)}$$

where I_{peak} is the peak current ($I_{peak} = I_{max} - I_\infty$), τ is the time constant for repassivation, t_0 is the time to produce the scratch, and I_∞ is the baseline current at $t = \infty$. If sample had Tafel behaviour, the following equation was then given:

$$I_{peak} = j_{crit} A_0 \exp\left[\frac{\eta_f}{\beta_a}\right] + \frac{\delta \rho Z F A_0}{M_w \tau} \qquad \text{(Eq.1.3)}$$

where j_{crit} is the critical current density for passivation, A_0 is the initial scratched area, η_f is the film overpotential (the difference between passivation and potentiosatically held potential) β_a is the Tafel slope, δ is the thickness of the repassived oxide film, ρ is the oxide film density, Z is charge per cation, F = Faraday's constant and M_w is oxide film molecular weight. It described the relationship between the peak current, oxide film thickness and exchange current density for ionic dissolution. It suggests that greater contact area resulting from plastic deformation (an application of high contact load) results in enhancing ion release.

Mischler *et al.* [28] presented a model to describe the corrosive behaviour of passive metals sliding against a hard insulating body. The anodic current under wear, $I_{a,w}$ was influenced by wear track length, l, sliding frequency, f, applied load, W, and hardness of the material, H. The model is expressed as :

$$I_{a,w} = K_w lf \left(\frac{W}{H}\right)^{\frac{1}{2}} \int_0^{\frac{1}{f}} idt \qquad \text{(Eq.1.4)}$$

where K_w is the proportionality factor.

However, in tribocorrosion systems, the material degradation is caused by the combination of tribological material removal and electrochemical processes. It is not simply sum of the two processes together. Synergistic effects play a very important role.

A number of works have been carried out in erosion-corrosion; another form of tribocorrosion process. Neville *et al.* [34] examined erosion-corrosion behaviour for materials under slurry conditions. Material degradation was evaluated. An equation (Eq. 1.5) was used to determine the mechanisms of material loss which caused by mechanical impact and electrochemical process.

$$T = E + C + S \quad \text{(Eq.1.5)}$$

$$S = E_C + C_E \quad \text{(Eq.1.6)}$$

$$T = E + C + E_C + C_E \quad \text{(Eq.1.7)}$$

T is the total weight loss in erosion corrosion environments. The pure material degradation caused by erosion is referred as E. C is the material loss due to electrochemical corrosion process only. S is the synergy effect which involves E_C (corrosion effect on erosion) and C_E (erosion effect on corrosion). Cathodic protection was employed to obtain erosion damage. Electrochemical tests and Faraday's Law were used to determine material loss due to corrosion effect. It can be expressed as :

$$C_E = \frac{QM}{nF} \quad \text{(Eq.1.8)}$$

where Q is the quantity of charge. M is the material molecular weight. The valence of the material is referred as n. F is Faraday's constant. Theoretically, the amount of charge can be calculated by the equation 1-9:

$$Q = \int_{t_0}^{t_f} (I - I_0)\,dt \quad \text{(Eq.1.9)}$$

where t_0 is the time at tribological movement starts and the current is I_0, t_f is the time at which the relative movement is completed and at t_f the current is I.

Many studies for investigations of wear and corrosion have used those equations [35-39]. A model was developed by Jiang *et al.* [40] to suit wear-corrosion under a aqueous sliding contact from Eqs. 1-4 and 1-7. It is able to simulate the synergy in a tribocorrosion system. The model is rather complex but covers the most important issues in a tribocorrosion system and provides a tool for further development. The effect of sliding distance, contact frequency, load and susceptibility of metals were related to this model and their effects on synergism were discussed. The low corrosion rate of orthopaedic materials relies on the protective passive film. When material under frequent mechanical motion, the protective barrier can be removed and re-form. In this sense, how fast the removal force is should be taken into account. Additionally, a wear map was constructed. A biphasic structure was concluded. One of them was called *wear-induced corrosion*, which is the equivalent to C_E in Eq. 1.7. It dominates the material degradation processes when low load and high contact frequency are applied. Another one was named *corrosion-induced wear*, which corresponded to E_C in Eq. 1.7. Considering dental or joint implants, material degradation is a consequence of combined attack by corrosion and wear. The two mechanisms do not proceed separately. Therefore, the conjoint processes of wear and corrosion are of great importance in the design of orthopaedic prostheses.

3. Tribology Effect on Corrosion

3.1. Effect of Tribology on Corrosion Potential

Clear passivity is shown for the three alloys in static conditions; expected due to the presence of a Cr_2O_3-containing film of several nm thickness spontaneously formed in air. One key issue with respect to tribo-corrosion is the behaviour of the film once a mechanical wearing/rubbing action is applied. One way to monitor the film (semi-quantitatively) is to measure E_{corr} (open circuit potential (OCP) or free corrosion potential) as the sliding action starts (and then stops).

In Figure 3.1 (a), for HC CoCrMo (High Carbon CoCrMo) the trend in E_{corr} as a function of time is shown. From point 1, the sample was immersed into the solution (50% serum) under static conditions. At this point the E_{corr} was stable. The reciprocating motion of the plate was started at point 2 but with no load added. Only the weight of the loading arm with the silicon nitride ball (<5N) in the holder acted on the plate. At this point a shift in E_{corr} in the active direction was recorded, consistent with removal of the passive film and an increase in the rate of the anodic reaction (as is expected when there is a removal of the charge transfer barrier). At position 3, the load on the silicon nitride ball was increased to 80N. At this point a significant shift in E_{corr} in the active (negative) direction was measured, followed by a transient period between points 3 and 4 during which the potential eventually stabilized. The steady state E_{corr} was reached when an equilibrium was reached between the rate of formation and the rate of removal of the passive film (depassivation versus repassivation). When load was added, plastic deformation occurred on the sample surface, which may contribute to the shift of potential. The initial transient period of potential moved at point 3 indicates the initiation of plastic deformation.

For 316L (a stainless steel used in biomedical devices) (Figure 3.1 (b)) in 50% serum, the same general trend of E_{corr} was seen but with more potential oscillations than for the CoCrMo alloys. Figure 3.2. and Figure 3.3. show the free corrosion potential and friction coefficient in DMEM and in 0.36% NaCl with the same experimental procedure as described above. Some similarities and differences can be seen.

- The friction coefficient and the free corrosion potential response correspond to each other very well.
- The transition period (when load was added) can be obtained for all tests. It indicates that an initial impact was occurred due to the load. Possibly there was an initial formation of tribofilm. However, the recovery from the impact in 50% serum seems quicker than in 0.36% NaCl. It may not be a simple passive (oxide) film recovery. Likely, it is an organic molecule-passive film interaction.
- 316L exhibited potential and friction coefficient oscillations in all three environments, especially in 50% serum. CoCrMo alloys, on the other hand, showed rather stable behaviour.

- In 50% serum, all materials showed an initial decrease of friction coefficient followed by an increase until the friction stabilized. In 0.36% NaCl, the friction coefficient increased from the start then reached a steady value. It may be caused by the removal of adsorbed proteins from the sample surface in 50% serum.
- In all solutions, after the load was removed, the removal of the passive film ceased and the protective barrier reformed. The final potential for CoCrMo was always more noble than the initial potential before load was applied. In contrast, the final recovery potential of 316L was more active (lower/more negative) than the stable free corrosion potential in the beginning of the tests (Figure 3.1. and Figure 3.2.).

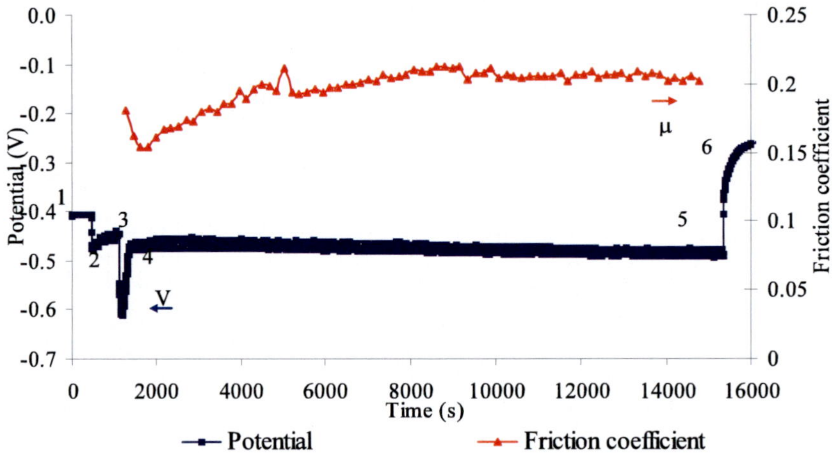

(a)

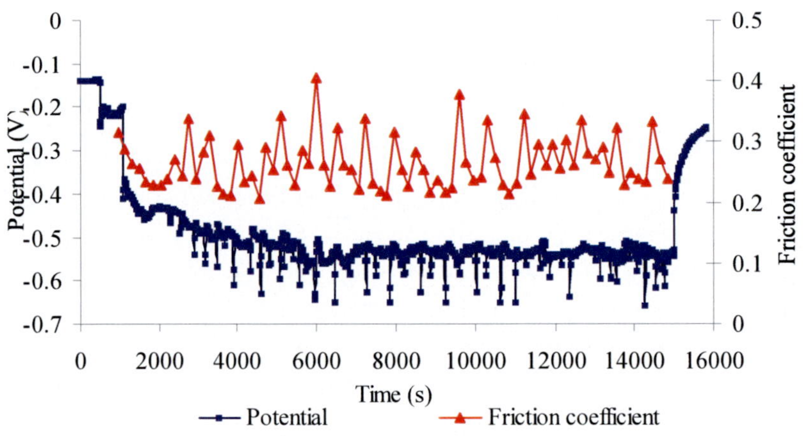

(b)

Figure 3.1. E_{corr} and μ versus time for (a) HC CoCrMo and (b) 316L in 50% serum for 4 hr tests.

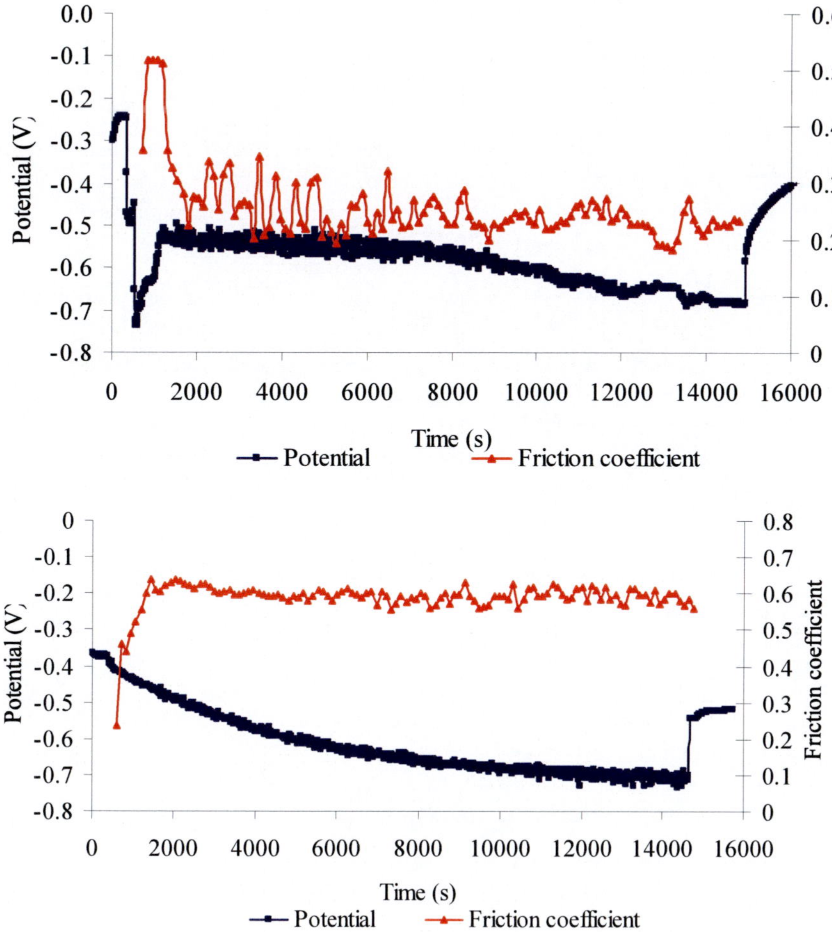

(b)

Figure 3.2. E_{corr} and μ versus time for (a) HC CoCrMo and (b) 316L in DMEM for 4 hr tests.

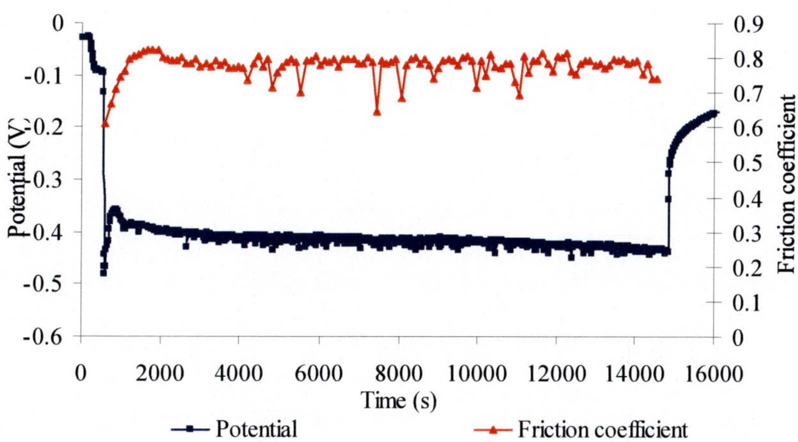

(a)

Figure 3.3 Continued on next page.

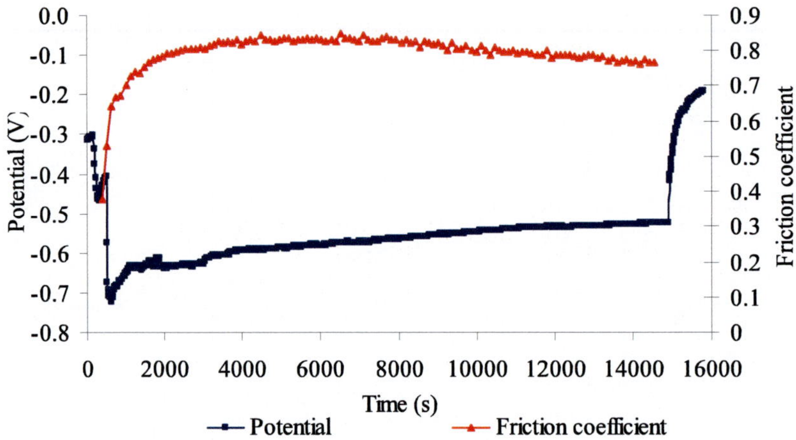

(b)

Figure 3.3. E_{corr} VS. time for (a) HC CoCrMo and (b) 316L in 0.36% NaCl for 4 hr tests.

Some observations of the friction coefficient from Figure 3.1., Figure 3.2. and Figure 3.3. can be summarized below:

- All three materials showed the lowest friction coefficient in 50% serum indicating that proteins in serum can lubricate the surface to some extent under the test conditions adopted.
- In addition, DMEM showed much lower friction coefficient values than 0.36%NaCl suggesting that the amino acids alone also have some friction reducing capability.
- HC CoCrMo appeared to have the lowest friction coefficient and the most noble OCP with organic species involved (50% serum and DMEM). It seems that HC CoCrMo received a better lubrication benefit from proteins and amino acids.
- In 50% serum and DMEM, a more noble free corrosion potential corresponded a lower friction coefficient. However, in 0.36% NaCl, 316L had an exception. The behaviour may be related to the adsorption of organic species and the passive film removal and re-formation mechanisms.

No carbide pluck out can be observed by optical microscopy and SEM. Figure 3.4. shows the wear groove for three materials in 50% serum. HC CoCrMo had a smooth groove (R_a 1.2μm) and the 316L wear scar was quite rough (R_a 6.3 μm). Figure 3.5. shows images taken from SEM. 316L appeared as a much rougher surface and cracks of the film can be seen (Figure 3.5. (b)). Images were also captured for the edge of the wear scar (Figure 3.6.). HC CoCrMo showed a smooth transition from the scar to the un-rubbed surface (Figure 3.6. (a)). For 316L, severe deformation was observed and some debris has been generated (Figure 3.6 (b)).

At the free corrosion potential (E_{corr}) the passive film is being continually removed and reformed on the wear track. Corrosion processes were occurring under these conditions – the ionic species were released due to the charge transfer at the interface and it can potentially interact with the biological fluid constituents and affect the friction and wear response of that interface.

(a)

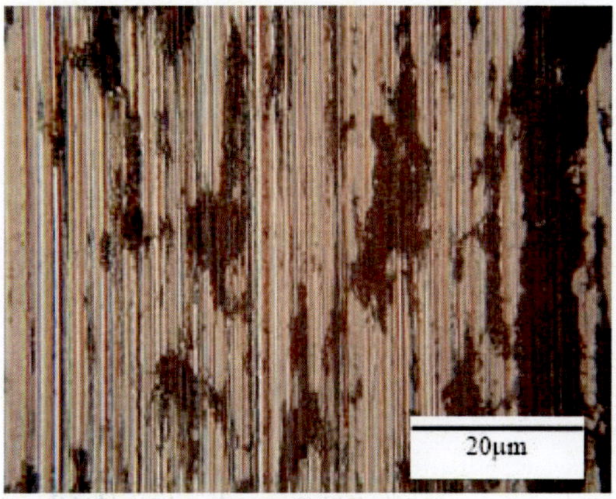

(b)

Figure 3.4. Optical images for (a) HC CoCrMo and (b) 316L taken after rubbing in 50% serum

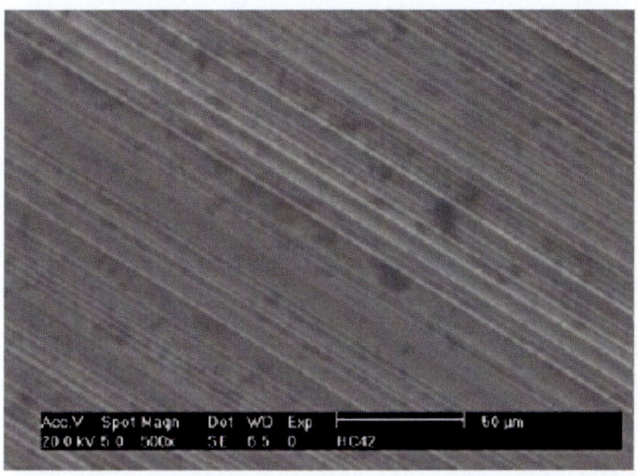

(a)

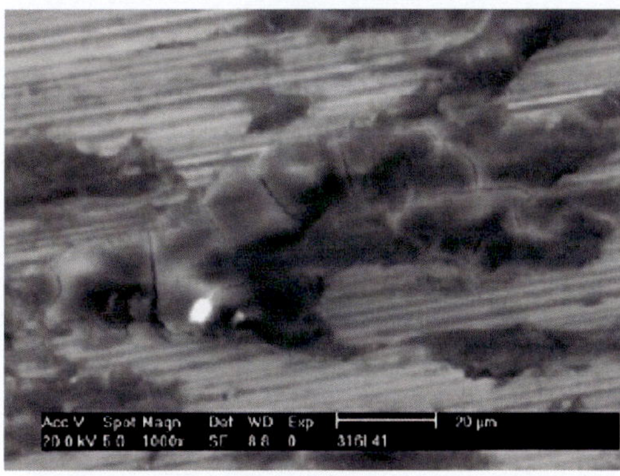

(b)

Figure 3.5. SEM images for (a) HC CoCrMo and (b) 316L taken after rubbing in 50% serum.

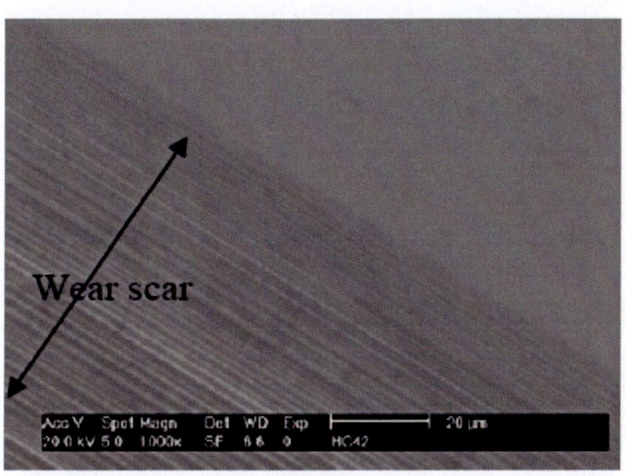

(a)

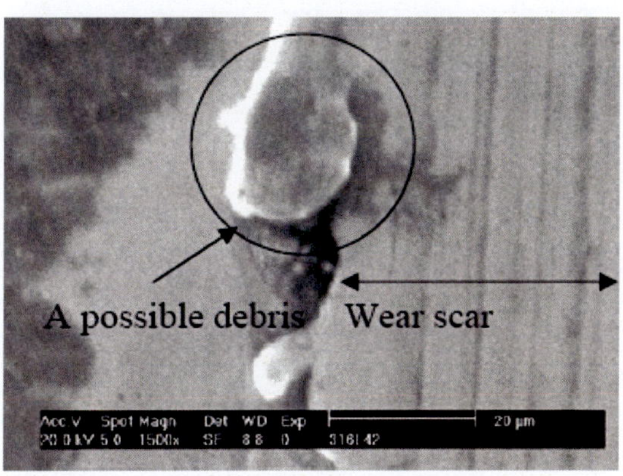

(b)

Figure 3.6. SEM images for the wear scar edge of (a) HC CoCrMo and (b) 316L taken after rubbing in 50% serum.

3.2 The Effect of Tribology on Corrosion Rate

Table 3.1. compares the corrosion current density for HC CoCrMo in two different conditions: in static and under rubbing. The Tafel extrapolation method was used to determine the corrosion current density value. From Table 3.1., some observations can be obtained:

- Tribological contacts certainly can accelerate the charge transfer and increase the corrosion current density. Due to the fact that the corrosion current density can represent the material degradation caused by corrosion, the current density here can be used to show some extent of the tribology effect on corrosion.
- The corrosion current density is about 15 times greater for HC CoCrMo in 50% serum under rubbing than in static conditions 316L showed an increase by a factor of 30 under the same conditions.
- 316L had the greatest corrosion current density in all three environments.
- It seems that the extent of wear effect on corrosion is very much material and solution dependent. With proteins involved, the corrosion current density under rubbing is less than in 0.36% NaCl for all materials.

Table 3.1. Results of Corrosion current density for materials in three environments

SOLUTION	CONDITION	CORROSION CURRENT DENSITY (μA/CM2)	
		HC COCRMO	316L
50% SERUM	IN STATIC	0.04±0.01	0.14±0.04
	UNDER RUBBING	0.68±0.03	6.10±0.07
DMEM	IN STATIC	0.03±0.01	0.06±0.02
	UNDER RUBBING	0.69±0.03	4.20±0.06
0.36% NACL	IN STATIC	0.03±0.01	0.04±0.01
	UNDER RUBBING	0.69±0.03	4.30±0.06

4. Corrosion Effect on Tribology

Figure 4.1. reveals several important aspects in relation to the wear and corrosion/wear interactions in sliding wear for these three materials in two solutions.

- Application of Cathodic Protection reduces the total damage (volume loss) in all cases indicating that corrosion or corrosion-related effects (effect of wear on corrosion or corrosion on wear) play an important role in degradation.
- In all solutions HC CoCrMo alloy shows the lowest volume and the highest volume loss is recorded for the 316L alloy.

- For 316L the higher total volume loss occurs for 50% serum and the 0.36% NaCl solution leads to the lower degradation. For HC CoCrMo alloy there is little difference between the two solutions.
- For 316L the highest wear volume loss (by applying CP) occurs for 50% serum and the lowest wear was found in 0.36%NaCl. For HC CoCrMo wear volume loss is the lowest in DMEM and in 50% serum and 0.36%NaCl the wear volumes loss was similar.

To date, there have been a few studies which attempt to quantify the contributions of material degradation due to corrosion, wear and their interactions in a biotribocorrosion environment where active biological solutions or saline solutions are used. This can be done by considering Eq.4.1 which allows the quantities of corrosion, wear and corrosion-wear to be analyzed.

$$T = W + C' + S \qquad \text{(Eq.4.1)}$$

where W is the material loss in the absence of corrosion, which was measured by applying CP. C' is the material loss due to electrochemical processes (corrosion) in the absence of wear. The synergy S includes two components which are

- the effect of wear on corrosion (C_w) and
- the corrosion effect on wear (W_c).

Corrosion-related damage (C'+S) can therefore be assessed for the three materials as can be seen in Figure 4.2. It should be remembered this is not only the corrosion rate – it is the damage that corrosion processes do when wear-corrosion processes occur in parallel. Hence it is referred to as *corrosion-related damage* rather than corrosion damage.

These results indicate that wear/corrosion interactions were involved and it is clear that in all cases in the present tests, corrosion and corrosion-related damage contributed to between 22% and 50% of the total damage, depending on the alloy and the solution. Figures 4.2. shows the components of total wear damage (W) and corrosion related damage (C'+S) for each alloy. Previously it was shown that the alloys exhibit passivity and so it can be concluded that the effect of corrosion in the absence of wear (C') can be assumed to be zero and that the synergy S (C_w and W_c) is the major corrosion-related component.

Figure 4.3. and Figure 4.4. assist in the understanding of the physical nature of wear/corrosion interactions. For the three alloys in 50% serum the physical differences in:

(i) the extent of wear
(ii) the wear mechanisms

can be observed.

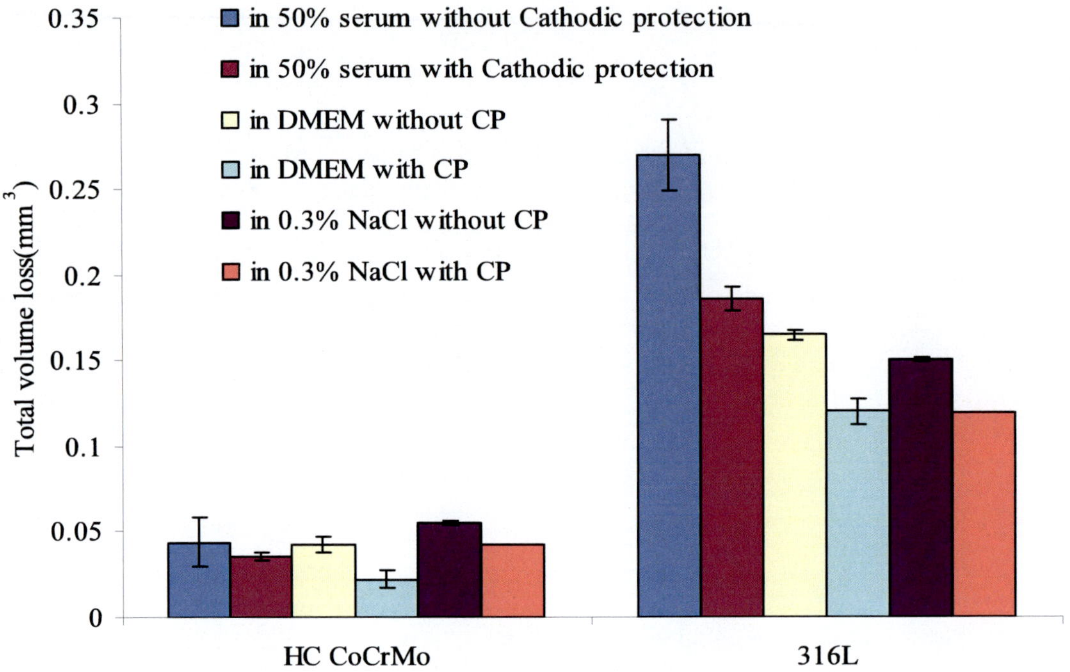

Figure 4.1. Material total volume loss without and with applied CP in three environments.

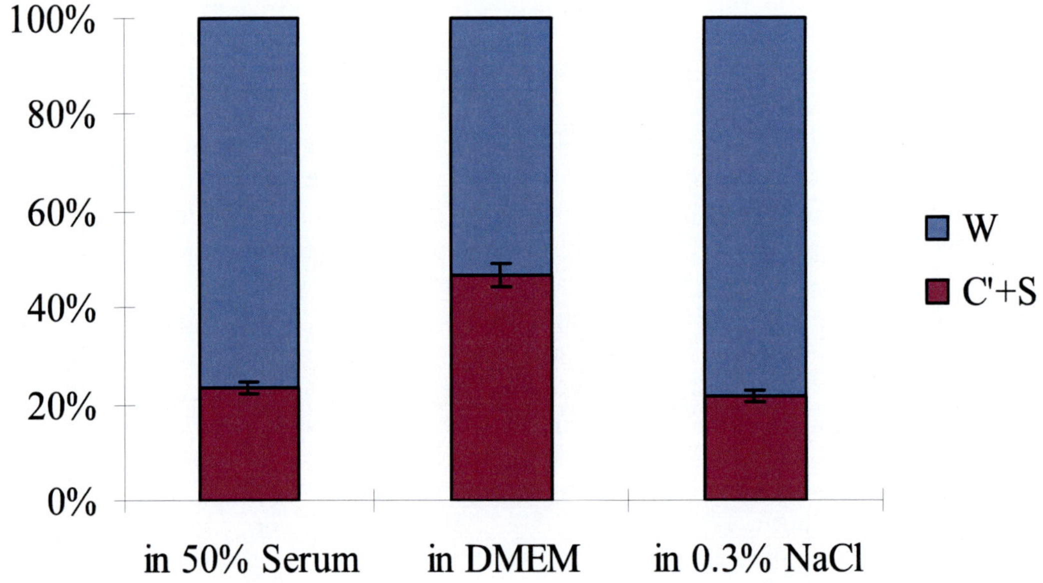

(a)

Figure 4.1 Continued on next page.

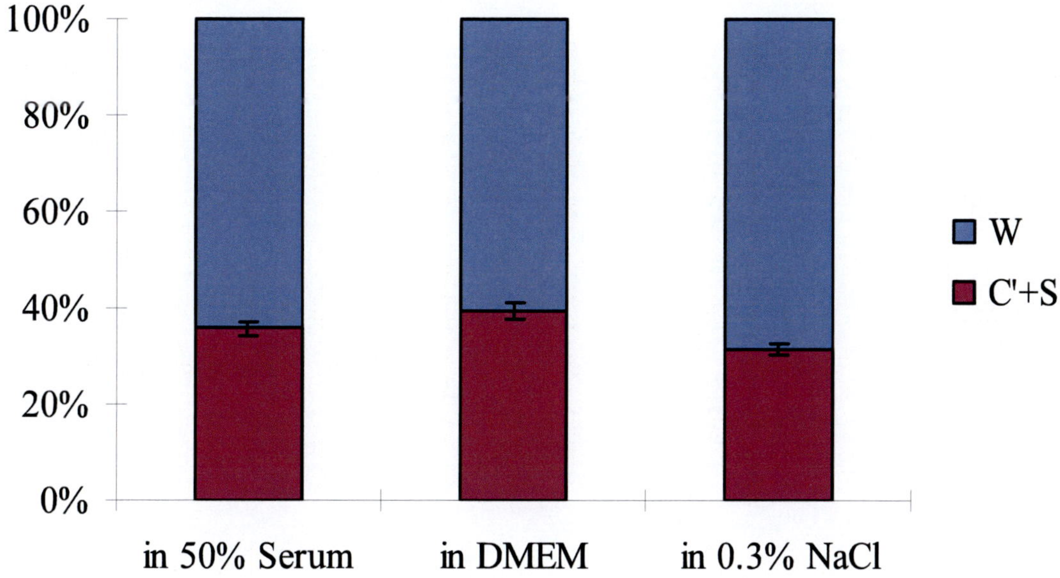

(b)

Figure 4.1. Components of volume loss for (a) HC CoCrMo (b) 316L in 50% serum, DMEM and 0.36% NaCl.

From the cross section of the wear scar taken by the interferometer (Figure 4.2.), by preventing corrosion, both the width and the depth of the wear scars were decreased in line with the volume loss reduction. Electrochemical processes occur there is evidence of pitting and gross adhesive wear. With applied CP only smooth grooves can be seen It has been reported that in wear-corrosion systems, damage consists of two processes:

(i) mechanical delamination of the passive layer in the wear track
(ii) a progressive electrochemical repassivation of that active wear track area (representing that part of the wear track that temporarily loses its passive character due to the mechanical interaction during sliding).

By applying CP, it can be seen that delamination is eliminated. No real change in surface structure on the HC CoCrMo alloy is observed – in agreement with the wear coefficient values. Both Cr-rich and Mo-rich carbides can be observed in the wear scar and out of the wear scar on HC CoCrMo. No evidence that carbides had been pulled out can be observed.

No evidence of any tribochemical reaction was obtained on the surface of the silicon nitride ball. This could be the low velocity (0.02m/s) and low temperature (37°C) applied in this chapter and also the high hardness of the silicon nitride ball (three times higher than HC CoCrMo). The ultra smooth surface finish (roughness less than 0.003 microns) and good density by HIPing (Hot Isostatic Pressing) for ball samples in this chapter can also minimize the possibilities of any tribochemical reaction on the ball surface.

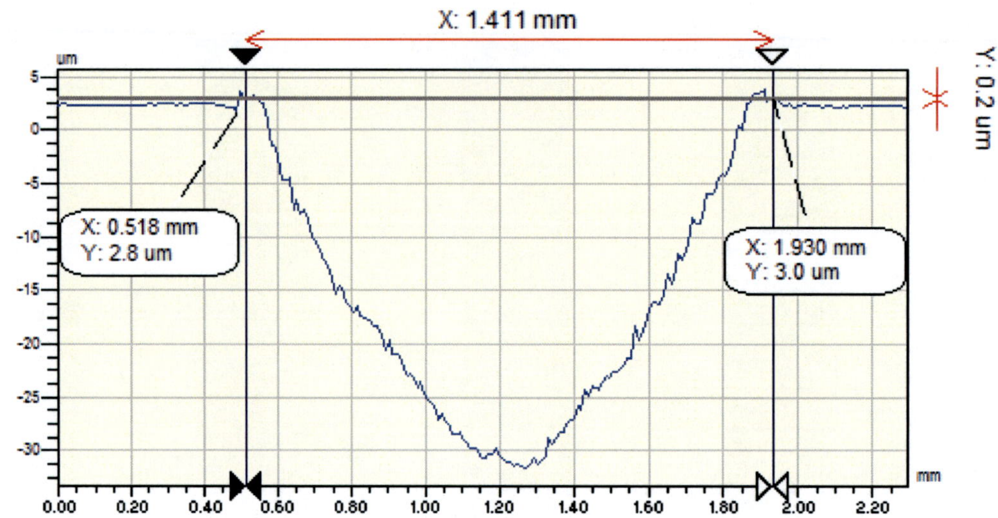
X: 1.411 mm
Y: 0.2 um
um
X: 0.518 mm
Y: 2.8 um
X: 1.930 mm
Y: 3.0 um
mm

(a)

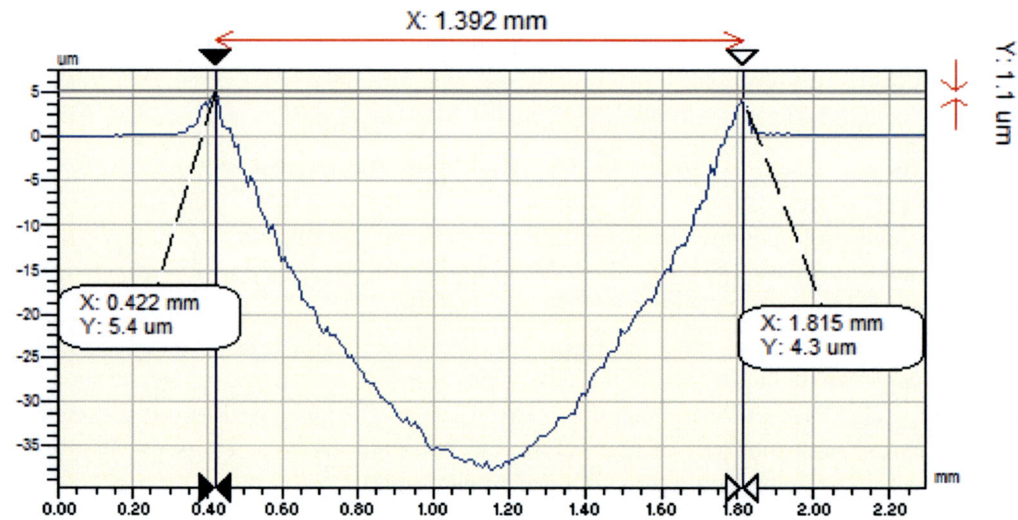
X: 1.392 mm
Y: 1.1 um
um
X: 0.422 mm
Y: 5.4 um
X: 1.815 mm
Y: 4.3 um
mm

(b)

- In 50% serum at open circuit conditions, all materials showed higher roughness. It indicates that the protein adsorption process presented.

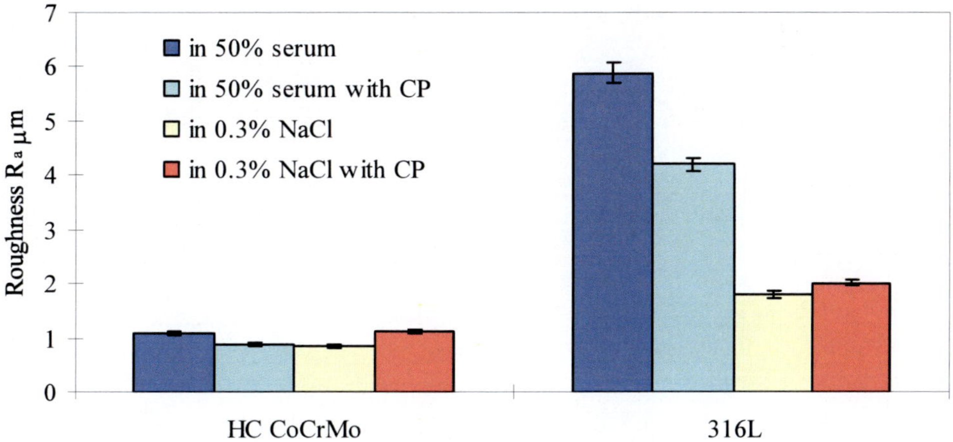

Figure 4.3. Roughness for wear scar of materials after 4 hours sliding tests in 50% serum and 0.36% NaCl with and with applied CP.

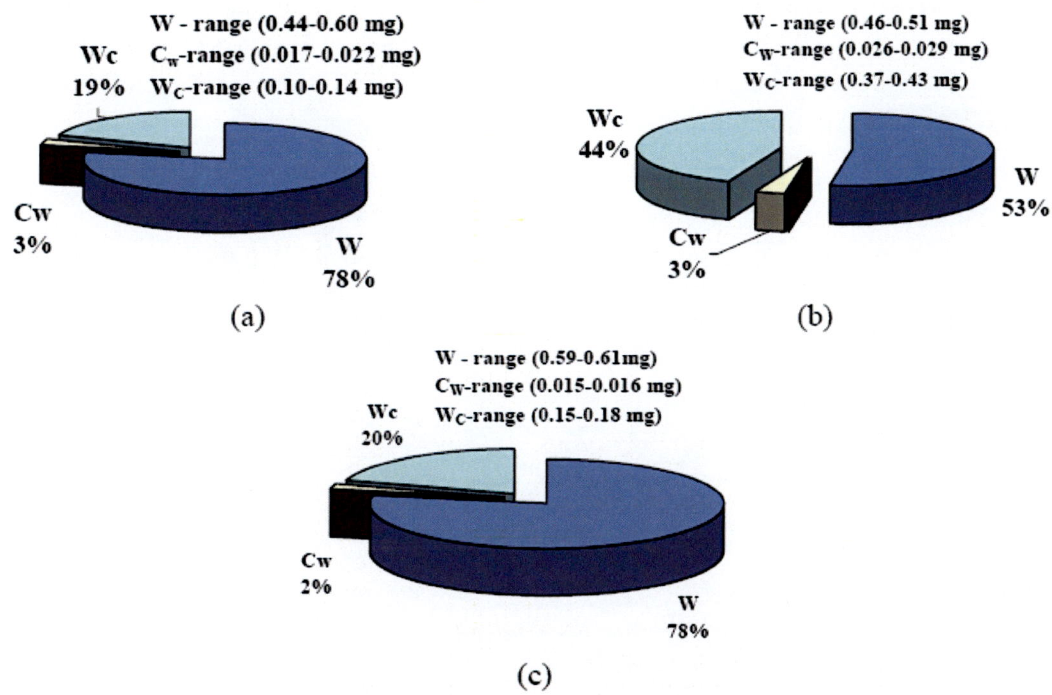

Figure 4.4. Percentages of volume loss for HC CoCrMo in (a) 50% serum (b) DMEM (c) 0.36% NaCl.

For all materials in three environments, mechanical damage is the major cause of the material loss (more than 50%). However, the synergy effect clearly contributes to the degradation. For HC CoCrMo, the effect of wear on corrosion is about 2% to 3%. It indicates that due to their good corrosion resistance, the mechanical damage did not affect the electrochemical processes massively. For 316L, 4% to 6% damage was from the wear effect on corrosion. The corrosion effect on wear is by far the largest component of damage accounting for corrosion/wear interactions (S) in the tribocorrosion system. Corrosion affects the surface properties then will influence the wear mechanisms abundantly. A detailed discussion is presented. However, an increase in error for W_c which results from combined data W and C_w can be gained. The errors for the combined data have been considered and the range of components for the material degradation processes is shown in Figure 4.4.

5. Tribo-Film on CoCrMo

Studies have been focused in the wear scar. It is the main contact area when friction and wear (mechanical damage) are evaluated. However, comparison is made with the film formation outside of the scar. Table 5.1. shows the concentration of elements in the wear scar after various etching times. Figure 5.1. compares the survey scans in and out of the wear scar. An apparent difference is that Co 2p was obtained in the top surface of the wear scar. It indicates that Co ions was released but still presented in the tribofilm. It gives the evidence that proteins bind with Co ions in the wear track. Proteins were adsorbed on the sample surface of unworn area, no Co 2p or Cr 2p was found on the top layer of the area outside the wear scar. Figure 5.2. and Figure 5.3. present the C 1s spectra inside and outside of the wear scar. The C 1s spectra observed outside the wear scar is very close to native albumin, which also suggests that an adsorption of proteins has occurred on the unworn surface. Detailed analysis for Co $2p_{2/3}$ is shown in Figure 5.4.

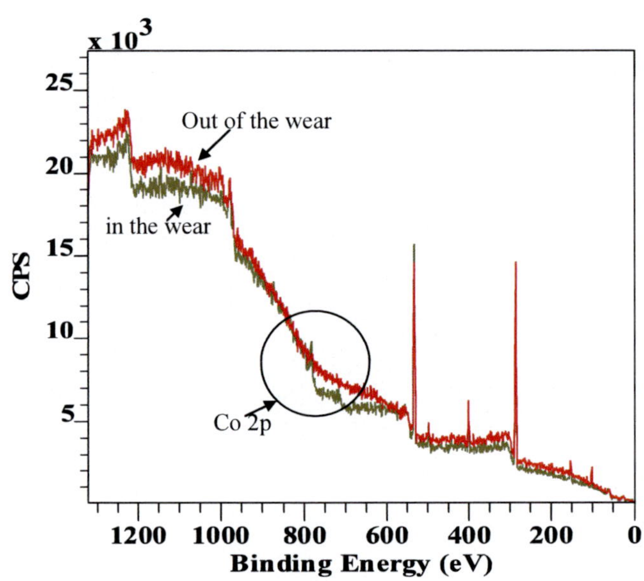

Figure 5.1. Survey scans for HC CoCrMo in 50% serum.

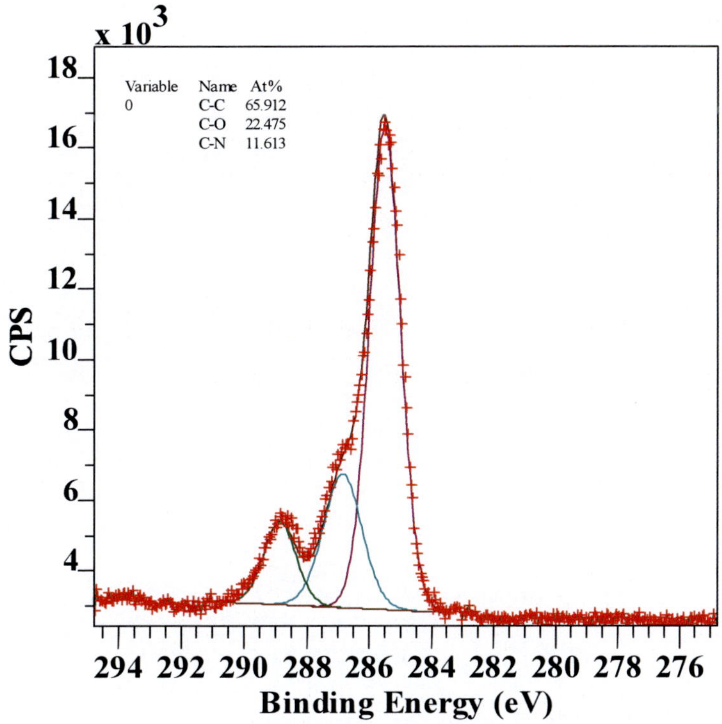

Figure 5.2. C 1s spectrum and fitted curves for HC CoCrMo wear scar in 50% serum.

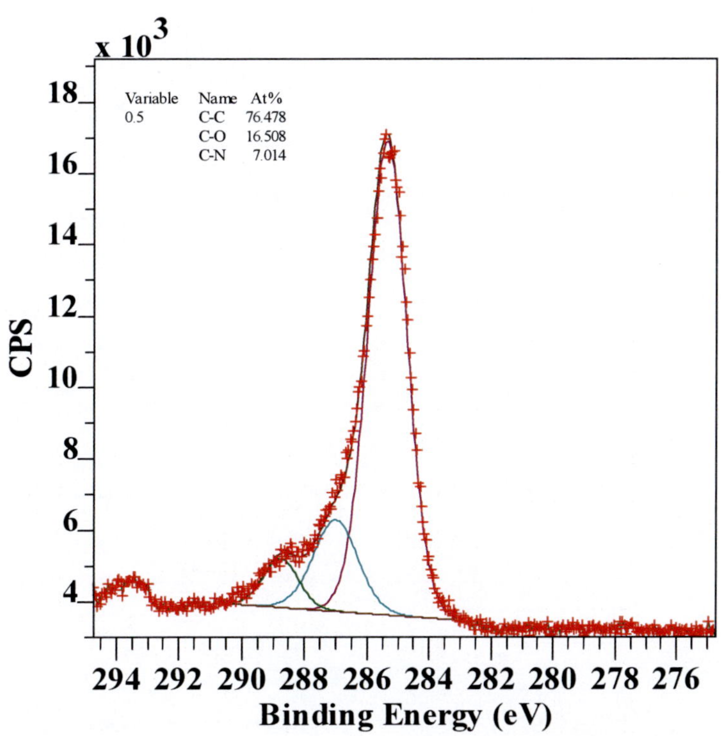

Figure 5.3. C 1s spectrum and fitted curves for out side of HC CoCrMo wear scar in 50% serum.

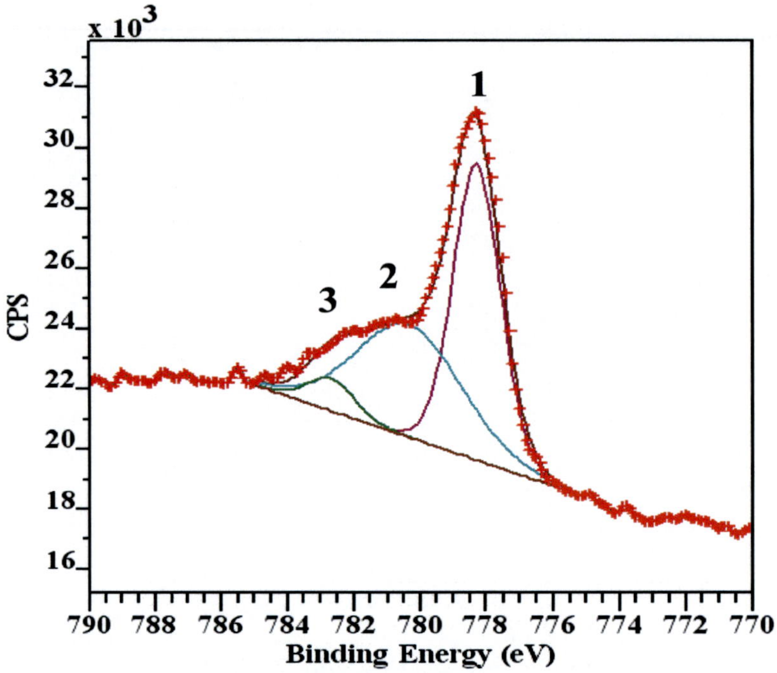

Figure 5.4. Co $2p_{3/2}$ peaks obtained from the wear scar of 4-hour rubbing HC CoCrMo in 50% serum.

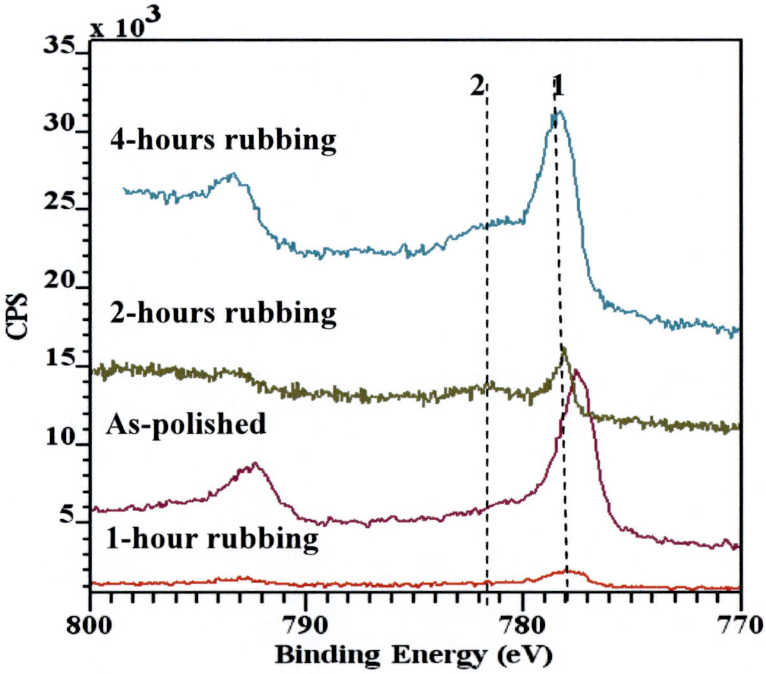

Figure 5.5. The change in the Co 2p spectrum as a function of rubbing time on HC CoCrMo samples.

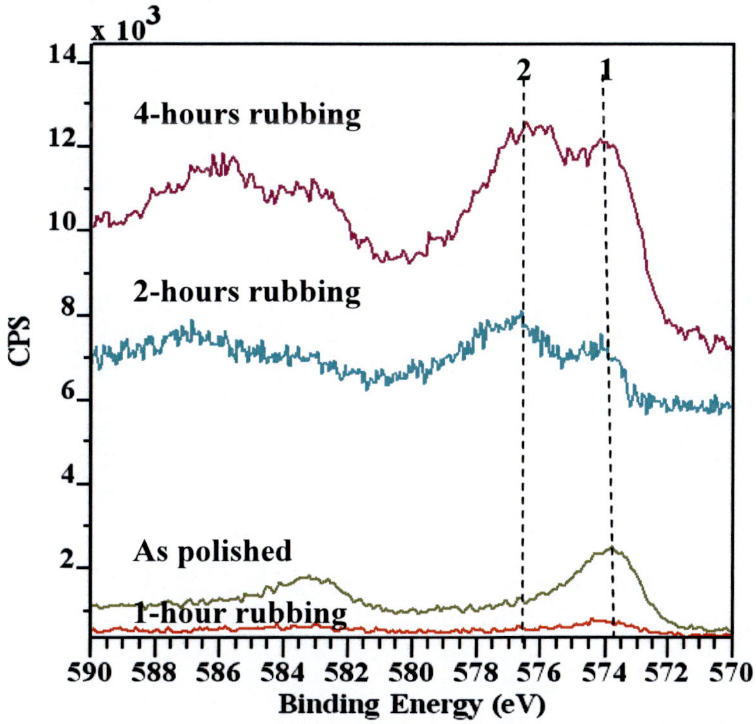

Figure 5.6.The change in the (b) Cr 2p spectrum as a function of rubbing time on HC CoCrMo samples.

Table 5.1. XPS results for HC CoCrMo in 50% serum after different etching duration

ETCHING TIME (MIN)	XPS (AT%)					
	C 1S	O 1S	N 1S	CO 2P	CR 2P	MO 3D
0	66	23	8	2	1	
0.5	62	14	8	8	7	1
5	33	11	6	28	18	4

A thin protein adsorption layer (3-4 nm) was observed on the sample surface for HC CoCrMo after 1, 2 and 4-hours sliding. Three main elements (C, N and O) were observed. All correspond to organic species. The C 1s peaks occur at 285.5, 287.1 and 289.0 eV (Figure 5.2.), indicating C bonds like C-C, C-H and C-O C-N, respectively. This can be compared with pure albumin where the range of binding energies for the C 1s peak is between 285.0 eV and 288.2 eV corresponding to various carbon species (aliphatic carbons, protein backbone and peptide carbons). Examining outside the wear scar, C 1s peaks (285.0-284.9 eV) (Figure 5.3.) are very consistent with C 1s peaks in the reference albumin. The shift of carbon binding energy and the broadening of carbon spectra in the wear scar are probably due to the denaturation of proteins resulting from the high pressure and instantaneous increase in temperature. The temperature rise in the contact area is defined by $\Theta = 0.946\frac{ha}{K}$ (the

Peclet number $P_e = \frac{ua}{2\kappa} << 1$, where $\kappa = \frac{K}{\rho c_p}$) (K=18 W $m^{-1}K^{-1}$ for 316L and 49 W $m^{-1}K^{-1}$ for CoCrMo), where $h = \mu p u$. The temperature rise for CoCrMo was 18°C at the very beginning of the test in 50% serum. However, as the tests progressed, in the steady state, the flash temperature was 4°C.

It is noticed that in the wear scar more C-C is observed than outside the wear scar. It may re-establish and result in a reduction of friction. This apparent change in layer structure, compared to the reference albumin is represented in Table 5.2. where the theoretical ratios of C/O and C/N in albumin are 3.2 and 3.7. These can be compared with the experimental results, after sliding, and it is clear the C/O and C/N ratios both increase. However, as sliding time increases, with no etching C/O and C/N ratios decrease and after 4 hours, the ratio of C/O and C/N are close to those of pure albumin. The increase of O 1s might be due to the absorption of water molecules or –OH from the environment.

Table 5.2. Binding energy for C 1s and C/N, C/O ratios

RUBBING TIME	ETCHING TIME	BE FOR C 1S (EV)[A]	C/N	C/O
THEORETICAL RATIO FOR ALBUMIN	0 MIN	285.0	3.7	3.2
1-HOUR	0 MIN	288.2	5.5	4.0
2-HOUR	0 MIN	288.6	5.0	3.9
4-HOUR	0 MIN	285.5	4.1	3.4

[a] BE for the main peak

By applying 0.5 minutes argon-ion etching, a passive film containing Cr oxides, Co oxides and Cr hydroxide was detected. It indicates that the protein-rich layer is about 3-4 nm thick. After 4 hours, a dense and thick tribo-film was found underneath the protein adsorption layer. Figure 5.4. shows different Co $2p_{3/2}$ states in the wear scar after 4 hours rubbing. Generally, there are three peaks. Peak 1 (778.4 eV) corresponds to typical Co in the CoCrMo alloy. Peaks 2 and 3 (780.0 eV and 782.8 eV) indicate that Co was present as different oxides (Co_2O_3/CoO) and as an organometallic form. It is well documented that Co, Cr, Fe and Ni can incorporate with Albumin at carboxylate and tyrosine positions or bind with the –SH group of a single cysteine residue. This formation could be responsible for the decrease of friction because of its polymer-like structure.

Figure 5.5. and Figure 5.6. shows the differences of tribo-film for Co 2p and Cr 2p peaks in the tribofilm after 1, 2 and 4 hours sliding compared with the as-polished sample. From Figure 5.5., it is apparent that in 50% serum, Cobalt changed from metallic form in the matrix (as-polished) to Cobalt oxides (Co_2O_3/CoO) and organometallic Co content in the tribo-film (peaks 1 and 2). The formation increased as sliding time increased. The Cr species also changed during tribological contact (Figure 5.5.). The Cr 2p spectrum changed from typical Cr formations in CoCrMo alloys (573.9 eV) to two species dominating the Cr content: peak 1 at 574.4 eV (Cr organometallic form) and peak 2 at 576.4 eV (Cr_2O_3/Cr_2O_4/organometallic).

These formations of organometallic Co/Cr and Co/Cr oxides are responsible for the reduction of friction and lower wear rate in the steady state regime. An indication of the thickness of the tribofilm was given, which was more than 25 nm with the thin protein adsorption layer (3-4 nm). Due to the dimension of albumin molecules (3×3×8), the layer seems to be a monolayer of proteins. Calculation of the thickness of the tribofilm was made according to the different rate of argon-ion etching process. For carbon specie, the rate of etching process is about 0.1 nm/s and for Cr_2O_3, the rate is approximately 0.08 nm/s [41-43].

6. Conclusion and Challenges

The subject of the tribocorrosion performance of implant materials in biological environments is challenging due to its complexity; many physical, chemical and electrochemical processes are involved. However, this chapter has made some progress towards the understanding of fundamental tribocorrosion processes and material degradation mechanisms in tribocorrosion systems. The link between tribological parameters with electrochemical response has been discussed. The performance of the three widely used implant materials has been evaluated in biological species-rich solutions. How materials are affected by service environment and vice versa have been addressed. The link between the tribofilms and tribocorrosion performance has been identified and discussed. The process – 'biotribocorrosion' is suggested to describe the phenomena of the joint effect of tribology and corrosion in a biological environment. Integrated *in-situ* electrochemical measurements appear to be very useful to determine corrosion and wear interactions. They can evaluate candidate materials in many respects (corrosion, wear and tribocorrosion).

Although the findings from this work are exciting and useful to the understanding and the development of implant materials in biotribocorrosion systems, several new questions have been initiated. Some directions are recommended to continue this study and a few are motioned below:

- An important concern of the use of metallic load bearing implants is the influence of released wear debris and metal ions to patients. In the current study, the awareness of wear debris to tribocorrosion behaviour presents, but further investigation of its exact effect on corrosion behaviour and tribological parameters is necessary.
- The current study is based on a ceramic-on-metal contact because it is the first attempt to understand the biotribocorrosion systems. In MoM joint replacement, the complexity of two metal components can make big difference. However, the principles of the reaction will not change. Future work can be established in this area.
- The binding of biological molecules (especially proteins) with metal ions is proved to be very important in biotribocorrosion systems. The nature of the film formation (tribofilm) under tribological contact still need more work and attention.
- The current study used a reciprocating ball-on-plate tribometer to investigate the tribocorrosion behaviour of different materials. The ultimate goal in this work should be to instrument a hip simulator to be able to superimpose electrochemical measurements during the movement cycle. As a step towards this the future work

will involve tribotesting of metal on metal systems using a multi-station pin-on-plate apparatus to obtain statistically relevant data.

- New materials, new treatments and new combinations and designs of joint replacements are developed fast. Evaluation of those factors is needed.

References

[1] Annual Report of The National Joints Registry (NJR), www.njrcentre.org.uk, 2003,

[2] Annual Report of The Swedish National Hip Arthroplasty Register, Swedish Orthopaedic Association, 2004,

[3] J. L. Tipper, E. Ingham, Z. M. Jin and J. Fisher, The science of metal-on-metal articulation, *Current Orthopaedics*, 2005, 19, pp.280-287

[4] C. B. Rieker, R. Schon and P. Kottig, Development and validation of a second-generation metal-on-metal bearing, *The Journal of Arthroplasty,* 2004, 19, pp.5-11

[5] D. W. Howie and M. A. McGee, Metal-on-metal resurfacing versus total hip replacement-the value of a randomized clinical trial., *The Orthopedic clinics of North America*, 2005, 36, pp.195-201

[6] D. F. Williams, Definitions in biomaterials, Proceeding of Consensus Conference of the European Society for Biomaterials, 1987, Amsterdam, Elsevier

[7] J. B. Park and Y. Kon Kim, Metallic Biomaterials, The Biomedical Engineering Handbook, 2000, Boca Raton, CRC Press LLC

[8] B. D. Ratner, A history of biomaterials, Biomaterials Science, 2004, London, Elsevier Academic Press

[9] M. Zhang, Biocompatibility of materials, Biomaterials and Tissue Engineering, 2004, Heidelberg, Spring-Verlag

[10] M. A. Jacobs, M. B. Schmidt and R. Farrar, Comparison of wear rates for metal-on-polyethylene and metal-on-metal hips, The AAHKS sixth annual meeting, 1998,

[11] A. Yew, M. Jagatia, H. Ensaff and Z. M. Jin, Analysis of contact mechanics in McKee-Farrar metal-on-metal hip implants, Proc Instn Mech Engrs Part H: *Journal of Engineering in Medicine,* 2003, 217, pp.333-340

[12] I. D. Learmonth, S. Gheduzzi and T. P. Vail, Clinical experience with metal-on-metal total joint replacements: indications and results, Proc. IMechE Part H: Engineering in Medicine, 2006, 220, pp.229-237

[13] D. Dowson and Z. M. Jin, Metal-on-metal hip joint tribology, Proc. IMechE Part H: Engineering in Medicine, 2006, 220, pp.107-118

[14] F. W. Chan, J. D. Bobyn, J. B. Medley, J. J. Krygier and M. Tanzer, Wear and lubrication of metal-on-metal hip implants, *Clinical Orthopaedics and Related Research*, 1999, 369, pp.10-24

[15] D. Dowson, A. A. J. Goldsmith, C. M. McNie and S. L. Simith, A tribological study of metal-on-metal total replacement hip joints, Friction, Lubrication and Wear of Artificial Joints, 2003, London, Professional Engineering Publishing

[16] C. Pabinger, R. Biedermann, B. Stocke, M. Fischer and M. Krismer, Migration of metal-on-metal versus ceramic-on-polyethylene hip prostheses, *Clinical Orthopaedics and Related Research*, 2003, 412, pp.103-110

[17] S. A. Catledge and Y. K.Vohra, Structural and mechanical properties of nanostructured metalloceramic coatings on cobalt chrome alloys, *Applied Physics Letters*, 2003, 82, pp.1625-1627

[18] L. L. Hench and S. Best, Ceramics, glasses, and glass-ceramics, Biomaterials Science, 2004, London, Elsevier Academic Press

[19] D. Dowson, History of Tribology, 1998, London, Professional Engineering Pubs

[20] V. Saikko, A multidirectional motion pin-on-disk wear test method for prosthetic joint materials, *Journal of Biomedical Materials Research*, 1998, 41, pp.58-64

[21] K. Kato, Wear in relation to friction - a review, Wear, 2000, 241, pp.151-157

[22] K. W. J. Wright, H. S. Dobbs and J. T. Scales, Wear studies on prosthetic materials using the pin-on-disc machine, *Biomaterials*, 1982, 3, pp.41-48

[23] D. Dowson, C. Hardaker, M. Flett and G. H.Isaac, A hip joint simulator study of the performance of metal-on-metal joint Part I: *The role of Materials, Arthroplasty*, 2004, 19, pp.124-130

[24] A. Wang, V. K. Polineni and a. et, Effect of femoral head surface roughness on the wear of UHMWPE acetabular cups, *Journal of Arthroplasty*, 1998, 13, pp.615-625

[25] Z. M. Jin, P. Firkins, R. Farrar and J. Fisher, Analysis and modelling of wear of cobalt-chrome alloys in a pin-on-plate test for a metal-on-metal total hip replacement, *Proc. Instn. Mech. Engrs. H*, 2000, 214, pp.559-568

[26] H. C. Amstutz, P. Campbell, H. McKellop, T. P. Schmalzried, W. J. Gillespie, D. Howie, J. Jacobs and J. Medley, Metal on metal total hip replacement workshop consensus document, *Clinical Orthopaedics and Related Research*, 1996, 319S, pp.297-303

[27] M. A. Wimmer, J. Loos, R. Nassutt, M. Heitkemper and A. Fischer, The acting wear mechanisms on metal-on-metal hip joint bearings: in vitro results, *Wear*, 2001, 250, pp.129-139

[28] S. Mischler, A. Spiegel and D. Landolt, The role of passive oxide film on the degradation of steel in tribocorrosion systems, *Wear,* 1999, 225-229, pp.1078-1087

[29] D. Landolt, S. Mischler and M. Stemp, Electrochemical methods in tribocorrosion: a critical appraisal, *Electrochemica Acta*, 2001, 46, pp.3913-3929

[30] Y. Okazaki, Effect of friction on anodic polarization properties of metallic biomaterials, *Biomaterials,* 2002, 23, pp.2071-2077

[31] P. Ponthiaux, F. Wenger, D. Dress and J. P. Celis, Electrochemical techniques for studying tribocorrosion processes, *Wear*, 2004, 256, pp.459-468

[32] I. Garcia, D. Drees and J. P. Celis, Corrosion-wear of passivation materials in sliding contacts based on a concept of active wear track area, *Wear*, 2001, 249, pp.452-460

[33] J. R. Goldberg and J. L. Gilbert, Electrochemical response of CoCrMo to high speed fracture of its metal oxide using an electrochemical scratch, *Journal of Biomedical Materials Research*, 1997, 37, pp.421-431

[34] A. Neville and T. Hodgkiess, Characterisation of high-grade alloy behaviour in severe erosion-corrosion conditions, *Wear*, 1999, 233-235, pp.596-607

[35] Y. Yan, A. Neville, D. Dowson, S. Williams and J. Fisher. Tribo-corrosion analysis of wear and metal ion release interactions from Metal-on-Metal and Metal-on-Ceramic contacts for the application of artificial prostheses, Proceedings of the Institution of Mechanical Engineers, Part J: Journal of Engineering *Tribology*. 2008, 222, pp. 483-492

[36] Y. Yan, A. Neville and D. Dowson. Biotribocorrosion of CoCrMo orthopaedic implant materials. *Tribology International*, 2007, 40, pp.1492-1497

[37] Y.Yan, A. Neville and D. Dowson. Tribo-corrosion properties of Cobalt-based medical implant alloys in simulated biological environments. *Wear*, 2007, 263, pp.1105-1111

[38] Y.Yan, A. Neville, D. Dowson and S.Williams. Tribocorrosion in implants—assessing high carbon and low carbon Co–Cr–Mo alloys by in situ electrochemical measurements. *Tribology International*. 2006, 39, pp 1509-1517

[39] Y.Yan, A.Neville and D. Dowson. Biotribocorrosion – an appraisal of the time dependence of wear and corrosion interactions Part I: *The role of corrosion. Journal of Physics* D: Applied Physics , 2006, 39, pp. 3200-3205

[40] J. Jiang, M. M. Stack and A. Neville, Modelling the tribo-corrosion interaction in aqueous sliding conditions, *Tribology International,* 2002, 35, pp.669-679

[41] C. D. Wagner, A. V. Naumkin, A. Kraut-Vass, J. W. Allison and C. J. Powell, NIST X-ray Photoelectron Spectroscopy Database, NIST Standard Reference Database 20, Version 3.4 (Web Version),

[42] http://lasurface.com/Data_base/Aw_test_princ_database.htm, La Surface Database

[43] T. Hanawa, S. Hiromoto and K. Asami, Characterization of the surface oxide film of Co-Cr-Mo alloy after being located in quasi-biological environments using XPS, *Applied Surface Science*, 2001, 183, pp.68-75

In: Tribology Research Trends
Editor: Taisho Hasegawa

ISBN: 978-1-60456-912-4

Chapter 4

AN INTEGRATED ADHESIVE WEAR TESTING METHODOLOGY

L. J. Yang *

Mechem Consultancy Services Pte Ltd, 127 Woodlands Industrial Park E5, Woodlands E-Terrace, Singapore 757500, Republic of Singapore

ABSTRACT

In adhesive (sliding) wear, a typical wear volume loss against sliding distance curve can generally be divided into three regimes: the transient, the steady-state and the severe wear regimes. Although the steady-state wear is usually linear, however both the transient wear and severe wear regimes are curvilinear. To solve this non-linear wear problem, an integrated adhesive wear model, in which the transient wear volume is described by an exponential equation while the steady-state wear by a revised Archard's equation, has been proposed by the author. With this integrated wear model, both the transient and the steady-state wear rate and wear coefficient can be modeled continuously. It is also possible to predict both the standard and net-steady state wear coefficients and wear rates with a suitable F_A value obtained from the transient wear data. Wear testing is a time consuming process, as the test has to be repeated with different sliding distances until a steady-state wear condition is achieved. During testing, it may also be difficult to judge correctly whether a steady-state wear condition has actually been attained. The standard wear coefficient value obtained would be higher if the sliding distance covered remains within the transient wear regime. On the other hand, excessive distance would also give a higher wear coefficient value as the wear might have occurred in the severe wear regime. Furthermore, different wear testing methods with different nominal specimen contact areas and testing parameters such as load and speed have also been used. It is therefore no surprising that wear coefficients as well as the wear rates obtained from different investigators have been found to vary significantly up to a

* Formerly School of Mechanical and Aerospace Engineering, Nanyang Technological University, Singapore. ylj@mechem.com.sg

deviation of 1,000%.

Since the wear testing methods currently used in the industry are 'non-standard' and inefficient, it is high time to find a more systematic one to determine the wear coefficient and wear rate more consistently, accurately and perhaps more economically. With these objectives in mind, an integrated adhesive wear testing methodology, based on the integrated wear model and the related wear equations developed by the author, has therefore been proposed. With this methodology, the wear testing will be divided into three stages:(i) to conduct the transient wear test; (ii) to predict the steady-state sliding distance, wear rate and wear coefficient; and (iii) to conduct the steady-state confirmation runs to obtain the measured steady-state wear rate and wear coefficient. This testing methodology will provide useful transient as well as the steady-state wear data, which will be valuable to meet different product design needs. It should be noted that the inclusion of transient wear in a test programme may not necessarily increase the wear testing time, as the transient wear tests are conducted with a shorter sliding distance. In reality, by doing away with the trial and error method for finding the steady-state wear regime, a lot of time will be saved. Based on the wear test data obtained previously, the proposed methodology was found capable of saving about 30-40% of testing time if only the confirmation run at the predicted sliding distance is chosen. Obviously this methodology will only work if the wearing pair has a sliding wear curve similar to that described in the integrated wear model; and it is assumed that no major change of wear mode is expected to happen in the wearing process.

This chapter will review the integrated wear testing methodology, the integrated wear model, the equations developed for the determination of steady-state sliding distance, wear coefficients and wear rates. Some wear data obtained previously from aluminium-based metal matrix composites will also be analyzed to support the proposed methodology.

Keywords: wear coefficient, wear rate, integrated wear model, adhesive wear, integrated wear testing methodology

Nomenclature

A, B	=	Experimental constants, Eq.8.
C	=	Constant for Eq.40, Eq.47. C=4.605, 6.908 and 9.210 for F_A=0.99, 0.999 and 0.9999 respectively.
d	=	Diameter of particles
D	=	Constant for Eq.41. D=0.215, 0.145 and 0.109 for F_A=0.99, 0.999 and 0.9999 respectively.
Dev(%)	=	Percentage of deviation
f_v	=	Volume fraction of particles.
g_1	=	Experimental constant, Eq.22
g_3	=	Experimental constant, Eq.25
F_A	=	Exponential function as described by Eqs.29, 35
F_B	=	Exponential function as described by Eqs.32,33, 36,
H	=	Brinell hardness
K_N	=	Net steady-state wear coefficient, Eq.15.
$K_{N(av)}$	=	Average net steady-state wear coefficient
K_P	=	Predicted wear coefficient, with Eq.38 and Eq.41
$K_{P0.99}$	=	Predicted steady-state wear coefficient, with Eq.41 and a D value of 0.215 to give a F_A value of 0.99.
$K_{P0.999}$	=	Predicted steady-state wear coefficient, with Eq.41 and a D value of 0.145 to give a F_A value of 0.999.

Nomenclature

$K_{P0.9999}$	=	Predicted steady-state wear coefficient, with Eq.41 and a D value of 0.109 to give a F_A value of 0.9999.
K_S	=	Standard steady-state wear coefficient, Eq.10
$K_{S(av)}$	=	Average standard steady-state wear coefficient.
L	=	Sliding distance
L_P	=	Predicted sliding distance, Eq.40.
L_{PN}	=	Net predicted sliding distance, Eq.50.
$L_{P0.99}$	=	Predicted sliding distance, with a F_A value of 0.99; C=4.605 in Eq.40.
$L_{P0.999}$	=	Predicted sliding distance, with a F_A value of 0.999; C=6.908 in Eq.40.
$L_{P0.9999}$	=	Predicted sliding distance, with a F_A value of 0.9999; C=9.210 in Eq.40.
L_t	=	Transient distance
L_{ta}	=	Approximate transient distance
L_q	=	Total sliding distance at point Q.
L_{qa}	=	Approximate total sliding distance
L_S	=	Steady-state sliding distance
L_{Sa}	=	Approximate steady-state sliding distance
m_A	=	Gradient of V_c curve at L=0, Eq.22.
m_B	=	Gradient of V_c curve at $L=L_t$, Eq.23.
P	=	Load
V	=	Volumetric (pin) material loss
V_q	=	Total volumetric (pin) material loss at point Q.
V_t	=	Transient volumetric (pin) material loss
V_{ta}	=	Approximate transient volumetric (pin) material loss
V_S	=	Steady-state wear volume
V_{Sa}	=	Approximate steady-state wear volume
(V/L)	=	Volumetric wear rate
$(V/L)_{M\text{-}Average}$	=	Average measured steady-state volumetric wear rate
$(V/L)_{M\text{-}Lowest}$	=	Lowest measured steady-state volumetric wear rate
$(V/L)_{M\text{-}LP}$	=	Measured steady-state volumetric wear rate at the predicted L_P
$(V/L)_M$	=	Measured steady-state volumetric wear rate
$(V/L)_N$	=	Net steady-state volumetric wear rate
$(V/L)_P$	=	Predicted steady-state volumetric wear rate, Eqs.44 and 47.
$(V/L)_{P0.99}$	=	Predicted steady-state volumetric wear rate with a F_A value of 0.99; C=4.605 in Eq.47.
$(V/L)_{P0.999}$	=	Predicted steady-state volumetric wear rate with a F_A value of 0.999; C=6.908 in Eq.47.
$(V/L)_{P0.9999}$	=	Predicted steady-state volumetric wear rate with a F_A value of 0.9999; C=9.210 in Eq.47.
$(V/L)_S$	=	Standard steady-state volumetric wear rate
W	=	Weight loss
ρ	=	Density

1. Introduction

To understand the wear behaviour of component parts, wear tests are often carried out with suitable wear testing techniques. There are different types of wear mechanisms involved, for example adhesive wear, abrasive wear or others. However, adhesive wear is by far the most dominant form of material loss among sliding components in machinery [1]. The pin-on-disc test is a classical method commonly used for adhesive wear experiments. During the experiment, the sliding between the pin and disc may result in wear on both contact surfaces of the pair. To facilitate measurement, the pin is generally the wearing member that has a lower hardness. Although weight loss and wear rate are often used for studying the wear characteristics of test specimens, recent investigations [2] have found that wear coefficient is a better parameter to be used instead. This is because the wear coefficient has taken into account not only the wear rate (V/L), the applied load, but also the hardness of the wear pin or counterface which can affect the wear rate significantly. Although there are some concerns in using the wear coefficient as a wear parameter, as variations by an order of 10 to 1 were observed [3], however it can be seen from the following sections that the variations can be minimized if suitable precautions are taken. Furthermore, more consistent wear coefficient values are also obtainable when a better analytical model, such as the integrated wear model, to be described in Section 2, is used in their computation [4-6].

For practical reasons, Fig.1(a) shows only the transient wear regime and the steady-state wear regime of a typical wear volume loss versus distance curve. The volume (or weight) loss is initially curvilinear and the rate of volume loss per unit sliding distance decreases until at P where it joins smoothly with the straight line PQ. The amount of volume loss in the regime given by OP is the transient (running-in or unsteady-state) wear and PQ is the steady-state wear. The slope of the linear steady state regime is used to express the wear rate of a material per unit sliding distance and, at a given load and speed, it is constant for a material depending on the nature (e.g. hardness) of the counter-surface.

The standard wear coefficient value obtained from a volume loss versus distance curve is a function of the sliding distance. Due to the higher initial running-in wear rates, it has a higher value initially and will reach a steady-state value when the wear rate becomes constant, as shown in Figure 1(b). This is because the standard method to calculate the wear coefficient is to make use of the total volume loss and the total sliding distance covered [7]. This practice would give a higher standard steady-state wear coefficient value since the higher wear rate from the transient wear is included in its computation. It is also obvious that variations of the standard wear coefficient values can occur since those determined near the transient wear regime will have higher values than those from the steady-state wear regime. Previous work has indicated that the ratio of the wear coefficient from the transient wear regime to that from the steady-state wear regime can vary from 5 to 10 [5]. Hence wear tests should be carried out to include both the transient wear and the steady-state wear. In this way, more useful wear data will be obtained to facilitate better product design to cater for different life cycles; and to avoid the problem associated with over-designing a component part. However, it should be noted that the term "standard steady-state wear coefficient" is used to differentiate it from another one, the "net steady-state wear coefficient" to be introduced later in this chapter. The introduction of the net-steady-state wear coefficient is to exclude the higher wear rate from the transient wear in its computation, to be described in Section 2.3.

Wear testing is a time consuming process, as the test has to be repeated with different sliding distances until a steady-state wear condition is achieved. Furthermore, it may also be difficult to judge correctly whether a steady-state wear condition has actually been attained. Common wear testing practice tends to perform the wear test near the transient wear regime with shorter sliding distances to save time; or in the so-called "steady-state" region without knowing exactly where it is located. Hence under-tested conditions or over-tested conditions were often used. This approach will lead to inaccurate results as higher wear coefficients and wear rates are likely to be obtained when wear tests are performed in the transient wear and in the severe wear regimes as mentioned earlier. Consequently a lot of time and resources are also wasted in the process. Although there are standard methods available to determine the common material properties such as tensile strength, yield strength, impact strength and many others, however there is no useful "standard method" available for the accurate determination of the steady-state wear rate and wear coefficient. In fact, different methods such as pin-on-disc, pin-on-cylinder and many others, with different nominal specimen contact areas, different testing parameters and even with different wear regimes, have been used. It is therefore no surprising that the wear rate and the wear coefficient values obtained from different investigators can vary significantly as mentioned earlier.

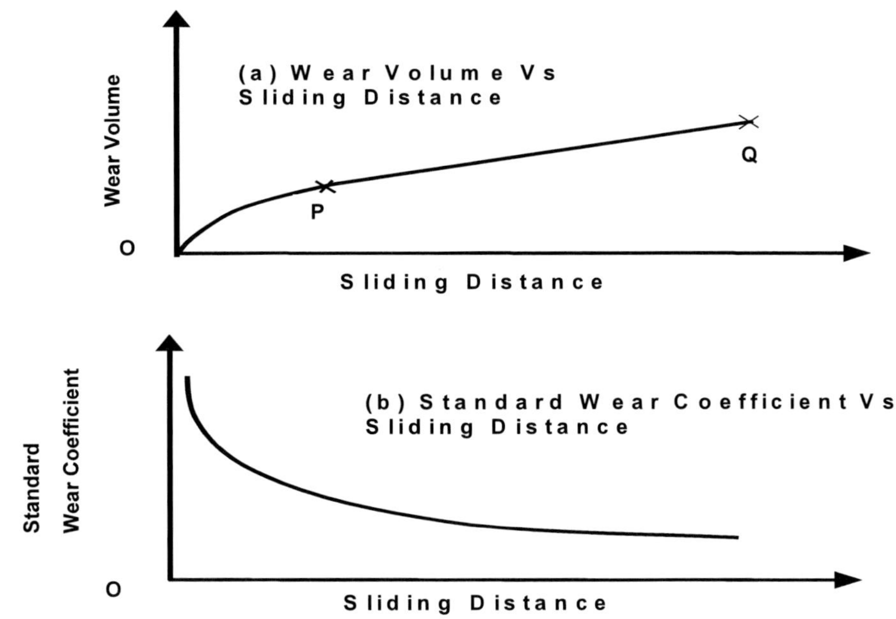

Figure 1. (a) Wear volume versus sliding distance curve; (b) Standard wear coefficient versus sliding distance curve

Recent work carried out by Yang [4-6] has found that, with the integrated wear model, it is possible to model the steady-state wear rate as well as the wear coefficient with the transient wear data only. It has been found that the F_A value, an exponential function, is a useful factor to be used to evaluate the wear test condition and to determine whether the steady-state wear regime has already been reached. Hence it is possible to predict both the standard and the net-steady state wear coefficients and wear rates with suitable F_A values, for example 0.99 to 0.999, together with the transient wear data (A and B values) obtained from the transient wear study [8-10]. Therefore with the integrated wear model, both the transient wear rate and the steady-state wear rate will be linked.

As currently all the wear data are obtained by using 'non-standard' methods, their usefulness is rather limited despite a lot of efforts and resources invested in carrying out the wear tests. It is high time to find some more useful ones to determine the wear coefficient and wear rate more consistently, accurately and perhaps more economically. With these objectives in mind, an integrated adhesive wear testing methodology, based on the integrated wear model and the related wear equations developed by the author, has therefore been proposed [11-12]. With this methodology, the wear testing process will be divided into three stages:(i) to conduct the transient wear test; (ii) to predict the steady-state sliding distance, wear rate and wear coefficient; and (iii) to conduct the steady-state confirmation run(s) to obtain the measured steady-state wear rate and wear coefficient. This testing methodology will provide useful and valuable transient as well as the steady-state wear data for different product design needs as mentioned earlier. More consistent steady-state wear coefficient and wear rate will also be obtained, since they are determined with the same F_A value. It will also save time and resources in conducting the wear tests. It should be noted that the inclusion of transient wear in a test programme may not necessarily increase the testing time, as the transient wear tests are conducted with a shorter sliding distance. In fact more time will be saved in wear testing as the exact sliding distance and wear coefficient value could be predicted from the transient wear test data instead of using the usual trial and error method. It will be shown later in this chapter that about 30-40% of testing time can be saved if only the confirmation run at the predicted sliding distance is chosen. Obviously this methodology of using limited confirmation runs will only work if the wear curve for the wearing pair has been well established and it is similar to that used in the integrated wear model as shown in Fig.3. Furthermore, it is also assumed that no major change of wear mode is expected to happen in the wearing process.

To facilitate understanding of the proposed integrated adhesive wear testing methodology, a review of the integrated wear model, the development of the various equations for modeling and predicting the standard and net steady-state wear rates and wear coefficients with their respective distances, will be carried out. It will describe the experimental methodology used in the previous studies. Some wear data obtained previously from aluminium-based metal matrix composites [4-5, 11-12] will also be analyzed and presented in this chapter to support the proposed methodology.

2. The Integrated Wear Model

For adhesive wear, the volume loss versus sliding distance curve can generally be divided into three regimes: the transient wear regime, the steady-state wear regime and the final rapid wear regime [13-14]. For practical applications, the transient wear regime and the steady-state wear regime are the two most important ones to be considered. Figure 2 shows the wear volume loss against distance curves plotted for MMC-A, MMC-B and MMC-C, with data obtained from both the moving-pin and the conventional pin-on-disc methods as shown in Table 1 [5]. In the figure, V(M)-A and V(C)-A indicate the wear volume loss values (V) of MMC-A tested under moving-pin and conventional pin-on-disc conditions respectively. Figure 2 indicates that the curves can generally be divided at least into two wear regimes, the transient wear regime and the steady-state wear regime. For each material, the actual wear volume loss curve is initially curvilinear and the rate of volume loss per unit sliding distance decreases until a point, i.e. at the transient distance, beyond which it will remain almost constant. By visual inspection of the curves, the end point of the transient region is about 2,000 m to 3,000 m. The wear loss values for the specimens tested with the conventional pin-on-disc method, indicated by the dotted lines, were found to be higher, as compared with those tested with the moving-pin method. It should be noted that the wear volume of a pin from the pin-on-disc test depends greatly on the condition of the wear track. The single wear track generated by the conventional pin-on-disc method can be damaged more easily by alumina particles or wear debris, to give a higher wear rate. A slightly increasing wear volume loss trend can also be observed on some specimens tested up to 12 km due to the same reason.

Figure 3 shows the proposed integrated mathematical wear model to cover the transient wear and the steady-state wear of a pin-on-disc system [6]. The wear model consists of two parts, the transient wear and the steady-state wear. Point P represents the end of transient wear and L_t is the transient wear distance. Point P also represents the beginning of the steady-state wear. Hence at point Q, the specimen will have covered a total sliding distance of L_q and the total volume loss will be V_q. An exponential equation is used to model the transient wear [15-16] while the Archard's wear equation [17] is revised to model the net steady-state wear. The following sub-sections will describe the wear model in details.

Table 1. Volume loss versus distance for MMC-A, MMC-B and MMC-C

Distance (m)	Volume loss (mm^3) (Moving-pin method)			Volume loss (mm^3) (Conventional-pin-on-disc method)		
	MMC-A	MMC-B	MMC-C	MMC-A	MMC-B	MMC-C
250	2.47	1.89	1.36	3.80	1.45	1.42
500	4.24	3.16	2.12	5.30	2.65	2.41
1000	6.42	4.24	3.52	7.32	4.42	4.62
1500	8.15	4.95	4.62	9.00	5.53	6.13
2000	9.17	5.69	5.18	10.13	6.11	7.03
2500	9.71	6.52	5.35	11.02	6.90	8.32
3000	10.40	7.08	5.60	12.12	7.57	9.04
6000	13.45	9.42	7.46	15.25	10.26	11.1
9000	17.54	12.99	9.81	17.89	13.59	13.8
12000	21.84	17.74	13.7	22.11	18.95	17.8

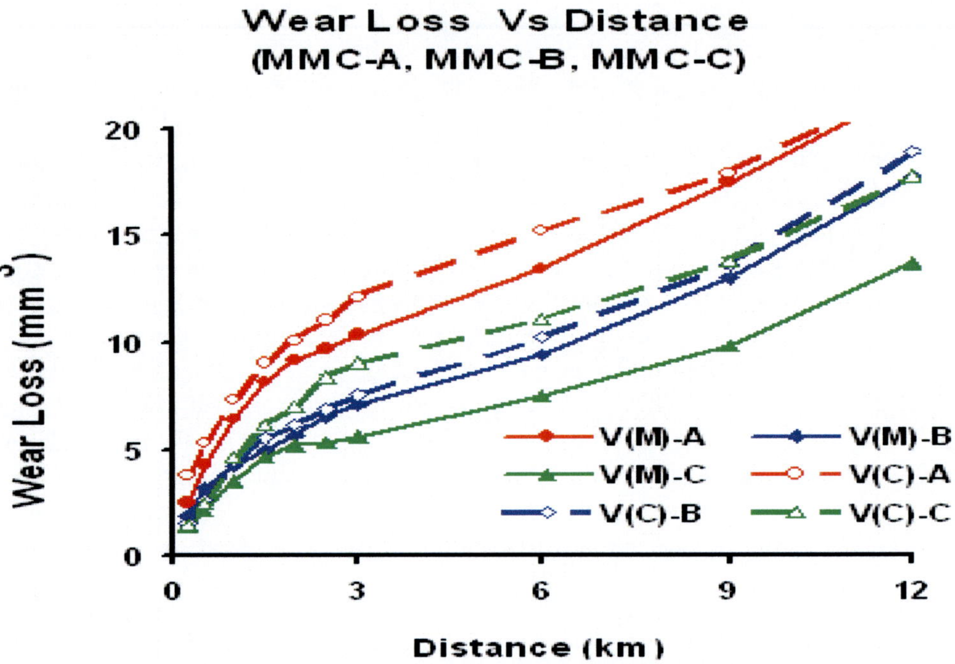

Figure 2. Wear volume loss versus distance curves for MMC-A, MMC-B and MMC-C

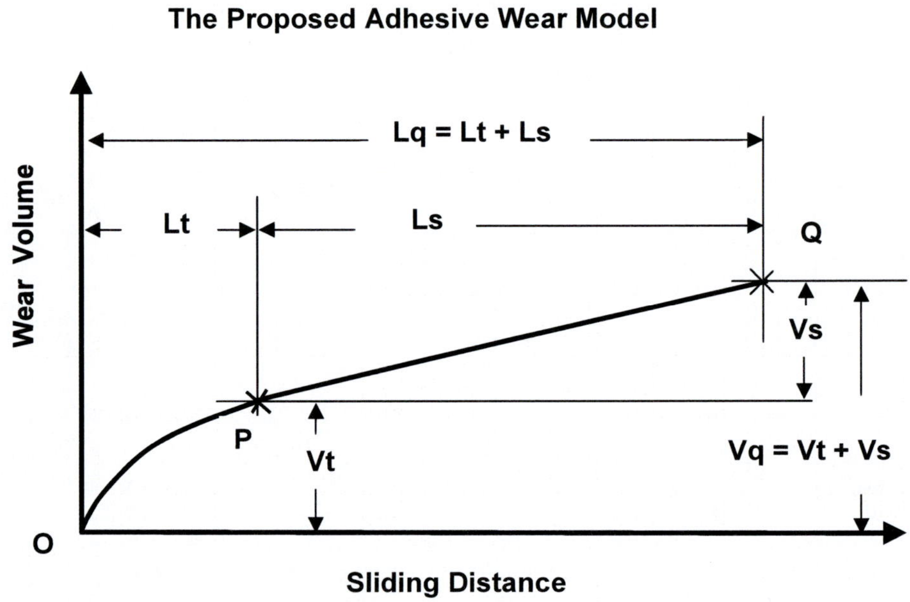

Figure 3. The integrated adhesive wear model

2.1 The Transient Wear

In the transient wear regime, the rate of volume removed per unit sliding distance is assumed to be a function of the volume of metal available at the junctions [15, 16].

$$\frac{dV}{dL} = -BV \qquad (1)$$

or re-arranging

$$\frac{dV}{V} = -BdL \qquad (2)$$

where the V term denotes the volume, L the sliding distance and B is a constant which depends on the applied load and the surface condition of the wear surface. The negative sign describes a situation where the original volume at the junctions diminishes with sliding distance.

Integrating Eq.2 gives

$$\ln V = -BL + C \qquad (3)$$

where C is the constant of integration.

Eq.3 can also be expressed as

$$V = \exp^{-BL+C} \qquad (4)$$

If A is the volume at the junctions at zero sliding distance, putting L =0 in Eq.4, then

$$A = \exp^{C} \qquad (5)$$

That is,

$$C = \ln A \qquad (6)$$

Hence

$$V = A\exp^{-BL} \qquad (7)$$

Thus the volume removed (V_t) which is the transient wear volume, at a sliding distance L is:

$$V_t = A - V = A - A\exp^{-BL} = A[1 - \exp^{-BL}]$$

that is,

$$V_t = A[1 - \exp^{-BL}] \qquad (8)$$

It should be noted that Eq.8 is an exponential equation in which L should be equal to or less than L_t, which is the transient wear distance. Obviously Eq.8 should be used only for modeling the transient wear up to the transient distance L_t. Fig.4 shows the measured and calculated volume loss against distance curves with Eq.8 for MMC-A, MMC-B and MMC-C, for data obtained with the moving-pin method. For the legends used in the figure, V(M), in solid lines indicates the measured volume while V(C), in dotted lines, the calculated volume. It can be seen from the figure that Eq.8 gives excellent agreement in the transient wear regime, beyond which each of them will give a horizontal line irrespective of the sliding distance.

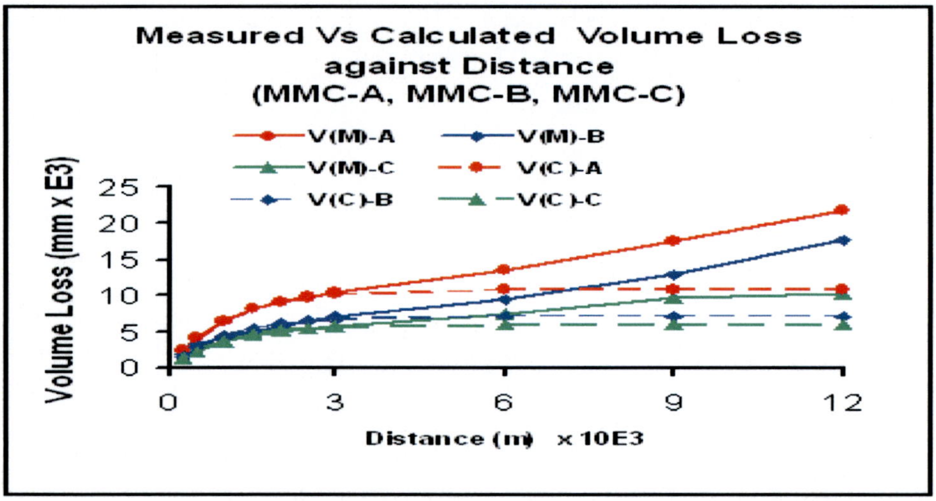

Figure 4. Measured and calculated volume loss against distance curves with Eq.8 for MMC-A, MMC-B and MMC-C. V(M) indicates the measured volume while V(C) the calculated volume

2.2 Standard Wear Coefficient (K_S)

Archard's wear equation [17], as shown in Eq.9, has been used to calculate the standard wear coefficient (K_S). As mentioned earlier, the term "standard wear coefficient" is used to differentiate it from the "net steady-state wear coefficient" to be introduced later in this chapter.

$$V = K_S \frac{PL}{3H} \qquad (9)$$

where V is the volumetric loss of the softer material after sliding for a distance L at load P normal to the wear surface. H is the Brinell hardness of the softer wearing material while K_S is a dimensionless standard wear coefficient characterizing the particular pin-disc interface. For known values of V, P, L and H, the standard wear coefficient can be calculated from Eq.10:

$$K_S = \frac{3HV}{PL} \qquad (10)$$

For a particular material, it should be noted that V can also be estimated from the weight loss W and the density ρ. Hence the standard wear coefficient can also be determined from Eq.11:

$$K_S = \frac{3HW}{PL\rho} \tag{11}$$

The standard wear coefficient value obtained from Eq.10 is a function of the sliding distance. Fig.5 shows the standard wear coefficient versus distance curves obtained for MMC-A, MMC-B and MMC-C, by using data from both the moving-pin and the conventional pin-on-disc testing methods. In the figure, K_S(M)-A and K_S(C)-A indicate the standard wear coefficient values (K_S) of MMC-A tested under moving-pin and conventional pin-on-disc conditions respectively. It can be seen from the figure that the standard wear coefficient (K_S) decreases more rapidly as the distance increases initially, but more gradually when approaching the steady-state wear and eventually becomes almost a horizontal line in the steady-state wear regime. The standard wear coefficient values for the specimens tested with the conventional pin-on-disc method, indicated by dotted lines, were found to be higher, as compared with those tested with the moving-pin method, due to their higher wear volumes caused by the more severely worn tracks. However, it should be noted that the standard method to calculate the standard wear coefficient is to make use of the total volume loss (V) and the total sliding distance (L) [7]. Consequently a higher steady-state wear coefficient value will be obtained since the higher wear rate from the transient wear is included in the computation. Hence there is a need to define the net steady-state wear coefficient (K_N), to be described in Section 2.3.

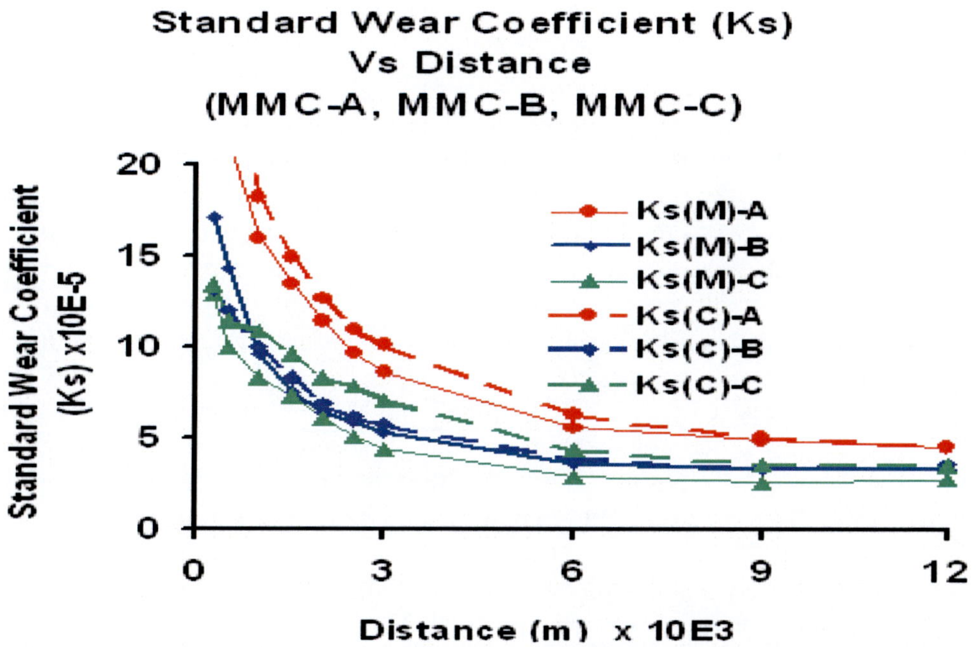

Figure 5. Standard wear coefficient versus distance curves for MMC-A, MMC-B and MMC-C

2.3 Net Steady-state Wear Coefficient (K_N)

To model the net steady-state wear, Archard's wear equation in its revised form as shown in Eq.12, is used to calculate the net steady-state wear volume, Vs.

$$V_S = K_N \frac{PL_S}{3H} \qquad (12)$$

where V_S and L_S are respectively the net steady-state wear volume and sliding distance, to be defined by Eq.13 and Eq.14 respectively. K_N is the dimensionless net steady-state wear coefficient.

Figure 3 shows the proposed integrated mathematical model to cover the transient wear regime and steady-state wear regime of a typical pin-on-disc system. Point P represents the end of transient wear and L_t is the transient wear distance. Point P also represents the beginning of the steady-state wear. Hence at point Q, the specimen will have covered a total sliding distance of L_q and the total volume loss will be V_q. The steady-state sliding distance (L_S) and the steady-state wear volume (V_S) can be calculated by using Eq.13 and Eq.14 respectively.

$$L_S = L_q - L_t \qquad (13)$$

$$V_S = V_q - V_t \qquad (14)$$

Hence the net steady-state wear coefficient (K_N) can be computed from Eq.15.

$$K_N = \frac{3HV_S}{PL_S} \qquad (15)$$

With the revised Archard's wear equation, Eq.15, a more precise net steady-state wear coefficient (K_N) can be obtained to facilitate a better comparison of the wear properties of different wearing pairs, since the higher wear rate from the transient wear is excluded. Furthermore, if the transient wear distance could be determined accurately, considerable time could also be saved in wear testing since a much longer sliding distance required by the conventional method would no longer be needed. Consequently, more consistent results would be obtained, as the wear track is less likely to be damaged by a shorter sliding distance.

Figure 6 shows the net-steady-state wear coefficients (K_N) plotted against the steady-state sliding distance (L_S) obtained for MMC-A, MMC-B and MMC-C, by using both the moving-pin and the conventional pin-on-disc testing methods. It should be noted that, in most cases, the net steady-state wear coefficient increases slightly with an increase of steady-state sliding distance. This was due to the occurrence of damaged wear tracks caused by fractured alumina particles when a longer sliding distance was used. However, the variations of the net steady-state wear coefficient values are confined to within a very much smaller range. The average net steady-state wear coefficient values (K_N), for MMC-A, MMC-B and MMC-C were respectively 2.95×10^{-5}, 2.86×10^{-5} and 1.66×10^{-5} when the moving-pin method was used; and 2.91×10^{-5}, 2.50×10^{-5} and 1.97×10^{-5} when the conventional pin-on-disc method was used.

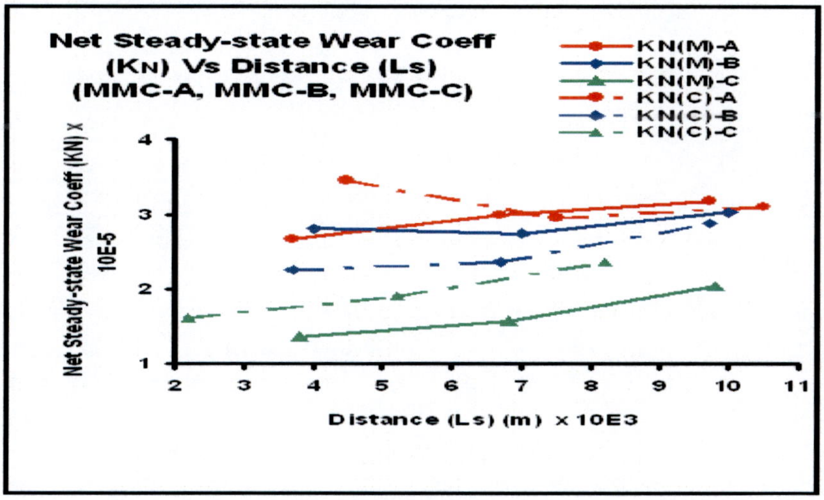

Figure 6. Net steady-state wear coefficient versus distance curves for MMC-A, MMC-B and MMC-C

2.4 Transient Distance

To determine the transient distance (L_t), it is assumed that the sliding distance L_t is the end of the transient wear and also the beginning of steady-state wear. From Fig.3, it is obvious that at L= L_t the gradient of transient wear is equal to the gradient of the steady-state wear.

Differentiate Eq.8 gives,

$$\frac{dV}{dL} = AB\exp^{-BL} \quad (16)$$

Differentiate Eq.15 gives

$$\frac{dV_s}{dL_s} = \frac{K_N P}{3H} \quad (17)$$

Equating Eq.16 and Eq.17 gives

$$AB\exp^{-BL_t} = \frac{K_N P}{3H} \text{ or } \ln\left[\frac{K_N P}{3HAB}\right] = -BL_t$$

Hence

$$L_t = \frac{-\ln[K_N P/3HAB]}{B} \quad (18)$$

Substitute Eq.15 into Eq.18 gives,

$$L_t = \frac{-\ln[V_S / ABL_S]}{B} \tag{19}$$

Eq.19 indicates that the transient distance L_t can be determined if the constants A and B as well as the L_S and V_S values are known. A and B values can be obtained from the transient data points from the respective volume loss versus distance curve by using standard softwares available from the market. However, both L_S and V_S values cannot be determined from Eqs.13 and 14 without a L_t value. To overcome this difficulty, a number of approximate transient distance values (Lt_a) are selected to work out some approximate L_{sa} and V_{sa} values. With these values, a number of transient distance (L_t) values can then be found from Eq.19. The average L_t value is then used to calculate the L_S and V_S values, by using Eqs.13 and 14 respectively. Consequently the net steady-state wear coefficient K_N can be computed from Eq.15. The choice of L_{ta} is not critical and it will not affect the calculated results significantly. This is because the net steady-state wear volume, as expressed by Eq.12, generally has a linear relationship. In the previous study, the L_{ta} values used were 2,000 m, 2,500 m and 3,000 m. The average transient distance values obtained, with Eq.19, for MMC-A, MMC-B and MMC-C were respectively 2,311 m, 1,999 m and 2,228 m when the moving-pin method was used; and 2,301 m, 2,334 m and 3,838 m when the conventional pin-on-disc method was used. The average value of the calculated transient distance was 2,502 m.

3. Review of The Wear Equations Proposed for the Integrated Wear Model

3.1 Equation for Modeling the Wear Coefficient of Particles Reinforced MMC Materials

The first standard wear coefficient equation proposed by Yang [4], Eq.27, was developed by using the mathematical model for the transient wear volume of an Al-SiC(P)/steel composite system postulated by Zhang, Zhang and Mai [18] as indicated in Eq.20.

$$V_c = \frac{g_1 P}{g_3 H} \frac{d(1-f_v)}{f_v}\left[1-\exp\left(\frac{-g_3 f_v L}{d(1-f_v)}\right)\right] \tag{20}$$

where V_c is the volumetric pin material loss, H is the pin hardness, P is the load, f_v is the particle volume fraction, d is the average particle size, L is the distance, g_1 and g_3 are experimental constants to be determined from Eq.22 and Eq.25 respectively. As will be mentioned later in this section, g_1 is a function of the applied load (P), the pin hardness (H) and the gradient (m_A) of the V_c curve at L=0. g_3 is a function of the average particle diameter (d), the volume fraction of particles (f_v), the transient distance (L_t), the gradients m_A and m_B of the V_c curve at L=0 and L=L_t respectively.

Differentiating Eq.20 with respect to L gives

$$\frac{dV_c}{dL} = \frac{g_1 P}{H} \exp\left[\frac{-g_3 f_v L}{d(1-f_v)}\right] \quad (21)$$

It should be noted that Eq.20 is valid in transient wear only and till the onset of steady wear. If it is allowed to progress from transient wear into steady wear, the curve will level off horizontally and this will violate Archard's steady state wear theory as shown in Fig.4. To determine g_1 and g_3, let the gradients of the V_c graph be m_A at L=0, and m_B at L= L_t, which is the onset of steady state wear. Invoking the first boundary condition when the first derivative of V_C equals to m_A and L= 0 in Eq.21, g_1 can now be determined from Eq.22.

$$g_1 = \frac{Hm_A}{P} \quad (22)$$

The second boundary condition is at L= L_t, which is the onset of steady state wear, where gradient of V_C is m_B.

$$m_B = \frac{g_1 P}{H} \exp\left[\frac{-g_3 f_v L_t}{d(1-f_v)}\right] \quad (23)$$

By employing natural logarithm on both sides of Eq.23,

$$\ln m_B = \ln\left(\frac{g_1 P}{H}\right) - \frac{g_3 f_v L_t}{d(1-f_v)} \quad (24)$$

Substituting Eq.22 into Eq.24 and rearranging gives

$$g_3 = \frac{d(1-f_v)}{f_v} \frac{(\ln m_A - \ln m_B)}{L_t} \quad (25)$$

With constants g_1 and g_3 determined, the transient wear volume loss (V_c) can be predicted from Eq.20. By substituting Eq.20 into Eq.10, one gets

$$K_S = \frac{3 g_1 d(1-f_v)}{g_3 f_v L}\left[1 - \exp\left(\frac{-g_3 f_v L}{d(1-f_v)}\right)\right] \quad (26)$$

It should be noted that the effects of load and pin hardness are taken into account in the determination of g_1, as indicated in Eq.22. Hence by substituting Eq.22 into Eq.26, one gets Eq.27 which shows clearly the effects of load and pin hardness on the wear coefficient. Table 2 shows the g_1 and g_3 values obtained from the previous study [4].

$$K_S = \frac{3Hm_A d(1-f_v)}{PLg_3 f_v}\left[1-\exp\left(\frac{-g_3 f_v L}{d(1-f_v)}\right)\right] \quad (27)$$

Table 2. g1 and g3 values for MMC-A, MMC-B and MMC-C

Material	Moving-pin		Conventional Pin-on-disc	
	g_1 (10^{-5})	g_3 (10^{-7})	g_1 (10^{-5})	g_3 (10^{-7})
MMC-A	8.2	2.14	12.63	2.09
MMC-B	5.65	1.56	4.37	1.20
MMC-C	4.25	0.547	4.47	0.525

3.2 Effects of Exponential Factors F_A and F_B

In order to make it easier to analyse the different behaviours of Eq.20 and Eq.27, it is useful to define two exponential factors F_A and F_B. Eq.20 can be re-written as

$$V_c = \frac{g_1 P}{g_3 H}\frac{d(1-f_v)}{f_v}[F_A] \quad (28)$$

where

$$F_A = \left[1-\exp\left(\frac{-g_3 f_v L}{d(1-f_v)}\right)\right] \quad (29)$$

Eq.20 was proposed for the transient wear volume only as it contains a function F_A as shown in Eq.29. F_A is an exponential function of L and it has the effect of increasing the volume rapidly to reach a steady state. Hence Eq.20 cannot be used for the modeling of the steady state wear volume as mentioned previously in Section 3.1. As Eq.27 also contains such a function, one may wonder why Eq.27 can be used successfully to determine the standard steady-state wear coefficient. To understand the rational, it is necessary to examine Eq.27 carefully. Eq.27 can be re-written as

$$K_S = \frac{3Hm_A d(1-f_v)}{Pg_3 f_v}\left[\frac{1}{L}\right]\left[1-\exp\left(\frac{-g_3 f_v L}{d(1-f_v)}\right)\right] \quad (30)$$

or

$$K_S = \frac{3Hm_A d(1-f_v)}{Pg_3 f_v}[F_B] \quad (31)$$

where

$$F_B = \left[\frac{1}{L}\right]\left[1 - \exp\left(\frac{-g_3 f_v L}{d(1-f_v)}\right)\right] \tag{32}$$

Comparing Eq.32 with Eq.29, it is obvious that

$$F_B = \frac{F_A}{L} \tag{33}$$

It is clear that Eq.27 contains factor F_B, which is related to F_A as shown in Eq.33. However, the behaviour of F_B is different from that of F_A. Table 3 shows some F_A and F_B values versus distance obtained in the previous study [4]. These values are also plotted in Fig.7 against distance, from 0.25 to 12 km, for MMC-A, MMC-B and MMC-C respectively.

It can be seen from Table 3 that the average value of F_A and F_B varies from about 0.19 to 1.0; and 7.6 x10^{-5} to 0.83 x10^{-5} respectively, from a sliding distance of 0.25 km to 12 km. F_A increases rapidly and reaches almost 100% of the maximum value at 9 km. However, F_B decreases more gradually, reaches to 1.1 at 9 km, or 83% reduction from its maximum value. Hence Eq.20, which contains factor F_A, is unsuitable for the determination of the wear volume in steady-state wear since it will give a horizontal line in the steady-state wear regime which is different from the actual volume loss versus distance line. On the other hand, Eq.27, which contains factor F_B, was found to be capable of predicting the standard steady state wear coefficients with a good accuracy.

Table 3. Values of factors F_A and F_B versus distance

Distance (m)	F_A						Av F_A	Av F_B (10^{-5})
	MMC-A		MMC-B		MMC-C			
	MP	CP	MP	CP	MP	CP		
250	0.209	0.231	0.212	0.170	0.206	0.113	0.190	7.60
500	0.374	0.409	0.379	0.310	0.370	0.213	0.343	6.85
1000	0.608	0.650	0.615	0.524	0.603	0.381	0.564	5.64
1500	0.754	0.793	0.761	0.672	0.750	0.513	0.707	4.72
2000	0.846	0.878	0.852	0.774	0.842	0.617	0.802	4.01
2500	0.904	0.928	0.908	0.844	0.901	0.699	0.864	3.46
3000	0.940	0.957	0.943	0.892	0.938	0.763	0.906	3.02
6000	0.996	0.998	0.997	0.988	0.996	0.944(a)	0.995(b)	1.64
9000	1.000	1.000	1.000	0.999	1.000	0.987	0.998	1.11
12000	1.000	1.000	1.000	1.000	1.000	0.997	1.000	0.83

Notes: (a) This value was too low and was omitted in the calculation of the average value as indicated in (b).
(MP= Moving-pin method; CP= Conventional pin-on-disc method)

It should be noted that the value of F_B is also equal to F_A/L, as indicated by Eq.33. The F_A value will only increase very slowly after the initial period. Furthermore, factor F_A can only have a maximum value of unity. Hence the value of F_B will decrease further as the distance (L) is increased, and will become zero when L is at infinity. Hence due care should be exercised to select the correct distance L. The best distance to be selected should give a F_A value that is close to, but slightly lower than unity, say 0.99 or 0.999.

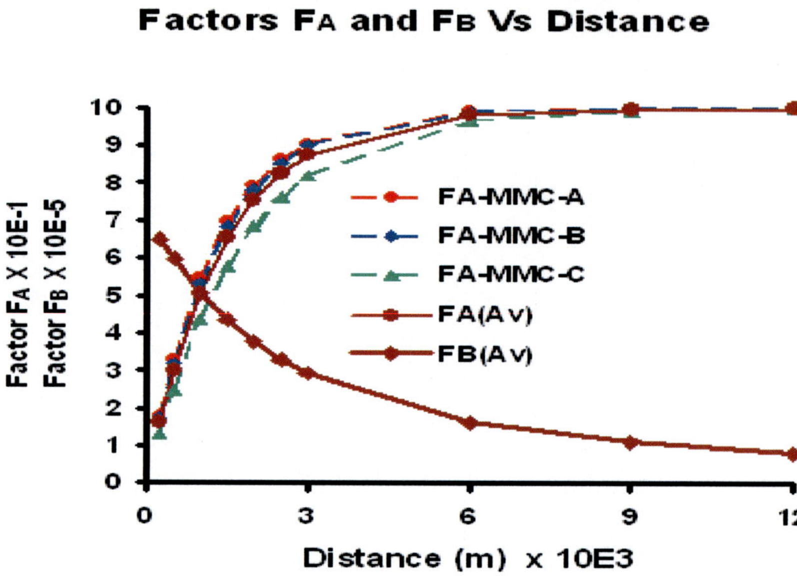

Figure 7. Characteristics of factors F_A and F_B versus distance

3.3 Equation for Modeling the Wear Coefficient of Other Materials

As mentioned earlier, Eq, 27 was developed to model the wear coefficient of an Al-SiC(P)/steel composite system, based on the mathematical model for the transient wear volume proposed by Zhang, Zhang and Mai [18]. However, computational experience with metal matrix materials have also indicated that there are some limitations in using Eq.27. First of all, the distribution of particles in a metal matrix composite is always non-uniform, to give different (f_v) values from one section to another. Next, different particle sizes (d) are present in a particular section. Furthermore, only two gradients, m_A and m_B are used to define the curve which may not turn up to be exact. In order to make modeling easier and with a better accuracy, and to be able to extend the same modeling technique to other metals and alloys, the second wear coefficient equation, as indicated in Eq.34, was proposed by Yang [5]. It was developed by using the same technique used to develop Eq.26, that is, by substituting Eq.8 into Eq.10.

$$K_S = \frac{3HA}{PL}\left[1 - \exp^{-BL}\right] \qquad (34)$$

It is obvious that with Eq.34, only two parameters (A and B) need to be determined, as the other parameters such as H, P, and L are known. With the use of established softwares, the values of A and B can be determined more accurately as all the transient data points are fully utilized to obtain the 'best-fit' curve for the transient wear volume to give more accurate results. Moreover, Eq.34 is also more generic in nature and can be used for modeling most of the materials and alloys, inclusive of particle reinforced metal matrix materials. Table 4 shows the A and B values, computed from wear volume loss data obtained by both the moving-pin and conventional pin-on-disc methods, and at distances of 500 m, 1,000 m, 1,500 m, 2,000 m, 2,500 m and 3,000 m, by using a commercial software DataFit Version 6 [19].

Table 4. Constants A and B, for transient wear equations for MMC-A, MMC-B and MMC-C

Material	Moving-pin method		Conventional pin-on-disc method	
	A	B (10^{-4})	A	B (10^{-4})
MMC-A	10.87	9.36	11.95	10.51
MMC-B	7.09	9.54	8.25	7.43
MMC-C	6.02	9.24	11.79	4.8

Note: The units for the volume and the sliding distance are in mm^3 and m respectively in the determination of A and B values.

Eq.34 has been found to be capable of modeling the wear coefficient for both the transient regime and the steady-state regime [8-10]. To make analyses easier, it is again useful to define two exponential functions, F_A and F_B as indicated in Eq.35 and Eq.36 respectively. Eq.34 will then be reduced to Eq.37, which indicates that the steady-state wear coefficient is a function of F_B, which is related to F_A, but will not behave like F_A as discussed in Section 3.2

$$F_A = \left[1 - \exp^{-BL}\right] \quad (35)$$

$$F_B = \frac{1}{L}\left[1 - \exp^{-BL}\right] \quad (36)$$

then Eq.34 becomes

$$K_S = \frac{3HA}{P}\left[F_B\right] \quad (37)$$

3.4 Equations for Predicting Steady-State Wear Coefficients

As mentioned earlier, the standard wear coefficient value will depend on F_B, which is also equal to F_A/L as indicated by Eq.36. Hence by selecting a suitable F_A value or a proper predicted sliding distance L_P, one should be able to predict both the standard and the net steady-state wear coefficients by using Eq.38.

$$K_P = \frac{3HA}{PL_P}\left[1 - \exp^{-BL_P}\right] = \frac{3HA}{PL_P}\left[F_A\right] \quad (38)$$

where K_P is the predicted wear coefficient and L_P is the predicted sliding distance, which is used to obtain a suitable F_A value and subsequently to determine the predicted wear coefficient value (K_P).

Previous works [8-9] have indicated that F_A values of 0.99 to 0.999 are useful for the prediction of standard steady-state wear coefficient and net-steady-state wear coefficient respectively. It is therefore obvious that Eq.38 can be simplified further. For example, to obtain a F_A value of 0.99, the sliding distance ($L_{P0.99}$) can be determined from Eq.39, which is derived by substituting F_A=0.99 into Eq.35 and re-arranging it. However Eq.40, in which C is a constant which is equal to 4.605, 6.908 and 9.210 for a F_A value of 0.99, 0.999 and 0.9999 respectively, is the general equation that can be used to determine the predicted sliding distance L_P.

$$L_{P0.99} = -\left[\frac{\ln(0.01)}{B}\right] = \frac{4.605}{B} \tag{39}$$

$$L_P = \frac{C}{B} \tag{40}$$

By substituting Eq.40 into Eq.38, one gets Eq.41, in which D is a constant, which is equal to 0.215, 0.145 and 0.109 for a F_A value of 0.99, 0.999 and 0.9999 respectively. It should be noted that D is approximately equal to 1/C as F_A is very close to unity. Eq.41 was found suitable for predicting the standard wear coefficient values and the net steady-state wear coefficient values for MMC-A, MMC-B and MMC-C, MMC-D and A6061-T6 with F_A values of 0.99 and 0.999 respectively [8-10]. It will be shown in Eq.47 in Section 3.5 that the term ABF_A/C or ABD in Eq.41 is actually the predicted steady-state wear rate $(V/L)_P$.

$$K_P = \frac{3HABF_A}{PC} = \frac{3HABD}{P} \tag{41}$$

3.5 Equations for Predicting the Steady-State Wear Rates

As mentioned in the previous section, the steady-state wear coefficient values can be predicted with Eq.41. Hence the standard steady-state wear rate $(V/L)_S$ can be predicted from the predicted standard wear coefficient values as indicated in Eq.42, which is derived from Eq.9. The net-steady-state wear rate $(V/L)_N$ can also be predicted similarly from Eq.43. Eq.44 is therefore the general equation which can be used for the prediction of the standard and net steady-state wear rates from their respective predicted standard and net steady-state wear coefficients. With this methodology, more consistent wear rates will be obtained since the same F_A value is to be used in their computations. Furthermore, the wear rates will also be linked automatically to the wear coefficients to facilitate wear design calculations [11-12].

$$\left(\frac{V}{L}\right)_S = \frac{PK_{P0.99}}{3H} \tag{42}$$

$$\left(\frac{V}{L}\right)_N = \frac{PK_{P0.999}}{3H} \quad (43)$$

$$\left(\frac{V}{L}\right)_P = \frac{PK_P}{3H} \quad (44)$$

Alternatively, the wear rate can also be predicted directly from the wear volume loss data by using Eq.45, which is obtained by dividing both sides of the transient wear equation (Eq.8) by L. Obviously the predicted wear rate $(V/L)_P$ can now be determined from Eq.46. This is because the wear rate (V/L) is again a function of F_B, and therefore the same methodology used previously to determine the wear coefficient is also applicable here. By substituting $L_P = C/B$ from Eq.40 into Eq.46, Eq.47 is obtained. As indicated previously, C is a constant, which is equal to 4.605, 6.908 and 9.210 for a F_A value of 0.99, 0.999 and 0.9999 respectively. Again D is a constant which is approximately equal to 1/C as F_A is very close to unity. However, F_A values of 0.99 and 0.999 are most suitable ones to be used for predicting the standard and net steady-state wear rates respectively. As these F_A values are exactly the same as those used previously for modeling and predicting the standard and net steady-state wear coefficients, the steady-state wear rates determined by using Eq.47 will also be identical to those obtained by using Eq.44.

$$\left(\frac{V}{L}\right) = \frac{A}{L}\left[1 - \exp^{-BL}\right] \quad (45)$$

$$\left(\frac{V}{L}\right)_P = \frac{A}{L_P}\left[1 - \exp^{-BL_P}\right] \quad (46)$$

$$\left(\frac{V}{L}\right)_P = \frac{AB}{C}\left[F_A\right] = ABD \quad (47)$$

4. Experimental Methodology Used in the Previous Studies

4.1 The Moving-Pin Technique

With the conventional pin-on-disc wear test, the pin specimen is held stationary on top of a precision ground rotating disc, with the required load applied through the pin. With the use of a variable speed motor, the rotational speed of the disc can be varied. The linear speed of the disc at the point where the pin is located is:

$$V_{disc} = 2\pi RN \quad (48)$$

where V_{disc} is the linear speed of the rotating disc, R is the distance from the centre of the pin to the centre of the disc, N is the rotational speed of the disc. The conventional pin-on-disc wear test is conducted with a fixed R and only a small portion of the disc is utilized. The continuous rubbing on the same surface can also damage the wear track to affect the test results.

In an attempt to overcome the above-mentioned shortcoming, the moving-pin technique with a spiral wear track was developed by Yang [20]. This technique was found particularly useful for testing hard pin material like cemented tungsten carbide. It was again used in the previous study [5], as the aluminium-based matrix composites containing alumina can be abrasive to cause a rapid wear of the disc material. It can be seen from Eq.48 that, for the same N, the linear speed of the disc is directly proportional to R. With the aid of computer numerical control (CNC) technology, R can be varied while the linear velocity of the disc is kept constant. The feed rate of the pin can also be specified. However, a high feed rate should be avoided in order not to affect the actual wear rate.

4.2 Experimental Set-up

Figure 8 shows the schematic experimental setup for carrying out the wear tests. The fixture is clamped onto the tool post of the CNC lathe and carries a square tool pin holder and a pneumatic cylinder which is used to provide the required load which is adjustable during the wear testing of the pins. A pressure gauge was used to monitor the load pressure and a throttle valve was used to adjust the required load. A Rikadenki Type R-63 multi-pen recorder and a Kistler Type 9121 force dynamometer were used to measure the force between the pin and the disc during wear testing [20]. The pin holder was designed to ensure that the pin specimen would be held firmly during the wear tests, with 20 mm of the pin held inside the holder and 5 mm of it protruding out for wear test purposes. The protruded pin surface must have full contact with the disc to obtain accurate experimental results. Thus close dimension tolerance was maintained when machining the pin holder and the pins.

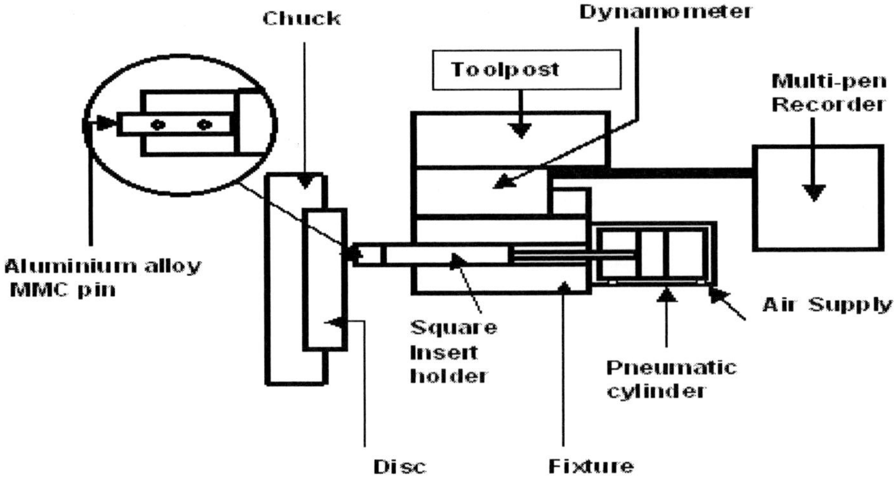

Figure 8. Schematic experimental setup for carrying out the wear tests

4.3 Disc and Pin Materials

The discs used in the previous study [5] were made of Assab DF2 tool steel (equivalent to AISI 01) hardened and tempered to 60 HRC (697Hv), and ground to a surface finish of 0.3

μm (R_a). The discs had a diameter of 180 mm and a thickness of 30 mm. The pins were machined accurately to a size of 10 mm x 10 mm and a length 25 mm by a wire-cut electro-discharge machining (EDM) process to fit the pin holder. The pin materials used in this experiment were A6061 aluminium-based matrix composites with different nominal volume fraction of alumina particles, MMC-A with 10% alumina, MMC-B with 15% alumina and MMC-C with 20% alumina. These materials were supplied by Comalco in the form of flat bars extruded from Duralcan billets. The density of the specimens was measured by using a Ultrapycnometer. The hardness of the three composite materials was tested using a Brinell hardness tester. Scanning Electron Microscopy (SEM) was used to determine the cross sectional area of the particles. Table 5 shows the density, hardness and average particle size of the pin materials.

Table 5. Density, hardness and average particle size of pin materials

Property	MMC-A	MMC-B	MMC-C
Density (g/cm^3)-Measured	2.817	2.895	2.973
Density (g/cm^3)-Calculated	2.823	2.885	2.946
HB (MPa)	611	554	579
Av. particle size (μm)	30	36	24

Ref [21] gives the density of A6061 as 2.70 g/cm^3 and 3.75 to 3.93 g/cm^3 for alumina. Based on the nominal volume fraction of alumina given by the manufacturer, the theoretical density was calculated, based on a value of 3.93 g/cm^3 for alumina, and listed in Table 5. It can be seen from the table that the deviations of both the measured and calculated density values are small, indicating the nominal volume fraction of alumina in each material is approximately correct. However, MMC-A contains a lower volume fraction of alumina and should give a lower hardness value. This does not seem to be the case, presumably due to measuring errors or due to the different processing parameters used in its manufacture. Table 5 shows the average diameter ranges from 24 to 36 μm for the three composite materials, which were obviously manufactured in different batches, with different alumina sizes.

4.4 Experimental Technique

Wear tests for the MMC-A, MMC-B and MMC-C were carried out at distances of 250 m, 500 m, 1,000 m, 1,500 m, 2,000 m, 2,500 m, 3,000 m, 6,000 m, 9,000 m and 12,000 m. Both the moving-pin and the conventional pin-on-disc techniques were used. A constant load of 7.5 kgf, linear velocity of 4.58 m/s and a feedrate of 0.05 mm/rev was used for the moving-pin technique. The pin would start at a radius of 87.5 mm on the rotating disc and would travel to a radius of 20 mm before returning back to the initial point. A stopwatch was used to time the cycle time needed based on the wear distance and constant linear velocity of 4.58 m/s for the collection of weight loss data. As for the conventional pin-on-disc experiment, the pin was held stationary at a radius of 87.5 mm of the rotating disc and a constant sliding speed of 500 rpm. The other factors remained the same as the moving pin technique. Three repetitions were carried out for each experiment [5].

5. The Proposed Integrated Wear Testing Methodology

As mentioned earlier, currently the wear coefficient as well as the wear rate are obtained from experiments conducted with different nominal specimen contact areas, different wear testing parameters, unknown wear regimes, or in the so called "steady-state wear regime" without knowing exactly its location. Consequently inaccurate test results are generally obtained and the test data are of limited use to the industry. Hence there is a need to develop a standard method for the determination of wear rates and wear coefficients, which should be measured at both the transient wear regime and the steady-state wear regime to meet the needs of modern product design. An integrated wear testing methodology is therefore suggested in this proposal. It will comprise the following steps.

Step (1) To perform transient wear tests

The unique feature of this methodology is to conduct a transient wear test first, with minimum number of steady-state wear test runs, to establish the transient distance and the A and B values. This is because the prediction of the steady-state sliding distance and the corresponding wear rate and wear coefficient in the next step will require the A and B values from the transient wear curve. In order to obtain a better accuracy of the A and B values, the transient wear curve should be divided into a number of small segments with a smaller increment of sliding distance between two experimental runs.

Step (2) To predict the steady-state sliding distances and their wear rates and wear coefficients from the transient wear data

With the A and B values obtained from Step (1), the second step is to predict the steady-state sliding distances with Eq.40, and their standard steady-state wear rate and wear coefficient and net steady-state wear rate and wear coefficient with Eq.47 and Eq.41; and with a F_A value of 0.99 and 0.999 respectively. It should also be noted that the steady-state wear rate and wear coefficient vary with the sliding distance used in their computation as discussed earlier. With a fixed or common F_A value, the steady-state wear rate and wear coefficient values will be more accurately determined, since the steady-state wear regime, where the test is to be conducted with a particular sliding distance, can now be defined precisely. Obviously different investigators will also be more likely to obtain comparable and consistent steady-state wear rate and wear coefficient values.

Step (3) To perform confirmation experimental runs to determine the standard steady-state wear rate and wear coefficient and the net-steady-state wear rate and wear coefficient.

To continue with the steady-state wear measurement from Step (1) until a horizontal steady-state wear rate or wear coefficient versus distance curve has appeared. Obviously the sliding distance to be covered should include the predicted sliding distance values calculated from Step (2). It should also be noted that a bigger increment of sliding distance between two experimental runs could be used in this state to save time, as it will not affect the accuracy of the transient wear curve anymore. Furthermore, if experimental time is a major concern, one might perform only confirmation runs to determine the wear rate and wear coefficient values at the predicted sliding distance(s). Obviously, in doing so, one should have tested the wearing pair with similar operating parameters many times previously to know their wear characteristics well. One should also bear in mind that the wear volume versus sliding distance curve obtainable from the testing pair is similar to that used in the integrated wear model as indicated in Fig.3 earlier. For new wearing pairs or testing conditions, it is always safer to carry out full transient and steady-state wear tests with sufficient repetitions to get meaning-full statistical values.

As both transient wear and steady-state wear need to be carried out with the proposed methodology, one might think that it would take a longer time to obtain the test data. However, it should be noted that the transient wear tests do not take a long time to do since the sliding distance involved is very much shorter. Moreover time will also be saved from doing away with the conventional trial and error technique used to determine the steady-state wear rate and wear coefficient. A shorter overall testing time can therefore be expected. Hence with the proposed methodology, the steady-state wear rate and wear coefficient of a wearing-pair will be determined more accurately and consistently, and with a significantly reduced testing time.

It should also be noted that the wear resistance of a component part depends on the composition, grain size or particle size, the manufacturing process and the heat-treatment or surface coating used, as well as its surface finish. Obviously the wear properties of wearing pairs may vary from one pair to another even though they may have made of the same materials. Hence it is always advisable to carry out full tests, to include both the transient wear regime and the steady-state wear regime as far as possible, until one is extremely confident about the wearing characteristics of the wearing pair before any decision is made to use the confirmation runs only.

6. Analysis of Wear Test Data

6.1 The measured volumetric wear rate and standard wear coefficient

Tables 6 and 7 show respectively the measured volumetric wear rate and the standard wear coefficient obtained for MMC-A, MMC-B and MMC-C, from data collected by both the moving-pin and the conventional pin-on-disc methods. The measured volumetric wear rate in Table 6 is determined by dividing the wear volume loss by the respective sliding distance. In the table, V/L(M) and V/L(C) indicate the volumetric wear rate obtained from the moving-pin and the conventional pin-on-disc method respectively. On the other hand, the measured standard wear coefficient (K_S) in Table 7 is computed from its measured volume loss against sliding distance by using Eq.10. In the table, again K_S(M) indicates standard wear coefficient obtained from the moving-pin testing method; while K_S(C) from the conventional pin-on-disc testing method. It should be noted that both the wear rate (V/L) as well as the standard wear coefficient (K_S) at the transient wear regime are higher than those at the steady-state wear regime for each case.

Table 6. The measured volumetric wear rate for MMC-A, MMC-B and MMC-C

	MMC-A		MMC-B		MMC-C	
Distance (km)	V/L(M)[a] (mm^3/km)	V/L(C) (mm^3/km)	V/L(M) (mm^3/km)	V/L(C) (mm^3/km)	V/L(M) (mm^3/km)	V/L(C) (mm^3/km)
0.25	9.88	15.20	7.56	5.80	5.44	5.68
0.5	8.48	10.60	6.32	5.30	4.24	4.82
1	6.42	7.32	4.24	4.42	3.52	4.62
1.5	5.43	6.00	3.30	3.69	3.08	4.09
2	4.59	5.07	2.85	3.06	2.59	3.52
2.5	3.88	4.41	2.61	2.76	2.14	3.33
3	3.47	4.04	2.36	2.52	1.87	3.01
6	2.24	2.54	1.57	1.71	1.24	1.85
9	1.95	1.99	1.44	1.51	1.09	1.53
12	1.82	1.84	1.48	1.58	1.14	1.48

Note: [a]V/L(M) indicates volumetric wear rate obtained from the moving-pin testing method; while V/L(C) from the conventional pin-on-disc testing method.

Table 7. The measured standard wear coefficient (K_s x10^{-5}) for MMC-A, MMC-B and MMC-C

Distance (km)	MMC-A		MMC-B		MMC-C	
	K_S(M) (10^{-5})	K_S(C) (10^{-5})	K_S(M) (10^{-5})	K_S(C) (10^{-5})	K_S(M) (10^{-5})	K_S(C) (10^{-5})
0.25	24.6	37.8	17.1	13.1	12.8	13.4
0.5	21.1	26.4	14.3	12	10	11.4
1	16	18.2	9.6	10	8.3	10.9
1.5	13.5	14.9	7.5	8.3	7.3	9.6
2	11.4	12.6	6.4	6.9	6.1	8.3
2.5	9.7	11	5.9	6.2	5.1	7.9
3	8.6	10.1	5.3	5.7	4.4	7.1
6	5.6	6.3	3.6	3.9	2.9	4.4
9	4.9	5	3.3	3.4	2.6	3.6
12	4.5	4.6	3.3	3.6	2.7	3.5

Note: [a]K_S(M) indicates standard wear coefficient obtained from the moving-pin testing method; while K_S(C) from the conventional pin-on-disc testing method.

6.2 The Predicted Steady-State Sliding Distance (L_P)

The steady-state predicted sliding (L_P) distance values for MMC-A, MMC-B and MMC-C, for both data obtained with the moving-pin and the conventional pin-on-disc methods, were calculated with Eq.40, with a C value of 4.605, 6.908 and 9.210 for a F_A value of 0.99, 0.999 and 0.9999 respectively; and listed in Tables 8 and 9. It can be seen from both tables that, for data obtained with the moving-pin method, the L_P values for MMC-A, are respectively 4.92 km, 7.38 km, and 9.84km. For data obtained with the conventional pin-on-disc method, the corresponding values for the three materials are 4.38 km, 6.57 km and 8.76 km respectively. Similarly, the L_P values for MMC-B are respectively 4.83 km, 7.24 km and 9.65 km for data obtained with the moving-pin method; and 6.20 km, 9.30 km and 12.40 km for data obtained with the conventional pin-on-disc method. The L_P values for MMC-C are respectively 4.98 km, 7.48 km and 9.97 km for data obtained with the moving-pin method; and 9.59 km, 14.39 km and 19.19 km for data obtained with the conventional pin-on-disc method. It should be noted that the L_P values of 14.39 km and 19.19 km obtained for MMC-C tested with the conventional pin-on-disc method are on the high side due probably to the irregularity of wear curve V(C)-C, which is a lot higher than V(M)-C, as shown in Fig.2. Furthermore, the steady-state L_P values to be used for predicting the wear coefficient and the wear rate are also identical. Hence the same wear rate will be obtained irrespective whether it is computed directly by using Eq.47; or indirectly through the computation of the wear coefficient, by using Eq.44.

Table 8. Predicted sliding distance (L_P) and predicted wear coefficient (K_P) at different F_A values (MMC-A, MMC-B, MMC-C)

F_A	MMC-A		MMC-B		MMC-C	
	L_P (km)	K_P (10^{-5})	L_P (km)	K_P (10^{-5})	L_P (km)	K_P (10^{-5})
Moving-pin method						
0.99	4.92	5.5	4.83	3.3	4.98	2.8
0.999	7.38	3.7	7.24	2.2	7.48	1.9
0.9999	9.84	2.8	9.65	1.7	9.97	1.4
Conventional pin-on-disc method						
0.99	4.38	6.7	6.20	3.0	9.59	2.9
0.999	6.57	4.5	9.30	2.0	14.39	1.9
0.9999	8.76	3.4	12.40	1.5	19.19	1.5

Table 9. Predicted sliding distance (L_P) and predicted wear rate $(V/L)_P$ at different F_A values (MMC-A, MMC-B, MMC-C)

F_A	MMC-A		MMC-B		MMC-C	
	L_P (km)	$(V/L)_P$ (mm^3/km)	L_P (km)	$(V/L)_P$ (mm^3/km)	L_P (km)	$(V/L)_P$ (mm^3/km)
Moving-pin method						
0.99	4.92	2.19	4.83	1.45	4.98	1.20
0.999	7.38	1.47	7.24	0.98	7.48	0.80
0.9999	9.84	1.10	9.65	0.73	9.97	0.60
Conventional pin-on-disc method						
0.99	4.38	2.70	6.20	1.32	9.59	1.22
0.999	6.57	1.82	9.30	0.89	14.39	0.82
0.9999	8.76	1.36	12.40	0.67	19.19	0.61

6.3 The Predicted Steady-State Wear Coefficients

Eq.41, with a D value of 0.215, 0.145 and 0.109 for a F_A value of 0.99, 0.999 and 0.9999 respectively, was used to calculate the predicted steady-state wear coefficient (K_P) for MMC-A, MMC-B and MMC-C. It should again be noted that D is approximately equal to 1/C. Eq.41 was found suitable for predicting the standard wear coefficient values and the net steady-state wear coefficient values for MMC-A, MMC-B and MMC-C, MMC-D and A6061-T6 with F_A values of 0.99 and 0.999 respectively [8-10].

It can also be seen from Table 8 that, for data obtained with the moving-pin method, the ($K_P x10^{-5}$) values for MMC-A are respectively 5.5, 3.7 and 2.8; and for data obtained with the conventional pin-on-disc method, the corresponding values are 6.7, 4.5 and 3.4 respectively. For MMC-B, the two sets of ($K_P x10^{-5}$) values are: 3.3, 2.2 and 1.7; and 3.0, 2.0 and 1.5 respectively. For MMC-C, the two sets of ($K_P x10^{-5}$) values are: 2.8, 1.9 and 1.4; and 2.9, 1.9 and 1.5 respectively. It should also be noted again that MMC-A has the highest wear coefficient ($K_P x10^{-5}$) value and MMC-C the lowest. This is because MMC-A is less wear resistant as it contains a lower volume fraction of alumina while MMC-C the highest. The respective sets of predicted steady-state wear coefficients ($K_P x10^{-5}$) for MMC-A, MMC-B and MMC-C are also plotted in Figs.9, 10 and 11 respectively. For the legends used in the figures, e.g. In K_M(M)-A and K_M(C)-A, K_M(M)-A and K_M(C)-A indicate the wear coefficients for MMC-A, based on data obtained from the moving-pin and the conventional pin-on-disc testing methods respectively. In K_P(M)-A-P0.99 and K_M(C)-A-P0.999, K_P(M)-A and K_P(C)-A again indicate the wear coefficients for MMC-A based on data obtained from the moving-pin and the conventional pin-on-disc testing methods respectively; and P0.99 and P0.999 indicate the predicted steady-state wear coefficients obtained with a F_A value of 0.99 and 0.999 respectively. It can be seen from the figures that the predicted steady-state wear coefficients all lie in the steady-state regimes and are close to their respective measured wear coefficient curves. Again, the last two ($K_P x10^{-5}$) data points from MMC-C tested with the conventional pin on-disc method are not shown in Fig.11 since the L_P values of 14.39 km and 19.19 km are outside the range of the figure. It should be noted from the figures that a F_A value of 0.99 will give a higher predicted wear coefficient which is also the predicted standard wear coefficient; while a F_A value of 0.999 will give the predicted net steady-state wear coefficient. Furthermore, it should also be noted that both the standard steady-state wear coefficient and the net steady-state wear coefficient can also be determined from the predicted standard steady-state wear rate and the net steady-state wear rate to be described in Section 6.5, with their deviations in Sections 6.6 and 6.7.

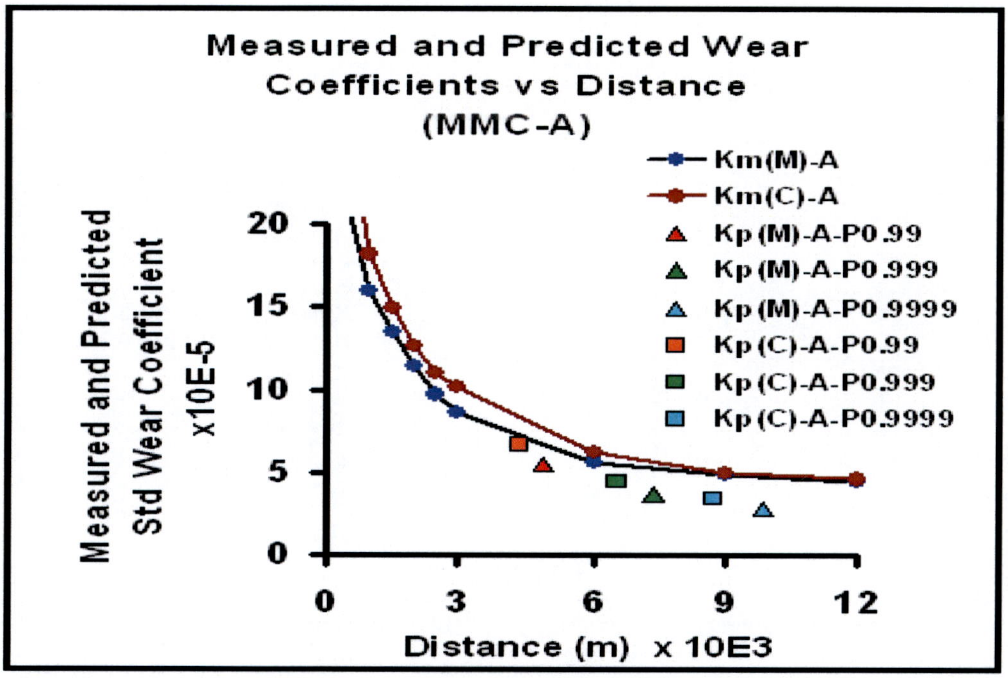

Figure 9. Measured and predicted wear coefficients versus distance curves for MMC-A

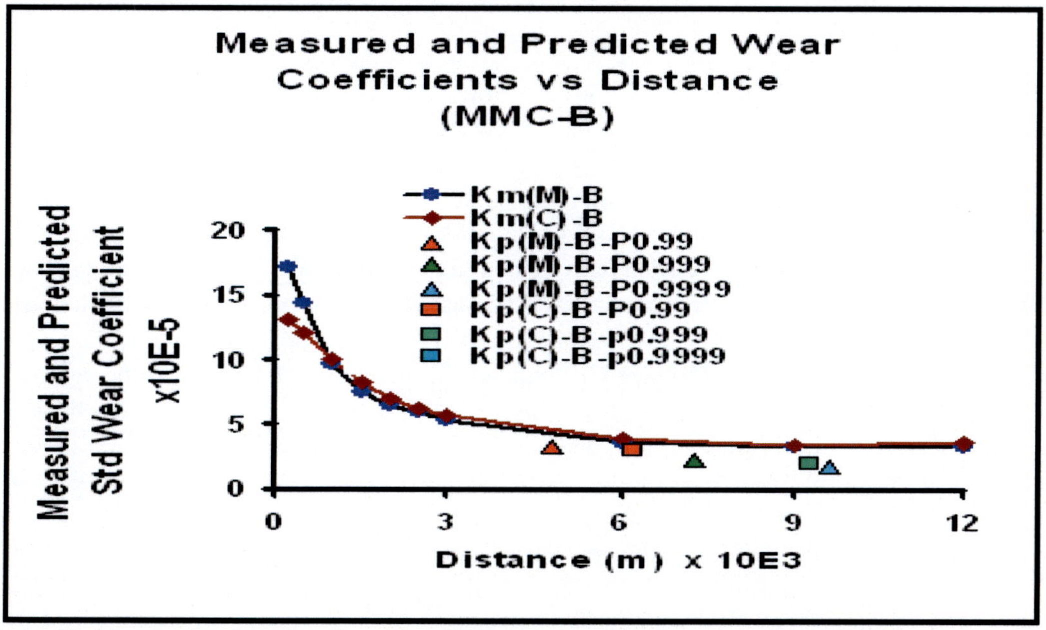

Figure 10. Measured and predicted wear coefficients versus distance curves for MMC-B

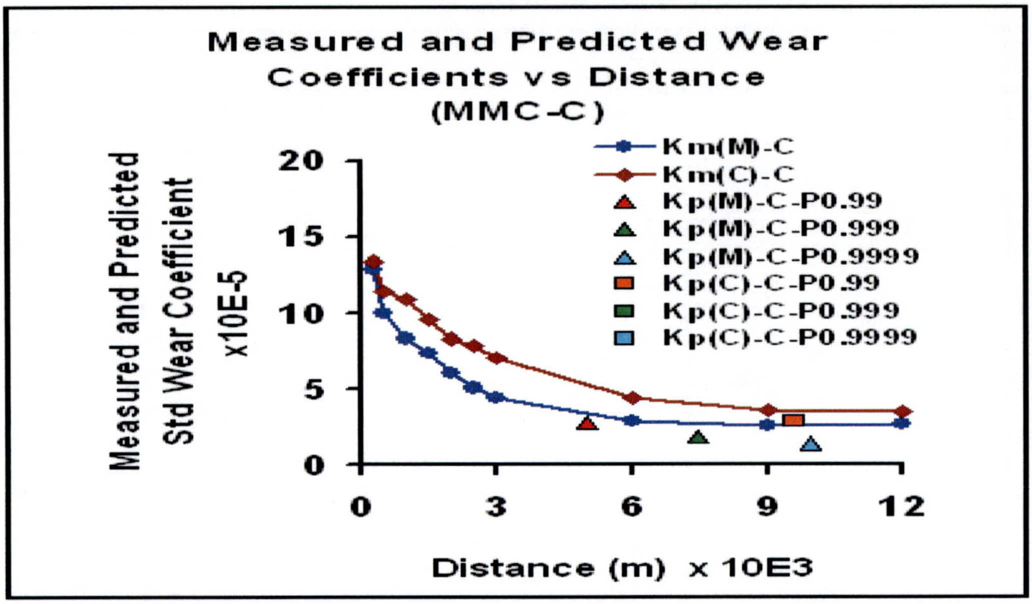

Figure 11. Measured and predicted wear coefficients versus distance curves for MMC-C

6.4 The Calculated Wear Rate

With the transient wear constants A and B values from Table 4, Eq.45 can be used to calculate the volumetric wear rates (V/L) for MMC-A, MMC-B and MMC-C, in both the transient wear regime and the steady-state wear regime. Figures.12, 13 and 14 show the measured and calculated wear rates, as well as the predicted steady-state wear rates (to be discussed in Section 6.5), plotted against sliding distance, for both data obtained from the moving-pin and the conventional pin-on-disc methods, for MMC-A, MMC-B and MMC-C respectively. For the legends used in the figures, e.g. In V/L(M)-A(M) and V/L (C)-A(C), V/L(M) and V/L(C) indicate wear rates based on data obtained from the moving-pin and the conventional pin-on-disc testing methods respectively; while A(M) and A(C) indicate the measured wear rate for MMC-A and calculated wear rate for MMC-A respectively. In V/L(M)-P0.99 and V/L(C)-P0.999, V/L(M) and V/L(C) again indicate wear rates based on data obtained from the moving-pin and the conventional pin-on-disc testing methods respectively; and P0.99 and P0.999 indicate the predicted steady-state wear rate obtained with a F_A value of 0.99 and 0.999 respectively. It can be seen from the figures that the calculated values generally agree well with the measured ones. However, the calculated values are slightly lower than the measured values when they reach the steady-state regime. Furthermore, it should be noted that the shape of the wear rate curves is similar to those of the wear coefficients, as shown in Figs.9 to11. The similarity can also be traced back to the equation for the standard wear coefficient, Eq.10, which contains the term (V/L), which is exactly the wear rate.

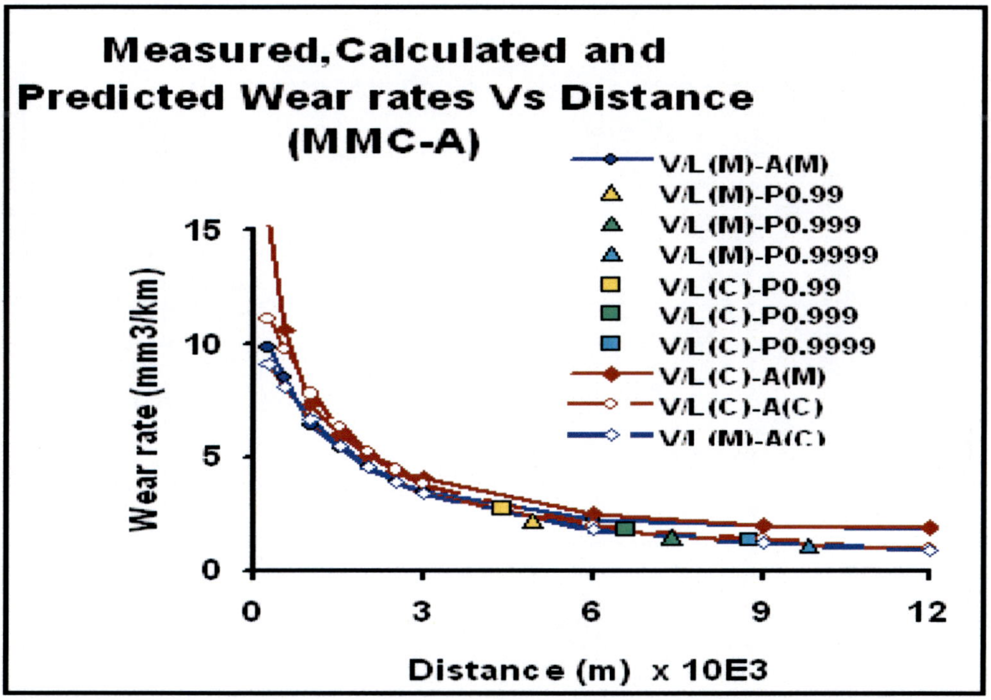

Figure 12. The measured, calculated and predicted wear rate versus distance curves for MMC-A

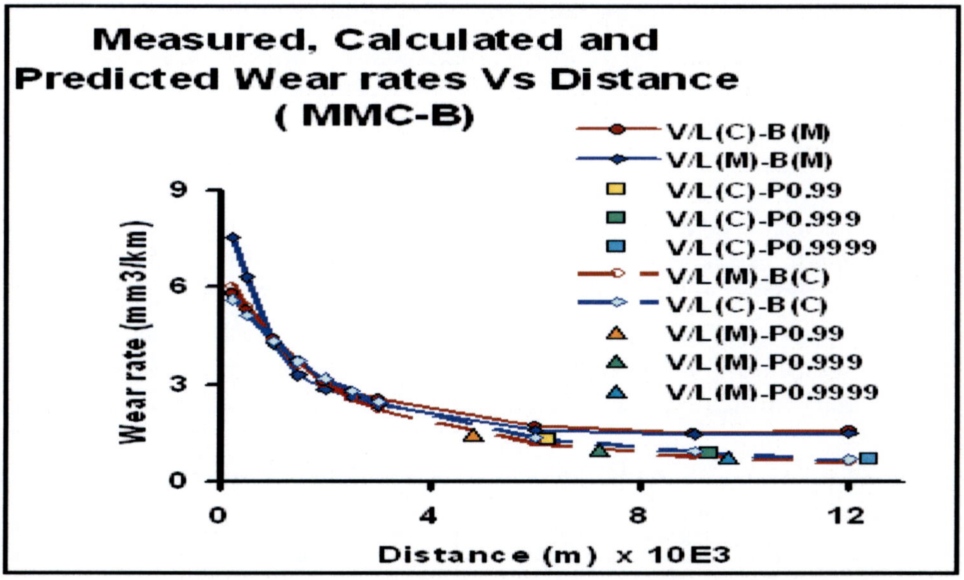

Figure 13. The measured, calculated and predicted wear rate versus distance curves for MMC-B

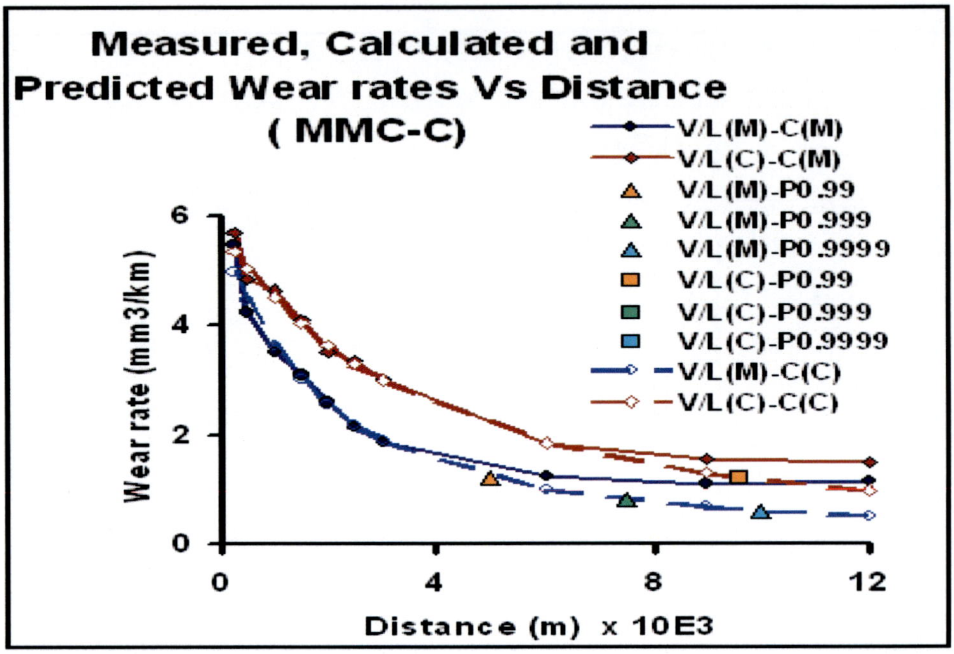

Figure 14. The measured, calculated and predicted wear rate versus distance curves for MMC-C

6.5 The Predicted Steady-State Wear Rate $(V/L)_P$

Eq.47, again with a C value of 4.605, 6.908 and 9.210 for a F_A value of 0.99, 0.999 and 0.9999 respectively were used to calculate the predicted steady-state wear rate $(V/L)_P$ for MMC-A, MMC-B and MMC-C. It can also be seen from Table 9 that, for data obtained with the moving-pin method, the $(V/L)_P$ values for MMC-A are respectively 2.19, 1.47, and 1.10 mm^3/km; and for data obtained with the conventional pin-on-disc method, the corresponding values are 2.70, 1.82 and 1.36 mm^3/km respectively. For MMC-B, the two sets of $(V/L)_P$ values are: 1.45, 0.98 and 0.73 mm^3/km; and 1.32, 0.89 and 0.67 mm^3/km respectively. For MMC-C, the two sets of $(V/L)_P$ values are: 1.20, 0.80 and 0.60 mm^3/km; and 1.22, 0.82 and 0.61 mm^3/km respectively. It should again be noted that MMC-A has the highest wear rate $(V/L)_P$ value and MMC-C the lowest. The respective sets of predicted steady-state wear rates $(V/L)_P$ for MMC-A, MMC-B and MMC-C are also plotted in Figs.12, 13 and 14 respectively. They all lie in the steady-state regimes and are close to their respective wear rate curves. Again, the last two $(V/L)_P$ data points from MMC-C tested with the conventional pin on-disc method are not shown in Fig.14. It should be noted from Figs.12-14 that the predicted net-steady-state wear rates given by a F_A value of 0.999 and 0.9999 are quite similar. As a longer distance will be required to achieve a F_A value of 0.9999, for practical reasons, it should be sufficient to conduct the tests up to a F_A value of up to 0.999. It will also be shown in Section 6.7 that, with such a F_A value, a more accurate net-steady-state wear rate will be obtained.

6.6 Deviation of Predicted Standard Steady-State Wear Rate $(V/L)_P$ versus Measured Standard Steady-State Wear Rate

In order to find the most suitable F_A value for the prediction of standard steady-state wear rate, a study was carried out to compare the predicted wear rates, obtained with F_A values of 0.99, 0.999 and 0.9999, with measured wear rates. Three types of measured wear rate were used, namely: (i) the average wear rate (V/L) obtained from the steady-state wear rates at 6 km, 9 km and 12 km; (ii) the lowest wear rate value obtained from 6 km, 9 km and 12 km; and (iii) the wear rate value obtained by interpolation or extrapolation, and by using the calculated predicted sliding distance (L_P). All the data obtained from both the moving-pin and the conventional pin-on-disc methods were used. Eq.49 was used to calculate the Dev (%) of each predicted $(V/L)_P$ value as compared with each of the measured $(V/L)_M$ value. Table 10 shows an example of calculation for the average Dev (%) for a F_A values of 0.99. The over-all average of the average Dev (%) obtained for $(V/L)_{P0.99}$, $(V/L)_{P0.999}$ and $(V/L)_{P0.9999}$ are, respectively, 18.8%, 50.4% and 96.7%. Hence $(V/L)_{P0.99}$ has given the least average deviation of about 19%. This also indicates that a F_A value of 0.99 should be used to predict the standard steady-state wear rate value and the required sliding distance. Fig.15 shows the deviations (%) of standard steady-state wear rates versus measured wear rates for F_A values of 0.99, 0.999 and 0.9999 respectively. Only absolute deviation values are used for plotting the figure. However, it is obvious from Fig.15 that F_A value of 0.99 gives the least deviation between the predicted wear rate and the measured one.

$$\text{Dev}(\%) = \frac{100\left[\left(\frac{V}{L}\right)_P - \left(\frac{V}{L}\right)_M\right]}{\left(\frac{V}{L}\right)_P} \tag{49}$$

Table 10. Deviation of predicted $(V/L)_{P0.99}$ values versus measured $(V/L)_M$ values

	K(M)-A	K(M)-B	K(M)-C	K(C)-A	K(C)-B	K(C)-C	Av. Dev.(%)
(I) $(V/L)_{P0.99}$ against $(V/L)_{M\text{-}Average}$							
$(V/L)_{P0.99}$	2.19	1.45	1.20	2.70	1.32	1.22	
$(V/L)_{M\text{-}Ave(6,9,12)}$	2.00	1.50	1.16	2.12	1.60	1.62	
Dev(%)	8.5	-3.3	3.5	21.3	-21.2	-33.0	15.1
(II) $(V/L)_{P0.99}$ against $(V/L)_{M\text{-}Lowest}$							
$(V/L)_{P0.99}$	2.19	1.45	1.20	2.70	1.32	1.22	
$(V/L)_{M\text{-}Lowest}$	1.82	1.44	1.09	1.84	1.51	1.48	
Dev(%)	16.9	0.7	9.2	31.9	-14.4	-21.3	15.7
(III) $(V/L)_{P0.99}$ against $(V/L)_{M\text{-}LP0.99}$							
$(V/L)_{P0.99}$	2.19	1.45	1.20	2.70	1.32	1.22	
$(V/L)_{M\text{-}LP0.99}$	2.68	1.88	1.47	3.38	1.7	1.52	
Dev(%)	-22.37	-29.66	-22.50	-25.19	-28.79	-24.59	-25.52
					Over-all Average		18.8

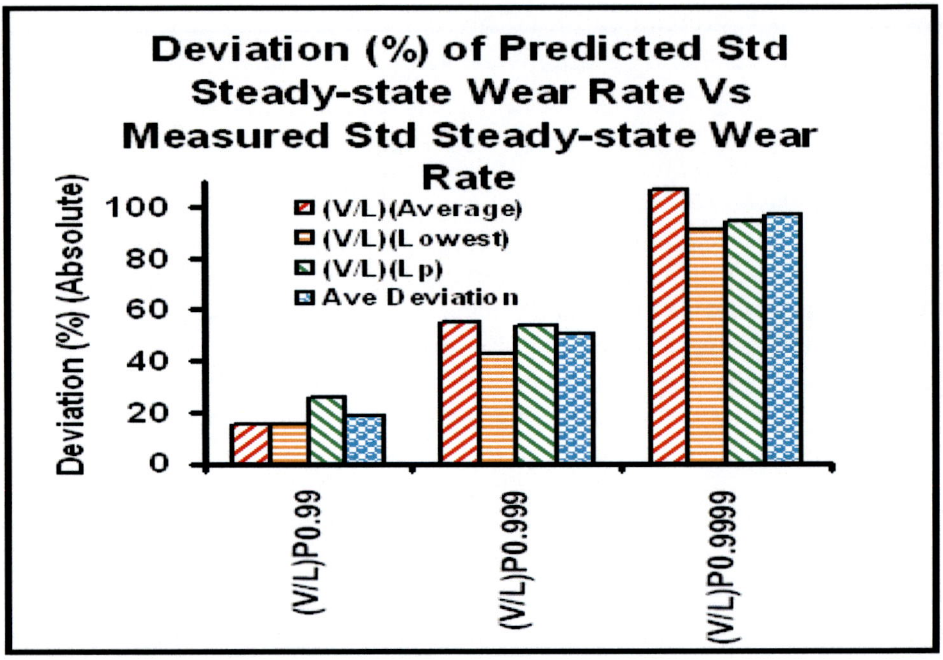

Figure 15. Deviations (%) of standard steady-state wear rate versus measure wear rates for F_A values of 0.99, 0.999 and 0.9999 respectively

6.7 Deviation of Predicted Net Steady-State Wear Rate $(V/L)_P$ versus Measured Net Steady-State Wear Rate

The deviation study carried out in Section 6.6 was repeated here in order to find the most suitable F_A value for the prediction of net steady-state wear rate. Three types of measured net steady-state wear rate were used, namely: (i) the average net steady-state wear rate (V/L); (ii) the lowest net steady-state wear rate; and (iii) the net wear rate obtained by interpolation or extrapolation, by using the net predicted sliding distance (L_{PN}) to be determined from Eq.50. To facilitate computation, the average transient distance (L_t) of 2,502 m, as indicated in Section 2.4, was used to calculate the net wear rates for all the six sets of data obtained from both the moving-pin and the conventional pin-on-disc methods. The same L_t value was also used in Eq.50, which is derived from Eq.13. Eq.49 was again used to calculate the Dev (%) of each predicted $(V/L)_p$ value as compared with each of the measured $(V/L)_M$ values. The overall average of the average Dev (%) for $(V/L)_{P0.99}$, $(V/L)_{P0.999}$ and $(V/L)_{P0.9999}$ are, respectively, 42.9%, 23.9% and 28.6%. Hence $(V/L)_{P0.999}$ has given the least average deviation of about 24%. This also indicates that a F_A value of 0.999 should be used to predict the net steady-state wear rate value and the required sliding distance. Fig.16 shows the deviations (%) of net steady-state wear rate versus measured wear rates for F_A values of 0.99, 0.999 and 0.9999 respectively. It is obvious from the Fig.16 that the F_A value of 0.999 has given the least deviations as compared with those obtained with F_A values of 0.99 and 0.9999.

$$L_{PN} = L_P - L_t \tag{50}$$

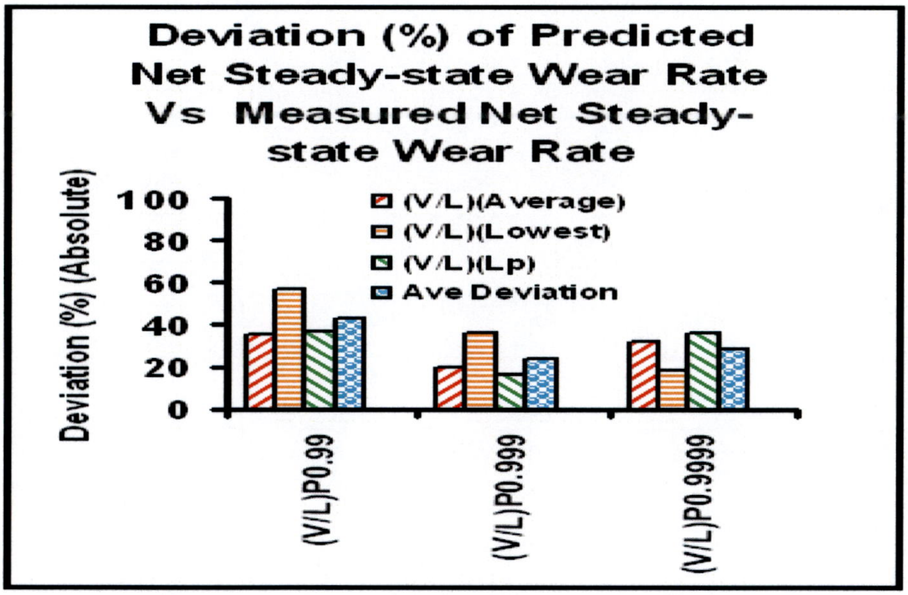

Figure 16. Deviations (%) of predicted net steady-state wear rate versus measured net steady-state wear rate for F_A values of 0.99, 0.999 and 0.9999 respectively

6.8 Testing Time Analysis

Table 11 shows the testing time analysis for three different cases. Case I is a full test consisting of seven transient wear tests and three steady-state wear tests with a total testing time of 137.38 min. Case II is a conventional test method consisting of four tests to be conducted in the unknown "steady-state regime". The total testing time required is 109.17 min. Case III is the proposed transient cum confirmation method, consisting of seven transient wear tests and one confirmation steady-state wear test. The total testing time required is 64.60 min. The testing times were calculated based on the linear testing speed of 4.58 m/s. Hence with the proposed transient cum confirmation testing method, the testing time saved is respectively 52.9% and 40.8% as compared with those needed by the first two methods. It should be noted that in the calculation of the testing times, only one test sample for each test is assumed to facilitate comparison. In actual testing, more than one test sample should be used for conducting each test. The sliding distance of 7,000 m used for the proposed method is based on the rounded average L_P value of 6,977 m.

7. Discussions

7.1 Significance of the Integrated Wear Testing Methodology

This chapter has introduced a new methodology for adhesive wear testing. An integrated wear model has been proposed to combine the transient wear curve with the steady-state wear

Table 11. Analysis of testing time

Testing method		Transient wear test		Steady-state wear test		Total testing time (min)
		Distance (m)	Testing time (min)	Distance (m)	Testing time (min)	
I	Full test (7 transient wear tests and 3 steady-state wear tests)	250	0.91	6,000	21.83	
		500	1.82	9,000	32.75	
		1,000	3.64	12,000	43.67	
		1,500	5.46			
		2,000	7.28			
		2,500	9.10			
		3,000	10.92			
		Sub-total	39.13		98.25	137.38
II	Conventional method (4 steady-state wear tests)			3,000	10.92	
				6,000	21.83	
				9,000	32.75	
				12,000	43.67	
			Sub-total		109.17	109.17
III	Transient cum confirmation method (7 transient wear tests and 1 steady-state confirmation wear test)	250	0.91	7,000	25.47	
		500	1.82			
		1,000	3.64			
		1,500	5.46			
		2,000	7.28			
		2,500	9.10			
		3,000	10.92			
		Sub-total	39.13		25.47	64.60

curve. With this integrated wear model, one is able to model and predict the steady-state wear rate and wear coefficient with the transient wear data, i.e. A and B values. This new powerful integrated methodology is capable to address some major problems encountered in the determination of standard wear rate and standard wear coefficient. First of all, very often standard wear rate and wear coefficient are measured in or near the transient wear regime, giving rise to higher values. By conducting the wear test in the transient wear and steady-state wear regimes, this problem can be avoided. The second difficulty facing scientists and engineers in the determination of steady-state wear rate and wear coefficient is to make sure that the wear has reached the steady-state regime. Very often, a lot of experimental runs need to be conducted. With the proposed methodology, it is possible to predict the steady-state sliding distance and the standard steady-state wear rate and wear coefficient with the constants A and B determined from the transient wear data. Consistent results will be obtained since a common F_A value, e.g. 0.99, will be used not only for the determination of the standard steady-state wear rate, but also for the standard steady-state wear coefficient, both of them are automatically linked. A lot of time will also be saved since one can now do away with the trial and error method to find the correct sliding distance. Hence the proposed methodology will greatly improve the testing efficiency, and to reduce significantly the time and other resources needed in the determination of standard steady-state wear rate and wear coefficient.

7.2 The Accuracy of this Methodology

As mentioned earlier in Sections 6.6 and 6.7, the average deviation of the predicted wear rates obtained with this methodology were about 19% and 24% respectively. These deviations may sound 'high' when compared with other engineering material testing practice. However, it should be noted that the wear data were obtained by two different sliding wear testing methods: the moving-pin method and the standard pin-on-disc testing method. Moreover, metal matrix composites reinforced with alumina particles, which could be very abrasive in nature, were used as the wearing pins. A very long sliding distance, up to 12 km, was also used in conducting the wear tests resulting in severe wear of the specimens as shown in their wear curves in Fig.2. Furthermore, one should also realize the fact that wear coefficient and hence the wear rate to be determined from a wear-pair can vary from 5 to 10 times by changing from the transient wear regime to the steady-state wear regime alone [5]. Hence it should be considered a significant achievement if one is able to obtain wear rates with such average deviations. Obviously, with more consistent wear data obtainable from testing less abrasive materials, better accuracy can be expected. Furthermore, since the same F_A value will be used by different investigators in the determination of the steady-state wear rates and wear coefficients, less variation in measured wear data will also be obtained.

7.3 The Effect of the Nominal Specimen Size

Common laboratory sliding wear devices are: block-on-ring, pin-on-disc, counter-rotating discs with adjustable slip, pin-in-v-block, and crossed cylinders [14]. The pin-on-disc is a very popular test method, while the block-on-ring (rotating cylinder) is also common. As different testing systems have been used, it is obvious that the nominal contact area of the wearing-pairs will also vary significantly from one system to another to give rise to different wear rate and wear coefficient values. Recent work by Yang [22] has indicated that different pin settings can change the contact geometry of a pin to give a significantly different wear rate. In another study [23], pin-on-disc experiments were carried out with Duralcan 284 MMC with two nominal specimen contact areas, 6 x 6 mm^2 and 10 x 10 mm^2, and tested with loads of 74 and 98 N, and speeds of 200 and 275 m/min. It was found that the wear coefficient values obtained for the specimens with a smaller nominal contact area were lower, by about 12 to 32% based on the measured and predicted values respectively. As indicated by Eqs.34 and 45, the wear coefficient and wear rate are proportional to A, the transient wear constant which is related to the wear asperity volume available. Obviously with a smaller nominal contact area, the A value would be smaller and so would be the wear rate and wear coefficient values. Hence 'A' might become a useful common link between wear test data obtained with different wear testing systems.

7.4 Selection of Testing Parameters

The testing parameters to be selected for the determination of the wear coefficient of a wearing-pair are the load (P) and sliding distance (L) as indicated by Eq.10. The load should be correctly chosen, as it may affect the wear mode, that is, whether it is mild wear or severe wear. In an un-lubricated wear test, as the load is increased, the wear rate will also increase. Eventually a point will be reached beyond which the wear rate will increase drastically. This increase in wear rate could be due to the high temperatures generated which in turn soften both the pin and disc materials; or it could also be due to the breakdown of the surface films, usually oxide, which could prevent surface damage and wear. On the other hand, the sliding distance is determined by the sliding speed and sliding time or the number of cycles. Although a high sliding speed will shorten the time in wear testing, however it may also increase the wear rate since more heat will be generated during testing to affect both the properties of the pin and the disc as mentioned earlier. Other factors, which may affect the wear rate in industrial applications, will be discussed in section 7.5.

7.5 Application of the Proposed Methodology in Industry

It should also be noted that in industry, in addition to unlubricated wear, wear can also occur under boundary lubricated condition and fluid film lubricated condition. Peterson [13] has indicated that, in addition to load, speed, sliding distance and geometry as mentioned earlier, the wear rate is also affected by the type of material, finish, film thickness, lubricant viscosity, lubricant quantity, ambient temperature and ambient atmosphere. However with a similar set of testing parameters, the shapes of the wear rate curves obtainable under unlubricated wear, boundary lubricated condition and fluid film lubricated condition are quite similar in a number of cases, but with different wear rate values of course. Hence there should be a good potential to apply the proposed methodology to wear testing under boundary lubricated condition and fluid film lubricated condition, if their volume loss versus distance curves are similar to that described in the proposed integrated wear model. However it should be noted that, with different testing parameters, some materials may exhibit a different type of volume loss versus distance curve [14].

8. Conclusion

This chapter has reviewed the integrated wear model, which comprises both the transient wear and the steady-state wear, as well as the wear equations developed to model the sliding distance, wear coefficient and wear rates, both in the transient wear regime and in the steady-state wear regime. Both the transient wear and the steady-state wear behaviour are closely linked. Hence with the transient wear data one is able to predict the steady-state wear rate and wear coefficient. An integrated adhesive wear testing methodology has also been proposed to determine the standard wear rate and wear coefficient of a wearing-pair by using the integrated wear model and the related wear equations developed by the author. With this methodology, the wear test will be conducted in three stages. First of all, transient wear test will be conducted to determine the constants A and B values. These values will then be used to predict the sliding distance required to attain the steady-state wear regime; and to predict the standard steady-state wear rates and wear coefficients. Steady-state confirmation runs will then be carried out. This methodology has many outstanding advantages. First of all, both the transient wear and the steady-state wear data will be obtained to facilitate product design with different life spans. Next, more consistent and accurate steady-state wear rates and wear coefficients will be obtained since a common F_A value will be used. Both the wear coefficient and wear rates will also be linked automatically to facilitate design calculations. It will also save significant time and labour in conducting the wear test, as the conventional trial and error method is no longer needed.

ACKNOWLEDGEMENT

The author would like to acknowledge the support provided by the School of Mechanical and Aerospace Engineering, Nanyang Technological University, Singapore, in carrying out all the experimental investigations.

REFERENCES

[1] Lansdown, A.R., Price, A.L. (1986). *Materials to resist wear*. Pergamon, Oxford.

[2] Yang, L.J., Loh, N.L. (1995). The wear properties of plasma transferred arc cladded stellite specimens. *Surface and Coatings Technology*, 71, 196-200.

[3] Godet, M. (1988). Modeling of friction and wear phenomena. In *Approaches To Modeling Of Friction And Wear;* Ling, F.F.;. Pan, C.H.T.; Ed.; Springer-Verlag.

[4] Yang, L.J. (2003). The transient and steady wear coefficients of A6061 aluminium alloy reinforced with alumina particles. *Composites Science and Technology*, 63, 3-4, 575-583.

[5] Yang, L.J. (2003). Wear coefficient equation for aluminium-based matrix composites against steel disc. *Wear*, 255, 1-6, 579-592.

[6] Yang, L.J. (2003). An integrated transient and steady-state adhesive wear model. *STLE Tribology Transactions*, 46, 3, 369-375.

[7] Rabinowicz, E. (1980). Wear coefficients-metals. In *Wear Control Handbook;* Peterson, M.B., Winer, W.O., Ed. The American Society of Mechanical Engineers, New York.

[8] Yang, L.J. (2004). Prediction of net steady-state wear coefficients in an Al_2O_3/steel system with an integrated wear model. *Tribology Lett*, 17, 2, 105-118.

[9] Yang, L.J. (2005). A methodology for the prediction of standard steady-state wear coefficient in an aluminium-based matrix composite reinforced with alumina particles. *Jnl of Materials Processing Technology*, 162-163, 138-148.

[10] Yang, L.J. (2004). Prediction of steady-state wear coefficients in adhesive wear. *STLE Tribology Transactions*, 47, 335-340.

[11] Yang, L.J. (2005). A test methodology for the determination of wear coefficient, *Wear*, 259, 7-12, 1453-1461.

[12] Yang, L.J. (2006). Determination of steady-state adhesive wear rate. *Journal of Tribology*, 128, 4, 725-734.

[13] Peterson, M. (1980). Design considerations for effective wear control. In *Wear Control Handbook;* Peterson, M.B., Winer, W.O., Ed. The American Society of Mechanical

Engineers, New York.
[14] Blau, P. (1989). *Friction and wear transitions of materials: break-in, run-in, wear-in.* Noyes Publications, Park Ridge, New Jersey, USA.
[15] Queener, C.A., Smith, T.C., Mitchell, W.L. (1965). Transient wear of machine parts. *Wear*, 8, 391-400.
[16] Sarkar, A.D. (1976). *Wear of metals*. Pergamon Press, Oxford, New York.
[17] Archard, J.F. (1953). Contact and rubbing of flat surfaces. *Jnl of Applied Physics*, 24, 981.
[18] Zhang, Z.F., Zhang L.C., Mai, Y.W. (1996). Running-in wear of steel/ SiCp-Al composite system. *Wear*, 194, 38-43.
[19] Oakdale Engineering, http://www.oakdaleengr.com.
[20] Yang, L.J. (1999). Pin-on-disc wear testing of tungsten carbide with a new moving pin technique. *Wear*, 225-229, 557-562.
[21] Mallick, P.K. (1980). *Composites Engineering Handbook.* Marcel Decker, New York.
[22] Yang, L.J. (2004). Wear coefficient of tungsten carbide against hot-work tool steel disc with two different pin settings. *Wear*, 257, 481-495.
[23] Yang, L.J. (2007). The effect of nominal specimen contact area on the wear coefficient of A6061 aluminium matrix composite reinforced with alumina particles. *Wear*, 263, 939-948.

In: Tribology Research Trends
Editor: Taisho Hasegawa
ISBN: 978-1-60456-912-4

Chapter 5

FRICTION FROM RECIPROCATING SLIDING OF DIFFERENT SCALES

Erjia Liu*
School of Mechanical and Aerospace Engineering, Nanyang Technological University

1. INTRODUCTION

Tribology is the science and technology of interacting surfaces in relative motion and related subjects and practices, which can be studied on macro (conventional), micro and nanoscales. For macrotribology, many friction and wear mechanisms have been proposed with different testing methods with or without lubrication under different environmental conditions. Large mass, heavy load, elastoplastic deformation, and significant wear have been characteristic of macrotribology. In macrotribology, the properties of bulk materials are dominating.

The effects of friction are due to physical interactions between bodies or objects moving relatively to each other. As a consequence of friction, the process of motion and the dynamic behavior of a system are influenced or disturbed and part of the energy of motion is dissipated. The friction force caused by interfacial adhesion between the asperities of mating surfaces is proportional to the real area of contact and the shear strength of the contact. The ratchet contribution to the coefficient of friction between two rough surfaces is dependent on the slope of the asperities of a surface having a smaller slope.

Nanotribology, brought about by magnetic recording technology, is to study interfacial phenomena in micro- and nano-structures used in magnetic storage systems, micro-electro-mechanical systems (MEMS), and nano-electro-mechanical systems (NEMS). Small mass, light load, elastic deformation, and slight wear or absence of wear are typical of nanotribology which is primarily concerned with the surface properties of materials.

* 50, Nanyang Avenue, Singapore 639798, Singapore
Tel: +65-67905504, Fax: +65-67924062, E-mail: mejliu@ntu.edu.sg

Experimental study of nanotribology was made possible by the advent of surface probing techniques such as scanning probe microscopy (SPM).

Reciprocating tribological testing with a ball-on-flat contact geometry (Figure 1) at a small displacement amplitude is suitable for locally identifying the tribological behaviour of materials in macrotribology, while lateral force microscopy (LFM), one of SPM techniques, is capable of assessing the nanotribological behaviour of materials.

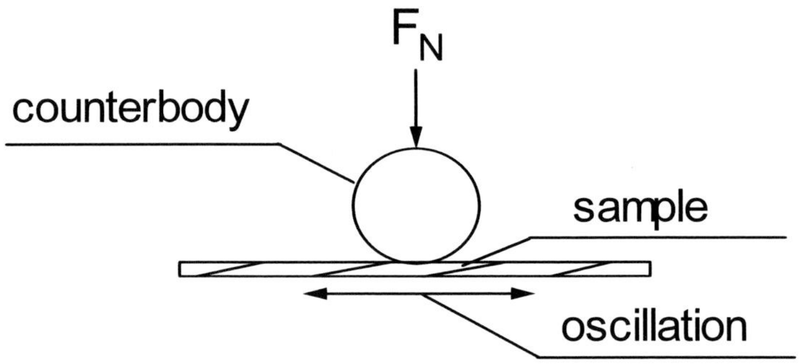

Figure 1. Schematic ball-on-flat contact geometry for reciprocating sliding. (From E. Liu et al, Tribo. Int. 40 (2007) 216.)

2. Friction with Reciprocating Testing

Principle of reciprocating testing

For a small tangential force applied across the contact surfaces of a pair of elastic bodies under a normal load, if there is no slip on the contacting surfaces, there will be no relative displacement of points on the contact surfaces. A contact surface traction, caused by the tangential force, increases from one half the average at the centre of the contact surface to infinity at the edge. The infinity of the traction leads to slip. For the contact of isotropic, linearly elastic, and perfectly smooth spheres and solids of revolution, from the Hertz theory, the contact area is circular and has the radius:

$$a = (3F_nR_0/4E_0)^{1/3} \text{ (m)} \tag{1}$$

with $1/R_0 = 1/R_1+1/R_2$ and $1/E_0 = (1-\nu_1^2)/E_1+(1-\nu_2^2)/E_2$

Here, F_n is the applied normal force (N); R_1 and R_2 are the respective radii of curvature of contacting surfaces before deformation (m); R_0 is the equivalent radius of curvature of the counterfaces before deformation (m); E_1 and E_2 are the respective elastic moduli of the two solids (Pa); and ν_1 and ν_2 are the respective Poisson ratios of the two solids.

For a ball-on-flat contact geometry as shown in Figure 1, the ball radius is R and the sample surface is flat, thus,

$$a = (3F_nR/4E_0)^{1/3} \text{ (m)} \tag{2}$$

The contact pressure is

$$p = p_0(1-r^2/a^2)^{1/2} \text{ (Pa)} \tag{3}$$

where $p_0 = 3F_n/2\pi a^2$ is the maximum contact pressure (Pa), and r is the radius of contact area (m) ($r \leq a$).

Figure 2 schematically shows three typical regimes in reciprocating tribological testing: (a) elastic recovery, (b) partial stick-slip, and (c) gross-slip.

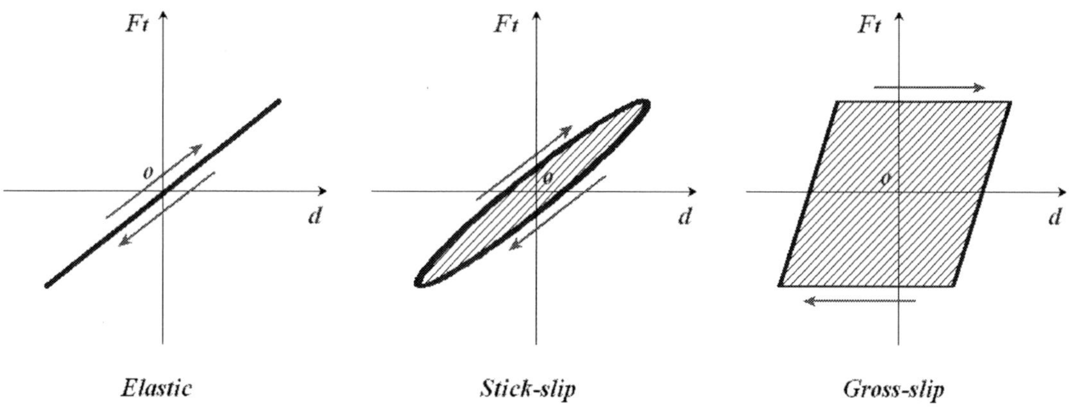

(a) *(b)* *(c)*

Figure 2. Friction regimes in reciprocating tribological testing: (a) elastic recovery, (b) partial stick-slip, and (c) gross-slip

For the ball-on-flat contact configuration with the counterbodies made of same material as shown in Figure 3, the ball is loaded by a tangential force F_t in the x direction of the x-y plane parallel to the sample surface in contact and a normal load F_n in the z-direction normal to the contact surface. A normal pressure built up over the circular contact area has a maximum at the centre and decreases towards zero at the rim of the contact circle. The contact surface traction has a minimum at the centre of the contact circle, which is counteracted by a friction proportional to the normal pressure. Thus, the friction has its maximum at the centre and approaches zero at the rim.

For a tangential force below a critical value, there is a central stick area inside which the (static) friction exceeds the traction and no slip takes place. Outside the stick zone there is a slip annulus. The contact area corresponding to the slip annulus is swept by those stress peaks of alternating signs every half-cycle during the oscillatory movement between contacting counterfaces.

The amount of slip in the slip annulus increases from zero at the stick-slip boundary to a maximum at the rim of the contact circle. In the case of stick-slip, the maximum pressure will no longer occur at the centre, but close to the boundary between the stick and slip zones. The stick-slip boundary circle shrinks with increasing tangential force, until it equals zero for the critical tangential force leading to a gross slip over the whole contact area. When plastic deformation is taken into account, the sharp traction peak at the stick-slip boundary of the elastic case is broadened into a yield annulus as shown in Figure 3.

In a reciprocating sliding test under gross-slip condition as shown in Figure 2c, the coefficient of friction corresponding to the maximum tangential force at the turnabout points is attributed to the static friction force, while the dynamic coefficient of friction can be determined from the area of the tangential force-displacement loop divided by twice the displacement stroke and by the normal load.

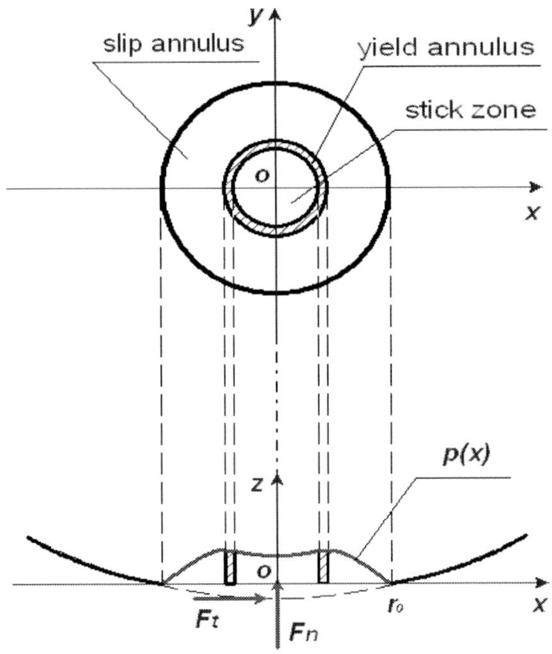

Figure 3. Schematic ball-on-flat contact geometry under mixed stick-slip conditions. (Adapted from O. Vingsbo and J. Schön, Wear, 162-164 (1993) 347.)

Case Study 1: Friction of carbon materials studied with reciprocating sliding

Diamond has a face-centred cubic crystal structure with eight carbon atoms in a unit cell (Figure 4a). In diamond, the tetrahedrally bonded C atoms are responsible for its unique properties such as the highest modulus of elasticity, hardness, and ambient thermal conductivity of any known solid materials together with low thermal expansion coefficient, large optical band gap, and excellent chemical inertness to most corrosive environments.

For a single crystal diamond with a smooth surface, its mechanical and tribological properties vary depending on its crystallographic planes and orientations. For a polycrystalline diamond coating, not only its crystal orientation but also its surface morphology, coating thickness, and the presence of non-diamond carbon phases it contains may influence its friction and wear behaviors.

Graphite has a hexagonal crystal structure with sp^2 carbon bonds in its basal plane and weak interlayer bonds like van der Waals forces between the stacked sheets as shown in Figure 4b.

Diamond-like carbon (DLC) materials are amorphous containing a mixture of sp^3 and sp^2 bonded carbon atoms. Hydrogenated DLC also contains a significant amount of hydrogen.

Diamond and DLC materials have many tribology related applications. For example, diamond and DLC can be coated on cutting tools, bearings, and mechanical components whose performances are largely determined by low friction and high wear resistance. DLC coatings have also been successfully applied to hard disks as overcoat, which can effectively protect the magnetic media underneath from wear and corrosion, and provide a very low coefficient of friction at the same time.

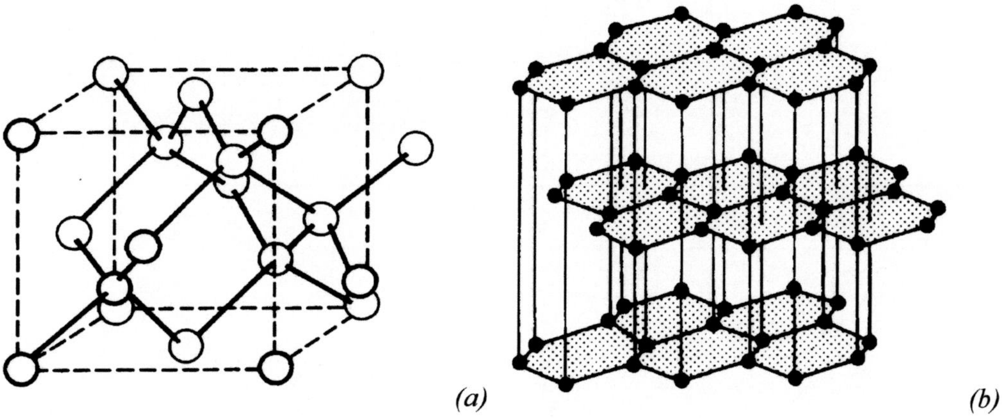

Figure 4. Spatial structures of (a) diamond and (b) graphite. (Figure a from T.J. Clark et al, in S.R. Lampman et al (eds.), Properties and Selection: Nonferrous Alloys and Special-Purpose Materials, Metals Handbook, (10th Ed.), ASM International, The Materials Information Society, Vol. 2 (1990) p.1009; Figure b from B.T. Kelly, Physics of Graphite, Applied Science Publishers, London, (1981).)

A number of carbon materials with a flat surface, such as a single crystal diamond with a polished (100) surface, a diamond coating by combustion flame (CF) chemical vapour deposition (CVD), a highly ordered pyrolytic (0001) graphite (HOPG), and three types of DLC coatings, namely *a*-C by cathodic vacuum arc without filtering, *a*-C:H by radio frequency (RF) CVD, and *ta*-C by filtered cathodic vacuum arc, were subjected to ball-on-flat reciprocating sliding testing. Linear contact displacement, normal contact force, and responding friction force were measured under the following testing conditions: Counterbody: corundum ball of 10 mm in diameter; Laboratory emvironment: 20°C, 50% RH, without lubrication; Normal load: 2 N; Oscillatory amplitude: 100 μm; Oscillatory frequency: 8 Hz; Number of reciprocating cycles: 10,000, 100,000, or 500,000. Under this set of testing conditions, a gross-slip regime was active for most samples.

Figure 5 shows the surface morphologies of all the samples imaged by atomical force microscopy (AFM). The polishing lines can be clearly seen on the (100) diamond surface (Figure 5a), while the CF diamond coating shows clear diamond crystallites (Figure 5b). The atomic image of the HOPG sample was measured on a freshly cleaved (0001) surface (Figure 5c). Of three DLC samples (Figs. 5d-f), the *a*-C coating has large graphitic particles, the *a*-C:H has a relatively smooth surface, and the *ta*-C contains very fine clusteres. Table 1 summarizes the surface roughnesses of all the samples and the thicknesses of the diamond and DLC coatings.

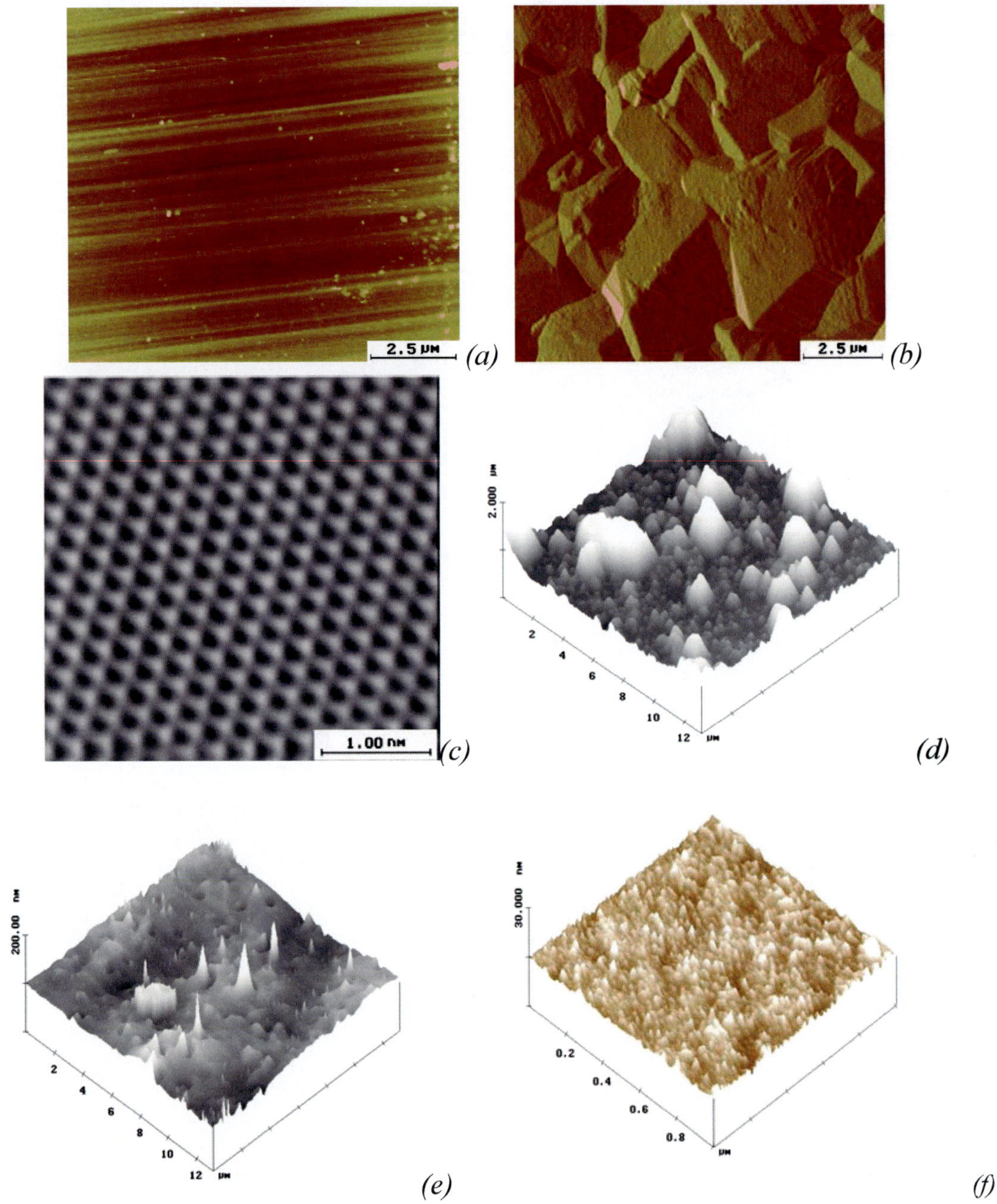

Figure 5. AFM images of carbon materials before ball-on-flat testing: (a) (100) diamond, (b) CF diamond coating, (c) (0001) graphite, (d) *a*-C coating, (e) *a*-C:H coating, and (f) *ta*-C film. (Figure b from E. Liu et al, Diam Relat. Mater. 5 (1996) 649; Figs. d&e from E. Liu et al, Surf. Coat. Tech. 106 (1998) 72; Figure f from E. Liu et al, J. Appl. Phys., 86 (1999) 6078.)

The Raman spectra of all the samples are illustrated in Figure 6. The (100) diamond shows a single peak at about 1332 cm^{-1} and the (0001) graphite peaks at about 1580 cm^{-1}. It is noticed that the CF diamond coating is contaminated by sp^2 cabon bonds. Both the *a*-C and the *a*-C:H coatings give a peak at around 1540 cm^{-1} (G-peak) overlapped with a broad band at around 1350 cm^{-1} (D-peak). The D-peak for the *a*-C:H is more significant than those of the *a*-

C and *ta*-C. The *ta*-C film generally shows weak features due to a much higher fraction of tetrahedral carbon bonds in the film.

Table 1. Surface roughnesses of all the samples and thicknesses of CF diamond and DLC coatings

Sample	Roughness (Ra / nm)	Thickness (μm)
a-C:H	7	~ 2
a-C	120	~ 2
ta-C	0.6	~ 0.1
CF diamond coating	290	~ 10
(100) diamond (45° to polish. lines)	2	-
corundum ball	6	-

The hardnesses/Young's muduli of the *a*-C, *a*-C:H and *ta*-C coatings are around 24.5 GPa/360 GPa, 8 GPa/100 GPa, and 54 GPa/420 GPa, respectively. In the *a*-C coating, the considerable sp^2 component is responsible for its lower hardness and E-modulus, while the hydrogen termination makes the sp^3 bonds in the *a*-C:H coating not that stiff as the sp^3 carbon bonding in the *ta*-C film.

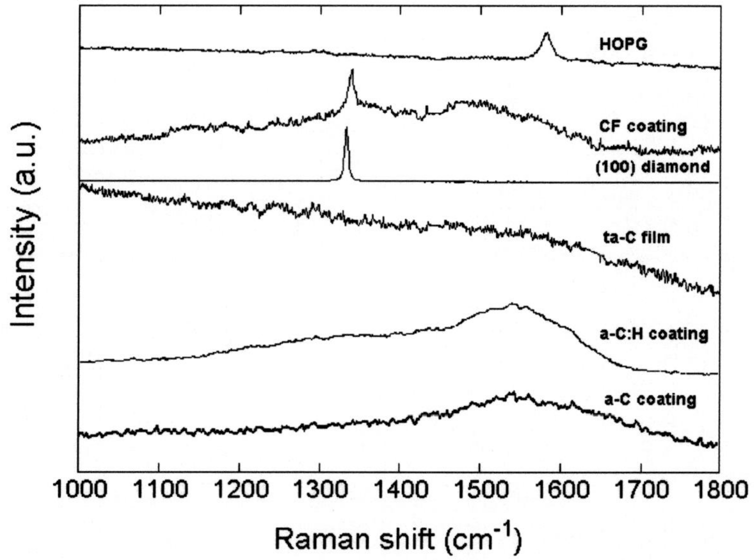

Figure 6. Raman spectra of carbon materials. (From E. Liu et al, Surf. Coat. Tech. 106 (1998) 72.)

For the (100) diamond and CF diamond coating, the coefficients of friction are highest during the first few cycles as shown in Figure 7, which correspond to the first surface conditions of the counterbodies in the contact. Then the coefficients of friction rapidly decrease to the steady state values due to wear occurring on either sample surface (e.g. worn diamond coating against corundum ball, Figure 8a) or counterbody surface (e.g. worn corundum ball surface against (100) diamond, Figure 8b) or both. Compared to the bulk diamond, the higher coefficient of friciton of the CF diamond coating at the steady regime is mainly due to the sp^2 contamination as well as the surface roughness as revealed in Figure 6.

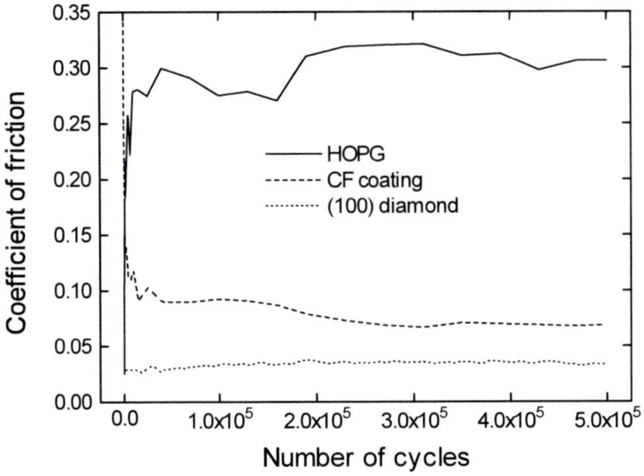

Figure 7. Coefficients of friction of (100) diamond, CF diamond coating, and (0001) graphite. (From E. Liu et al, Surf. Coat. Tech. 106 (1998) 72.)

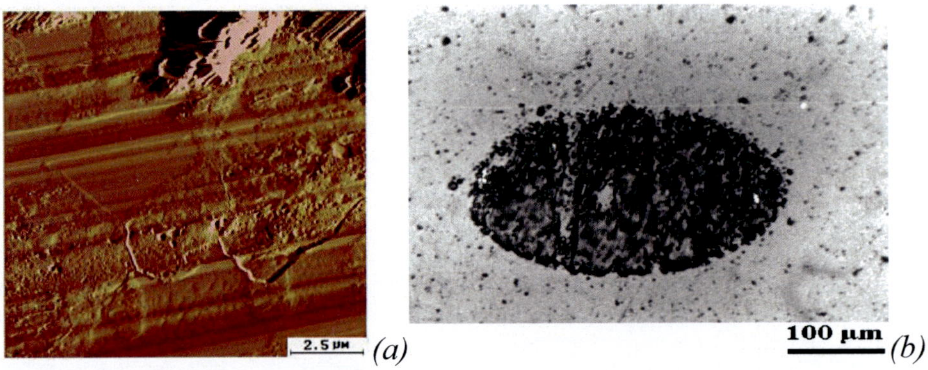

Figure 8. (a) AFM image of worn CF diamond coating and (b) optical micrograph of worn surface of corundum ball against (100) diamond. (Figure a from E. Liu et al, Diam Relat. Mater. 5 (1996) 649.)

The coefficients of friction of the DLC samples are shown in Figure 9. For the *a*-C and *a*-C:H coatings, the coefficient of friction is also the highest at the beginning of each test due to the same reason as for the diamond materials. Then the coefficient of friction rapidly decreases to the minimum value of about 0.1 for the *a*-C:H coating or to a steady state value for the *a*-C coating, due to wear debris lubricating effect and smoothened contacting surfaces as shown in Figs. 10a and b. At the steady regime, the coefficients of friction of the *a*-C:H and *a*-C coatings are about 0.14 and 0.08, respectively. Hydrogen terminations in the *a*-C:H coating are expected to be responsible for its higher friction than that of the *a*-C coating while the *a*-C coating behaves like a polycrystalline diamond material with considerable sp^2 contamination such as the combustion flame diamond coating as shown in Figure 7. Under the same test conditions, the coefficient of friction of the *a*-C coating is higher than that of the (100) diamond.

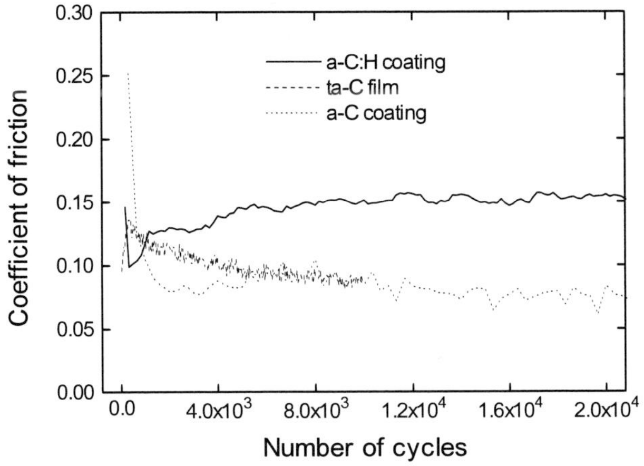

Figure 9. Friction coefficients of DLC coatings. (From E. Liu et al, Surf. Coat. Tech. 106 (1998) 72.)

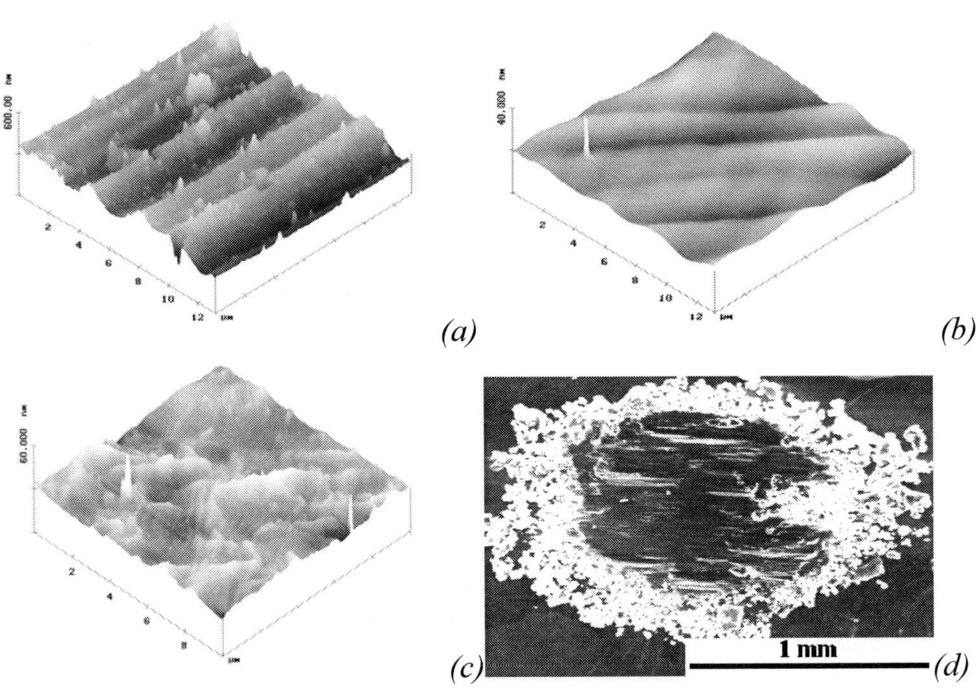

Figure 10. AFM images of worn *a*-C coating (a), *a*-C:H coating (b) and *ta*-C film (c), and SEM micrograph of worn (0001) graphite (d). (Figs. a-c from E. Liu et al, Surf. Coat. Tech. 106 (1998) 72; Figure d from E. Liu et al, J. Appl. Phys. 84 (1998) 4859.)

The friction behavior of the *ta*-C film is some different from that of the *a*-C and *a*-C:H coatings. After the first few cycles, the coefficient of friction increases to the maximum value, then decreases with the increase of sliding cycles. The *ta*-C film has a smoother surface and is expected to have a very thin top layer of sp^2 bonded carbon, which plays a lubricating role, so a lower coefficient of friction of the *ta*-C film than that of the *a*-C coating at the very beginning of each test is reasonable. With the removal of the sp^2 layer, the contribution from the surface roughness becomes significant, leading to a gradual increase in coefficient of friction. A continuous decrease of coefficient of friction till the steady state at the following cycles for the *ta*-C film is owing to the accumulation of wear debris (Figure 10c) and a continuous increase in true area of contact between the counterfaces.

The freshly cleaved (0001) HOPG is tested to reveal the effect of sp^2 bonded carbon on the friction behaviour of diamond and DLC materials. The general tendency of the friction behaviour of the (0001) graphite is significantly different from those of the diamond and DLC samples as depicted in Figs. 7 and 9. At the beginning of the test, the friction coefficient of the HOPG is the lowest and continues increasing with the number of oscilation cycles. In the steady regime, a coefficient of friction of about 0.35 is maintained, which is much higher than those of all the diamond and DLC samples.

The lowest coefficient of friction of the HOPG at the beginning of the test is due to the atomically smooth low-index basal plane with low surface energy. When the basal plane is damaged (Figure 10d), the edge sites with high indices can come into contact with the counterbody, which, together with surface roughening, contributes to the increasing coefficient of friction.

Both the CF diamond coating and the DLC coatings were worn during the ball-on-flat tests as revealed in Figs. 8a and 10a-c, where the sp^2 bonded carbon in the coatings is responsible for such damage.

From Eq 3, p_0 can exceed the elastic limits of the DLC coatings at the asperities even under a relatively low load. In fact, considerable wear on these materials indeed occurs. For the *a*-C coating, the wear debris can be attributed to the graphitic particles trapped in the coating during deposition. For the *a*-C:H coating, no clear evidence of wear debris has been found, though a worn surface is observed in Figure 10b. The oxidation of carbon and hydrogen could be the reason for the little wear debris produced on the *a*-C:H coating. For the graphite, significant material removal and transfer to the corundum counterface are evident. Since the *ta*-C film is quite thin and the wear test duration is shorter, little wear debris from the film can be resolved. The wear on the CF diamond coating (Figure 8a) can be considered a polishing process, causing the decrease of friction force, while no wear is found on the (100) diamond surface though slight wear is observed on the corundum counterface (Figure 8b).

The tangential force-displacement loops corresponding to the beginning and the end of each reciprocating sliding test for all the samples are summarized in Figure 11. Higher tangential forces at the 2nd cycles for the CF diamond and *a*-C coatings are due to their original rough surfaces (Table 1). The (100) diamond, *ta*-C film, (0001) HOPG, and even *a*-C:H coating show relatively low friction at the beginning which is attributed to their initial smooth surfaces.

It can be clearly seen that the friction of the CF diamond and *a*-C coatings significantly decreases and the changes of friction of the (100) diamond, *ta*-C film, and even *a*-C:H coating are insignificant, while the friction abruptly increases to a very high value for the HOPG in the ends of the respective tests. That different numbers of test cycles were used is because of the consideration of different coating thicknesses.

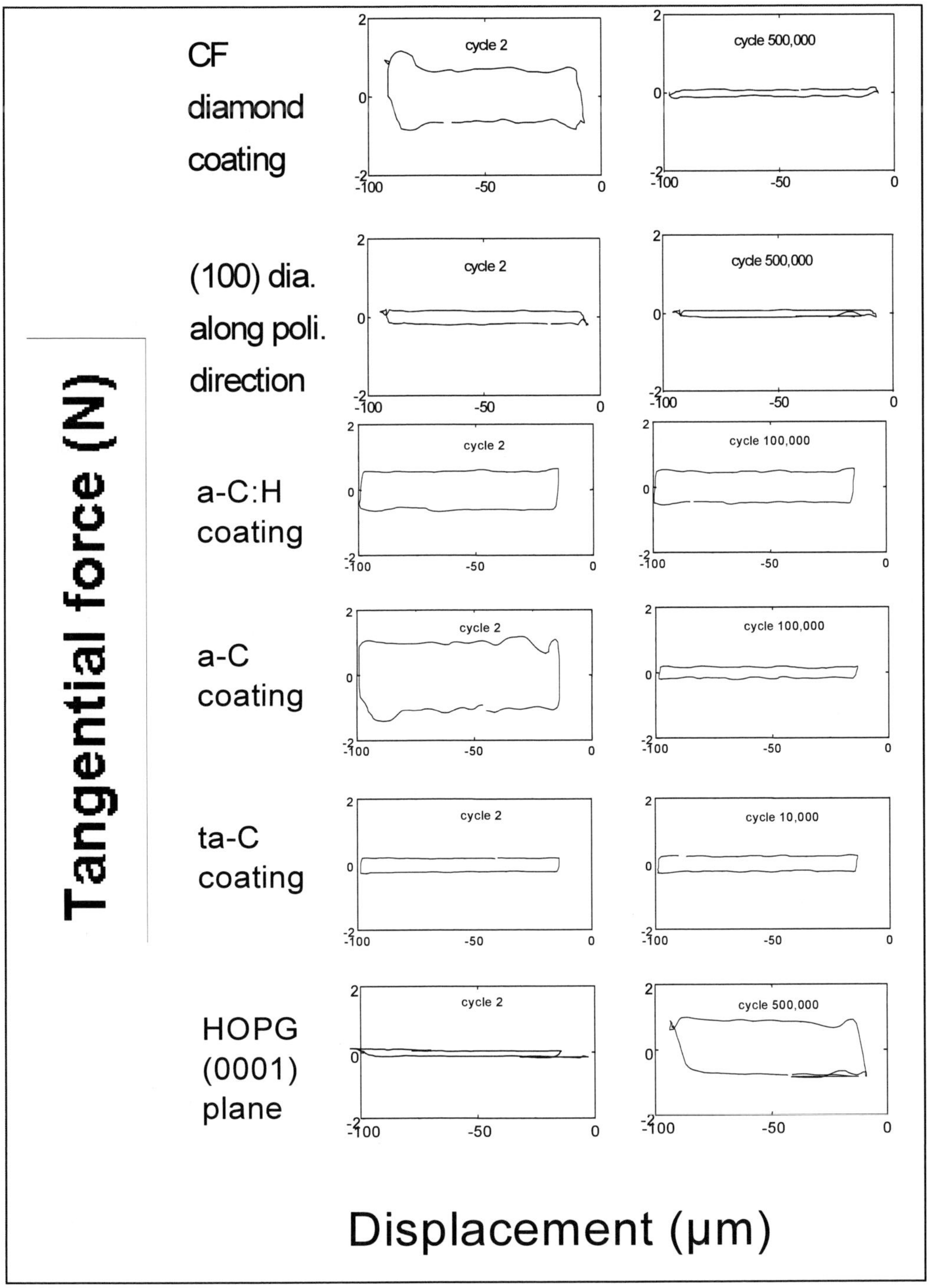

Figure 11. Tangential force-lateral tangential displacement loops corresponding to the beginning and the end of each reciprocating test for the carbon materials used. (From E. Liu et al, Surf. Coat. Tech. 106 (1998) 72.)

3. Friction with Lateral Force Microscopy (LFM)

Principle of LFM

Lateral force microscopy (LFM) is a variation of atomic force microscopy (AFM). AFM is a scanning probe technique to image a sample surface in three dimensions with atomic resolutions, based on measuring by a displacement sensor the deflection of a cantilever that supports a sharp tip at its free end. The cantilever deflection is caused by the intermolecular forces between the tip and the sample surface when the tip is brought into close proximity (nm to Å) to the surface. AFM can be operated in contact, tapping and non-contact modes.

In contact mode AFM, two force regimes can be distinguished: repulsive and attractive. When AFM is operated in the attractive regime, forces like van der Waals, electrostatic, magnetic and capillary forces can be sensed and give information about surface topography, distribution of charges, magnetic domain wall structure or liquid film distribution. At small separations of the order of angstrom, repulsive forces allow the sample surface topography to be traced with atomic resolutions.

A lateral force microscope (LFM) having a four-sector photodiode detector can simultaneously detect both torsion signals (lateral forces) and deflection signals (normal forces) of the cantilever based on the optical sensing set-up of the LFM as schematically shown in Figure 12. A photodiode voltage resulting from torsion is converted to a lateral tangential displacement of the cantilever tip and then the lateral tangential force is the product of the torsion elastic constant of the cantilever and the lateral tangential displacement of the tip.

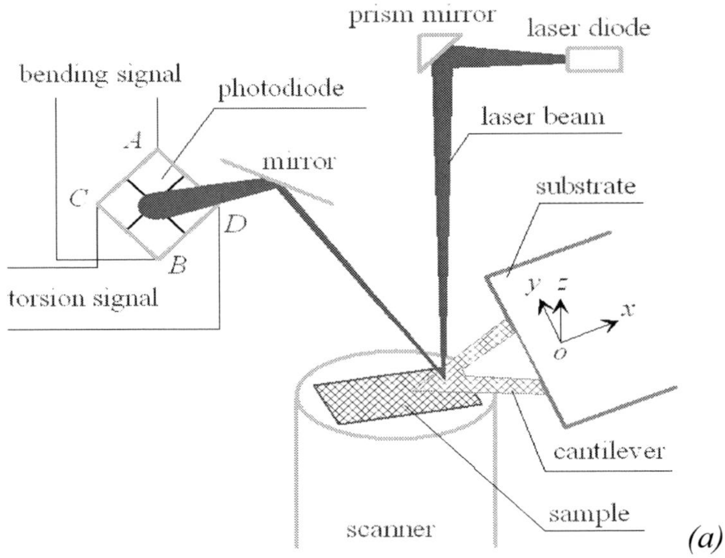

Figure 12. Continued on next page.

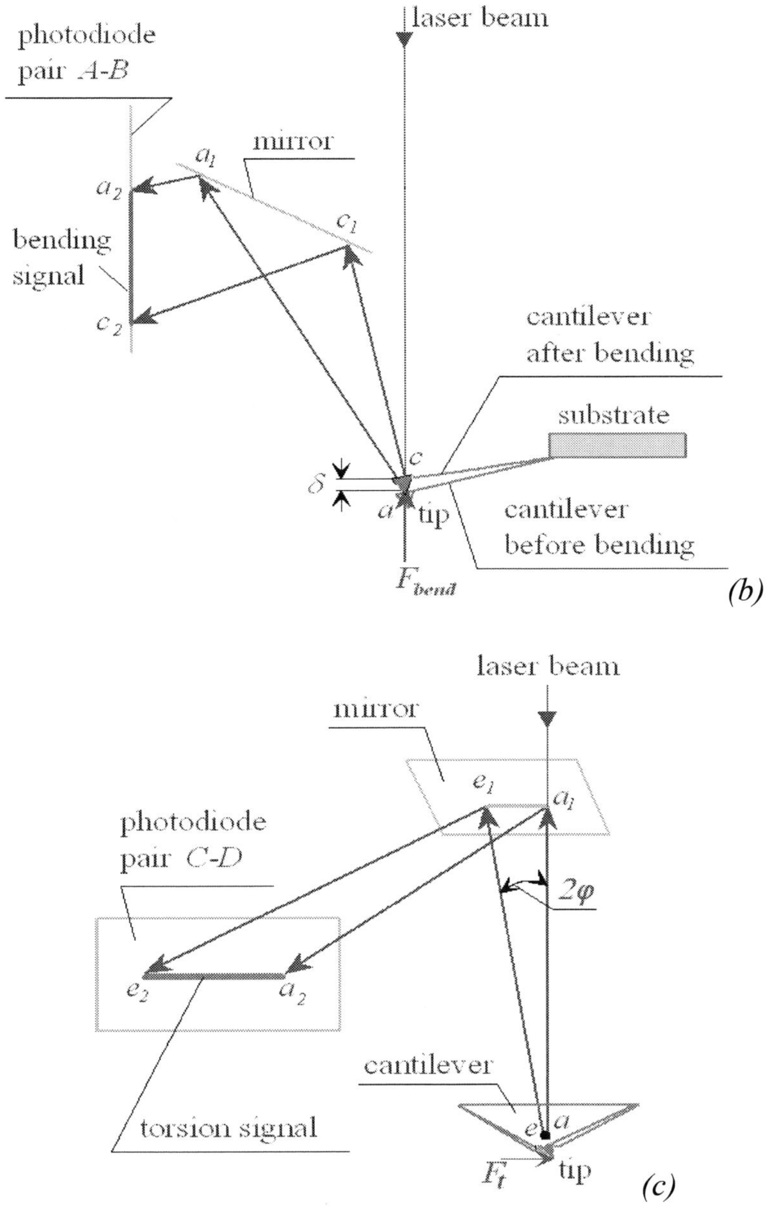

Figure 12. Schematic diagrams of (a) LFM optical path, (b) cantilever normal bending signals and (c) cantilever torsion signals. (Adapted from E. Liu et al, Wear 192 (1996) 141.)

In principle, LFM is more versatile than AFM. LFM is used for the fundamental understanding of adhesion, friction, wear and indentation processes of materials in contact at nano or atomic scales. For example, friction force acting on a tungsten tip sliding along the basal plane of a graphite under a small load in ambient air shows features with the atomic periodicity of the graphite surface (Figure 13). In magnetic storage systems and MEMS, friction and wear of lightly loaded micro/nanocomponents are highly dependent on surface interactions (few atomic layers).

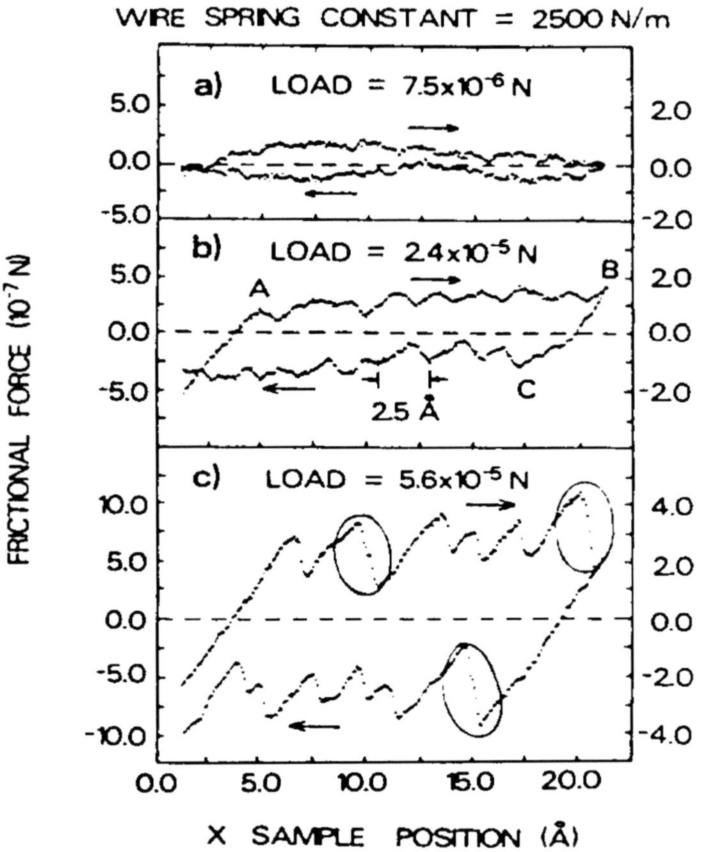

Figure 13. Frictional force applied to cantilever tip as a function of sample surface positions for three different loads. The circled sections in (c) indicate where double slips occur. (From C.M. Mate et al, Phys. Rev. Lett., 59 (17) (1987) 1942.)

Cantilever of LFM

Cantilever in LFM is a critical component which should meet the following criteria: small mass (e.g. $< 0.1\ \mu$g), short lever length (e.g. $< 200\ \mu$m), small beam thickness (e.g. $<$ 0.5 μm), low bending stiffness (flexible and sensitive to the interactions between cantilever tip and sample surface), high thermal stability, high mechanical stability (e.g. resonant frequency > 2 kHz), high torsion elastic constant, and incorporated with a sharp protruding tip at the free end as well as a mirror or electrode for reflecting deflection and torsion signals.

The thermal distortion of cantilever beam is

$$\frac{d\varepsilon}{dx} = \frac{\alpha dT}{dx} = \left(\frac{\alpha}{\lambda}\right)q \qquad (4)$$

where ε: strain; α: thermal expantion coefficient (1/K); T: temperature (K); λ: thermal conductivity (W/mK); and q: heat input per unit area (J/m^2), $q = -\lambda(dT/dx)$.

The thermal distortion of the beam can be minimized by maximizing $\left(\frac{\lambda}{\alpha}\right)$.

The mass of a light and compliant cantilever beam with a uniform cross-section is

$$m = Al\rho \text{ (kg)} \tag{5}$$

where m: mass (kg); A: cross-section area (m^2); l: length (m); and ρ: density (kg/m^3).

As the deflection of the beam caused by a bending force F_{bend} is

$$\delta = F_{bend} l^3 / C_1 EI \text{ (m)} \tag{6}$$

the stiffness of the beam is the ratio of bending force to beam deflection:

$$k = F_{bend} / \delta \text{ (N/m)} \tag{7}$$

with $I \propto A^2$ (m^4) (8)

Here δ: beam deflection (m); C_l: constant; E: Young's modulus (Pa); I: second moment of area (m^4); and k: stiffness (N/m).

Thus, the mass of the beam can be expressed as

$$m = \left(\frac{12k}{C_1 l}\right)^{1/2} l^3 \left(\frac{\rho}{E^{1/2}}\right) \text{ (kg)} \tag{9}$$

The mass of the cantilever beam can be minimized by maximizing $\left(E^{1/2}/\rho\right)$ and minimizing k.

The most commonly used LFM cantilevers have either rectangular or triangular shapes.

Bending of rectangular cantilever

Figure 14 schematically shows a rectangular cantilever. When the beam is deflected by a bending force F_{bend} as shown in Figure 15a, the curvature of the cantilever at any point along the x-direction is

$$\frac{d^2 z}{dx^2} = \frac{M_y}{EI_y} \tag{10}$$

where M_y: bending moment (Nm), $M_y = F_{bend}x$, $(0 \le x \le l)$; l: beam length (m); I_y: 2nd moment of area (m^4), E: modulus of elasticity (Pa); $I_y = \frac{bt^3}{12}$; b: beam width (m); and t: beam thickness (m).

Integrate Eq 10,

$$\begin{aligned} z' &= \int \frac{M_y}{EI_y} dx = \int \frac{F_{bend} x}{EI_y} dx \\ &= \frac{F_{bend}}{EI_y}\left(\frac{x^2}{2} + C_1\right) \end{aligned} \tag{11}$$

For $z'|_{x=l} = \frac{F_{bend}}{EI_y}\left(\frac{l^2}{2} + C_1\right) = 0$,

$$C_1 = -\frac{l^2}{2} \tag{12}$$

Thus,

$$z' = \frac{F_{bend}}{2EI_y}\left(x^2 - l^2\right) \tag{13}$$

Integrate Eq 13,

$$\begin{aligned} z &= \int z' dx = \int \frac{F_{bend}}{2EI_y}\left(x^2 - l^2\right) dx \\ &= \frac{F_{bend}}{2EI_y}\left(\frac{x^3}{3} - l^2 x + C_2\right) \end{aligned} \tag{14}$$

For $z|_{x=l} = \frac{F_{bend}}{2EI_y}\left(\frac{l^3}{3} - l^3 + C_2\right) = \frac{F_{bend}}{2EI_y}\left(-\frac{2l^3}{3} + C_2\right) = 0$,

$$C_2 = \frac{2l^3}{3} \tag{15}$$

Therefore,

$$z = \frac{F_{bend}}{2EI_y}\left(\frac{x^3}{3} - l^2 x + \frac{2l^3}{3}\right) \text{ (m)} \tag{16}$$

The deflection of the cantilever beam at the tip position (x = 0) is

$$\delta = z\,|_{x=0} = \frac{F_{bend}}{2EI_y}\left(\frac{2l^3}{3}\right) = \frac{F_{bend} l^3}{3EI_y} \text{ (m)} \tag{17}$$

Thus, the bending elastic constant of the rectangular cantilever is

$$k = \frac{F_{bend}}{\delta} = \frac{3EI_y}{l^3} = \frac{3Ebt^3}{12l^3} = \frac{Ebt^3}{4l^3}\left(\frac{\text{N}}{\text{m}}\right) \tag{18}$$

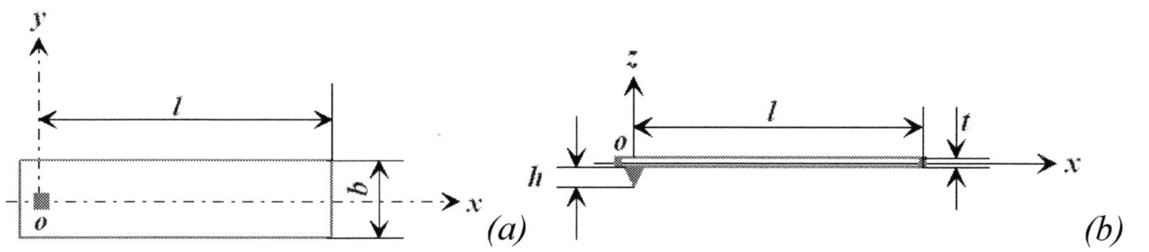

Figure 14. Schematic diagrams of rectangular cantilever

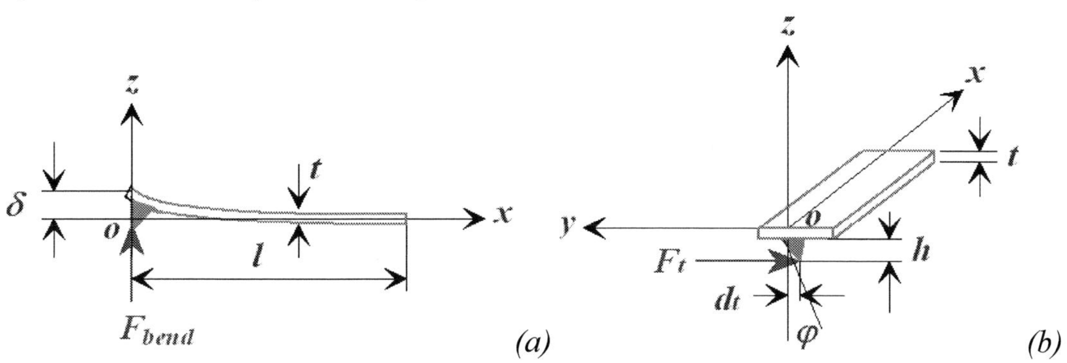

Figure 15. Schematic diagrams of rectangular cantilever: (a) deflected by normal bending force F_{bend} and (b) distorted by lateral force F_t.

Torsion of rectangular cantilever

For the rectangular beam distorted by a lateral force F_t (Figure 15b), the first derivative of torsion angle is

$$\frac{d\varphi}{dx} = \frac{T}{GJ}, \ (0 \le x \le l) \tag{19}$$

where: φ: torsion angle (rad); T: torsion moment (Nm), $T = F_t(h+t/2)$; h: tip height (m); G: shear modulus (Pa); and J: polar moment of inertia (m^4), $J = b^3t^3/3.6(b^2+t^2)$.

By integrating Eq 19, the torsion angle of the cantilever beam at the tip position ($x = 0$) is

$$\varphi = \int_l^0 \frac{T}{GJ} d(l-x) = \int_0^l \frac{F_t(h+t/2)\times 3.6(b^2+t^2)}{Gb^3t^3} dx$$
$$= \frac{3.6F_t l(h+t/2)(b^2+t^2)}{Gb^3t^3} \text{ (rad)} \tag{20}$$

The tip lateral tangential displacement is

$$d_t = \left(h+\frac{t}{2}\right)\sin\varphi \approx \left(h+\frac{t}{2}\right)\varphi$$
$$= \frac{3.6F_t l(h+t/2)^2(b^2+t^2)}{Gb^3t^3} \text{ (m)} \tag{21}$$

assuming that φ is small.

Thus, the torsion elastic constant of the rectangular cantilever is

$$k_t = \frac{F_t}{d_t} = \frac{Gb^3t^3}{3.6l(h+t/2)^2(b^2+t^2)} \left(\frac{\text{N}}{\text{m}}\right) \tag{22}$$

Bending of triangular cantilever

Figure 16 schematically illustrates a triangular hollow cantilever. When the beam is deflected by a bending force F_{bend} as shown in Figure 17a, the curvature of the cantilever at any point along the x-axis is

$$\frac{d^2z}{dx^2} = \frac{M_y}{EI_y} \tag{23}$$

where M_y: bending moment (Nm), $M_y = F_{bend}(x-l_3)$, $(0 \le x \le l_1)$; E: Young's modulus (Pa); I_y: 2nd moment of area (m^4), $I_y = I_{y1} = \frac{(b_1-b_2)t^3}{6}$ for $l_2 \le x \le l_1$ and $I_y = I_{y2} = \frac{b_1t^3x}{6l_1}$ for $l_3 \le x \le l_2$; b_1, b_2: beam widths of different segments (m); l_1, l_2, l_3: beam lengths of different segments (m); and t: beam thickness (m).

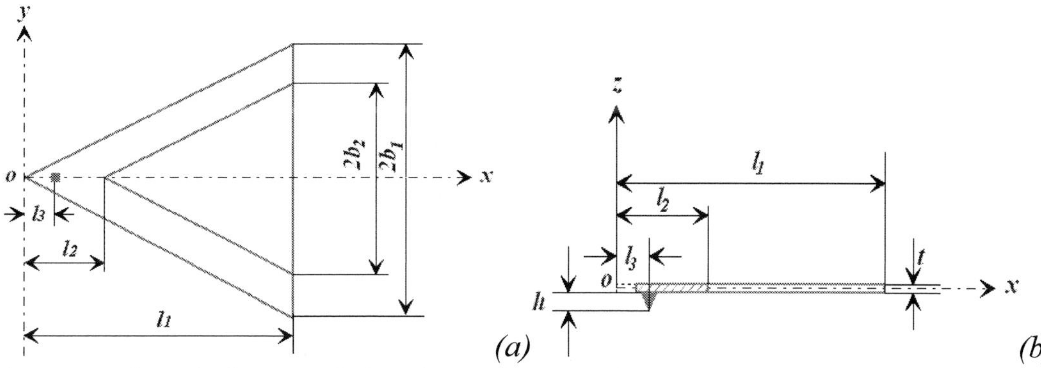

Figure 16. Schematic diagrams of triangular cantilever. (Adapted from E. Liu et al. Wear 192 (1996) 141.)

Integrate Eq 23,

$$\begin{aligned} z &= z_1 + z_2 \\ &= \iint_{(l_2 \le x \le l_1)} \frac{M_y}{EI_{y1}} dx^2 + \iint_{(l_3 \le x \le l_2)} \frac{M_y}{EI_{y2}} dx^2 \\ &= \frac{6F_{bend}}{E(b_1 - b_2)t^3} \iint_{(l_2 \le x \le l_1)} (x - l_3) dx^2 + \frac{6F_{bend} l_1}{Eb_1 t^3} \iint_{(l_3 \le x \le l_2)} \frac{(x - l_3)}{x} dx^2 \ \text{(m)} \end{aligned} \tag{24}$$

Applying the boundary conditions: $z_1 = z_1' = 0$ at $x = l_1$, $z_1 = z_2$ at $x = l_2$, and $z_1' = z_2'$ at $x = l_2$, to Eq 24, the deflection of the cantilever beam at the tip position ($x = l_3$) is

$$\begin{aligned} \delta = z|_{x=l_3} &= \frac{F_{bend}}{Et^3} \left\{ \frac{3l_1 \left[l_2^2 - 4l_2 l_3 + 2l_3^2 \ln(l_2/l_3) + 3l_3^2 \right]}{b_1} \right. \\ &\left. + \frac{2\left[l_1^3 - l_2^3 - 3l_3 \left(l_1^2 - l_2^2 \right) + 3l_3^2 (l_1 - l_2) \right]}{b_1 - b_2} \right\} \ \text{(m)} \end{aligned} \tag{25}$$

Thus, the normal bending elastic constant of the triangular cantilever is

$$\begin{aligned} k = \frac{F_{bend}}{\delta} &= Et^3 \left(\left\{ \frac{3l_1 \left[l_2^2 - 4l_3 l_2 + 2l_3^2 \ln(l_2/l_3) + 3l_3^2 \right]}{b_1} \right. \right. \\ &\left. \left. + \frac{2\left[l_1^3 - l_2^3 - 3l_3 \left(l_1^2 - l_2^2 \right) + 3l_3^2 (l_1 - l_2) \right]}{b_1 - b_2} \right\} \right)^{-1} \left(\frac{\text{N}}{\text{m}} \right) \end{aligned} \tag{26}$$

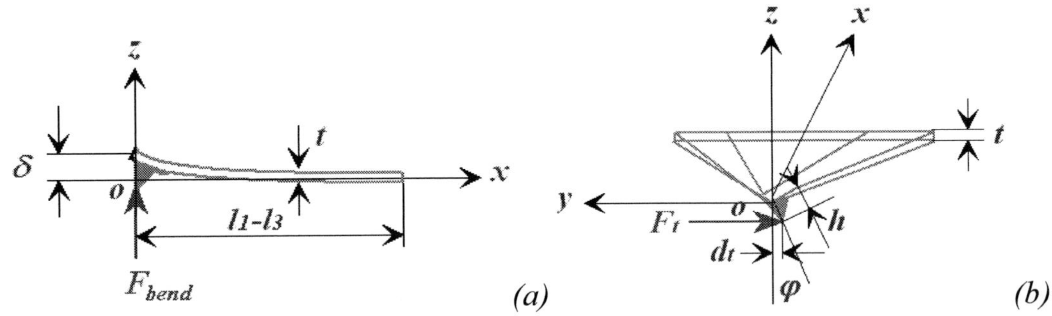

Figure 17. Schematic diagrams of triangular cantilever: (a) deflected by normal bending force F_{bend} and (b) distorted by lateral force F_t.

Torsion of triangular cantilever

For the triangular beam distorted by a lateral force F_t (Figure 17b), the 1st derivative of torsion angle is

$$\frac{d\varphi}{dx} = \frac{T}{GJ} \tag{27}$$

where T: torsion moment (Nm), $T = F_t\left(h + \frac{t}{2}\right)$; G: shear modulus (Pa); and J: polar moment of inertia (m^4), $J = J_1 = \frac{2t^3(b_1 - b_2)}{3}$ for $l_2 \le x \le l_1$ and $J = J_2 = \frac{2t^3 b_1 x}{3l_1}$ for $l_3 \le x \le l_2$.

By integrating Eq 27, the torsion angle of the cantilever beam at the tip position ($x = l_3$) is determined as

$$\begin{aligned}
\varphi &= \int_{l_2}^{l_1} \frac{T}{GJ_1}dx + \int_{l_3}^{l_2} \frac{T}{GJ_2}dx \\
&= \int_{l_2}^{l_1} \frac{3T}{2Gt^3(b_1 - b_2)}dx + \int_{l_3}^{l_2} \frac{3Tl_1}{2Gt^3 b_1 x}dx \\
&= \frac{3T}{2Gt^3(b_1 - b_2)} \int_{l_2}^{l_1} dx + \frac{3Tl_1}{2Gt^3 b_1} \int_{l_3}^{l_2} \frac{1}{x}dx \\
&= \frac{3F_t(h + t/2)}{2Gt^3}\left[\frac{1}{(b_1 - b_2)} \int_{l_2}^{l_1} dx + \frac{l_1}{b_1} \int_{l_3}^{l_2} \frac{1}{x}dx\right] \\
&= \frac{3F_t(h + t/2)}{2Gt^3}\left[\frac{l_1 - l_2}{b_1 - b_2} + \frac{l_1}{b_1}\ln\left(\frac{l_2}{l_3}\right)\right] \text{ (rad)}
\end{aligned} \tag{28}$$

The tip lateral tangential displacement is expressed as

$$d_t = \left(h + \frac{t}{2}\right)\sin\varphi \approx \left(h + \frac{t}{2}\right)\varphi$$
$$= \frac{3F_t(h+t/2)^2}{2Gt^3}\left[\frac{l_1 - l_2}{b_1 - b_2} + \frac{l_1}{b_1}\ln\left(\frac{l_2}{l_3}\right)\right] \tag{29}$$

assuming that φ is small.

Thus, the torsion elastic constant of the triangular cantilever is

$$k_t = \frac{F_t}{d_t} = \frac{2Gt^3}{3(h+t/2)^2\left[\frac{l_1 - l_2}{b_1 - b_2} + \frac{l_1}{b_1}\ln\left(\frac{l_2}{l_3}\right)\right]} \tag{30}$$

Case Study 2

A rectangular AFM cantilever, made from silicon (E = 180 GPa, ρ = 2300 kg/m^3), has the length l = 100 μm, width b = 10 μm and thickness t = 1 μm. (i) Estimate the bending spring constant and resonant frequency of the cantilever. The concentrated mass of the tip is assumed to be $m_{tip} \approx$ 4 ng. (ii) Calculate the sensitivity of the displacement sensor (at least 5 times smaller than the deflection of the cantilever, which is required to measure the deflection of the above cantilever). The interaction force between the tip and the sample surface is typically F_{bend} = 1 nN.

Solution

(i) For the cantilever, E = 180 GPa, ρ = 2300 kg/m^3, l = 100 μm, b = 10 μm, and t = 1 μm, according to Figure 15a and Eq 18, the bending spring constant of the cantilever is

$$k = \frac{F_{bend}}{\delta} = \frac{Ebt^3}{4l^3} = \frac{180\times10^9\times10\times10^{-6}\times(10^{-6})^3}{4\times(100\times10^{-6})^3} = 0.45\ \text{N/m}$$

As the mass of the cantilever is $m_{beam} = lbt\rho$ and the effective mass of the cantilever is $m_{eff} = m_{beam}/3$, the resonant frequency of the cantilever with the tip ($m_{tip} \approx$ 4 ng) is

$$f = \frac{1}{2\pi}\sqrt{\frac{k}{m_{tip} + m_{eff}}} = \frac{1}{2\pi}\sqrt{\frac{k}{m_{tip} + m_{beam}/3}} = \frac{1}{2\pi}\sqrt{\frac{k}{m_{tip} + btl\rho/3}}$$
$$= \frac{1}{2\pi}\sqrt{\frac{0.45}{4\times10^{-12} + 10^{-5}\times10^{-6}\times10^{-4}\times2300/3}} \approx 48.9 \text{ (kHz)} \tag{31}$$

(ii) Under $F_{bend} = 1$ nN, the deflection of the cantilever is

$$\delta = \frac{F_{bend}}{k} = \frac{10^{-9}}{0.45} \approx 2.2 \text{ (nm)}$$

Thus, the sensitivity of the displacement sensor (at least 5 times smaller than δ) is

$$s = \frac{\delta}{5} = \frac{2.2}{5} = 0.44 \text{ (nm)} \tag{32}$$

Mechanisms of Micro/Nanoscale Friction

There are three dominant mechanisms in friction measurement with LFM: adhesive, ratchet and plowing. Adhesive friction alone cannot explain the local variations in friction. Ratchet friction is dominating for the local variations in friction in order to relate coefficient of friction to the plowing component and the slope of local asperities. The ratchet mechanism combined with collision is valid provided that the tip radius of cantilever is smaller than the size of asperities. The tip's collision with an asperity on its ascending portion causes an additional torsion of the cantilever, which results in a higher coefficient of friction (Figure 18a), while the tip has almost no collision with the asperity on its descending portion with a negative slope (Figure 18b).

Supposing that the adhesive friction mechanism has no change during sliding, the coefficient of friction between cantilever tip and sample surface is

$$\mu_0 = S/N \tag{33}$$

where S is the shear force along the local surface and N is the normal force perpendicular to the local surface. The local normal and friction forces are measured with respect to the global horizontal and vertical directions.

The local coefficient of friction on ascending having an angle of α with the global surface as schematically illustrated in Figure 18a is expressed as

$$\mu_1 = \frac{F_t}{F_n} = \frac{S\cos\alpha + N\sin\alpha}{N\cos\alpha - S\sin\alpha}$$
$$= \frac{\mu_0 + \tan\alpha}{1 - \mu_0\tan\alpha} \approx \mu_0 + \tan\alpha \tag{34}$$

assuming $\mu_0 \tan\alpha << 1$.

Here, F_t is the friction force along the global surface, F_n is the normal force perpendicular to the global surface, and tan α is the uphill plowing component.

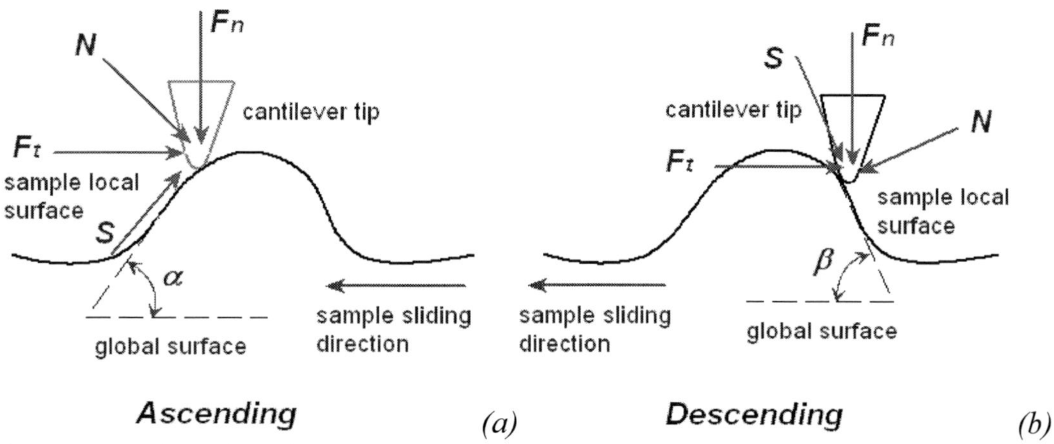

Figure 18. Cantilever tip sliding over an asperity. (Adapted from B. Bhushan, Tribology and Mechanics of Magnetic Storage Devices; 2nd ed.; Springer-Verlag: pp 883-900.)

Similarly, the local coefficient of friction on descending having an angle of with the global surface as schematically shown in Figure 18b is

$$\begin{aligned}\mu_2 &= \frac{F_t}{F_n} = \frac{S\cos\beta - N\sin\beta}{N\cos\beta + S\sin\beta} \\ &= \frac{\mu_0 - \tan\beta}{1 + \mu_0\tan\beta} \approx \mu_0 - \tan\beta\end{aligned} \tag{35}$$

assuming $\mu_0 \tan\beta << 1$.

Tan β in Eq 35 is the downhill plowing component. Note that the downhill plowing component has a negative sign.

The average local coefficient of friction across a symmetric asperity when $\beta = \alpha$ is

$$\begin{aligned}\mu_{ave} &= \frac{\mu_1 + \mu_2}{2} \\ &= \frac{\mu_0\left(1 + \tan^2\alpha\right)}{1 - \mu_0^2\tan^2\alpha} \\ &\approx \mu_0\left(1 + \tan^2\alpha\right)\end{aligned} \tag{36}$$

assuming $\mu_0^2 \tan^2\alpha << 1$.

In this condition, the plowing component of the local coefficient of friction in either direction is $\pm\tan\alpha$.

Case Study 3: Friction of (0001) HOPG studied with LFM

A commercial LFM (NanoScope III, Digital Instruments) with a Si_3N_4 triangular cantilever is used to measure the (0001) HOPG sample freshly cleaved under the following experimental conditions: ambient atomosphere: 20°C and 50% RH; normal force: several nN to several hundreds nN; lateral scan rate: 0.5 - 61 Hz; lateral scan size: 10 nm - 12 μm; and without lubrication.

A total normal force acting on the cantilever tip is composed of two parts: cantilever normal bending force and adhesive force between the tip and the sample surface in contact mode, which are determined from the relationship between the beam deflection and the travel distance of the *z*-scanner of the LFM from the sample surface as shown in Figure 19. In the diagram, the cantilever is positively deflected in the repulsive regime (segment between 4 and 5 in extending or between 5 and 6 in retracting) and negatively bent in the attractive regime (segment between 6 and 7). The adhesive force is taken as constant and measured as the maximum negative bending force of the cantilever at the pull-off point (point 7 in Figure 19). A cantilever normal bending force is calculated by multiplying the normal bending elastic constant of the cantilever by a cantilever deflection.

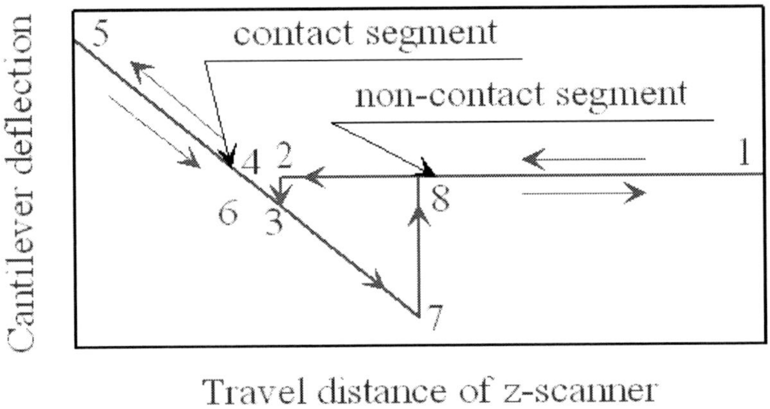

Figure 19. Schematic diagram of cantilever deflection vs. travel distance of *z*-scanner of LFM. (Adapted from E. Liu et al. *J. Appl. Phys.* 84 (1998) 4859.)

A friction force acting on the cantilever tip is determined from a cantilever torsion-tip lateral scan size loop as shown in Figure 20 when measured at a scan angle of 90° from the lengthwise direction of the cantilever (Figure 17b). The coefficient of friction is defined as

$$\mu = \frac{F_t}{F_{total}} = \frac{F_t}{\left(F_{bend} + F_{ad}\right)} = \frac{k_t d_t}{\left(k\delta + F_{ad}\right)} \tag{37}$$

depending on the setpoint position in the normal force curve measurement with LFM.

Here, F_t: lateral tangential force; F_{bend}: normal bending force; F_{ad}: adhesive force; and d_t: tip lateral tangential displacement.

Features on the sample surface can cause the fluctuations of the cantilever torsion leading to an increase in average friction force as indicated by Eqs 34-36.

In macrotribology with a ball-on-flat contact geometry, adhesive forces can be ignored in most cases, because they are generally small compared to an applied load. When the normal load is reduced to a very small value, the effect of adhesive forces on friction gradually becomes significant. However, in micro/nanotribological study with LFM, the adhesive forces can be comparable with or even higher than an applied normal force.

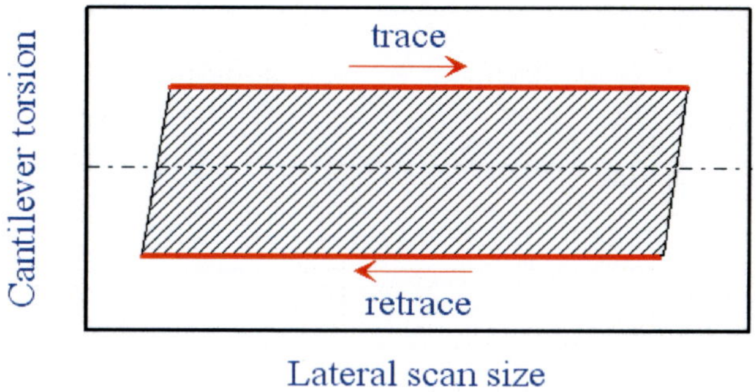

Figure 20. Schematic diagram of cantilever torsion-tip lateral scan size loop

When the friction force is plotted against the total normal force (sum of adhesive force and cantilever normal bending force), a linear relationship is observed in Figure 21 and the slope of the fitting line is taken as the coefficient of friction. The linear relationship between the friction force and the total normal force is due to the assumption that only the elastic contact between the sample surface and the cantilever tip has been maintained throughout the tests. The contribution of the cantilever normal bending alone to the friction is also overlaid in the graph under the assumption that a constant adhesive force has been maintained throughout the measurements. The offset of friction force at zero total normal force in the graph is due to uncertainties or scattering. The coefficient of friction of the basal plane of the HOPG sample measured in such a way is about 0.009.

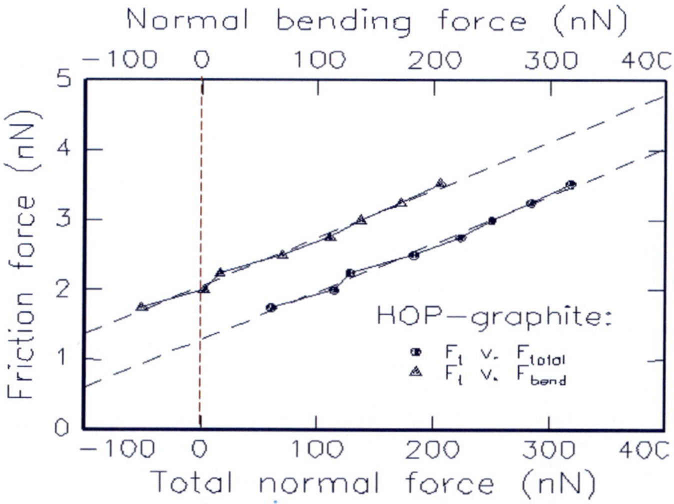

Figure 21. Friction force vs. total normal force or cantilever normal bending force measured on (0001) graphite with LFM. The dashed lines represent the linear least square fitting. (From E. Liu et al. *J. Appl. Phys*. 84 (1998) 4859.)

The coefficients of friction of the graphite sample (0.001 - 0.01) measured at different lateral scan sizes (Figure 22a) and scan rates (Figure 22b) under a total normal force of 374.4 nN are consistently lower than or comparable with the initial value of friction coefficient of the same sample (around 0.03) acquired from a ball-on-flat reciprocating testing (Figure 7). The initial low coefficient of friction of the graphite sample obtained from the ball-on-flat test is corresponding to its atomically smooth (0001) surface. However, the coefficient of friction of the graphite from the ball-on-flat test after the (0001) surface is worn may not be comparable with the LFM results obtained under the loading conditions used here.

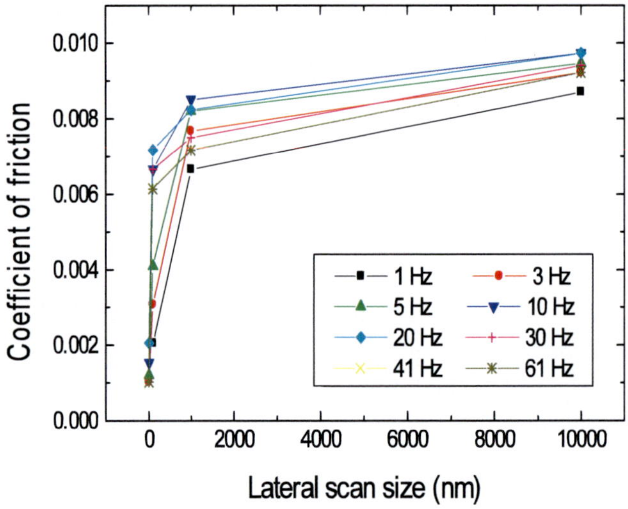

(a)

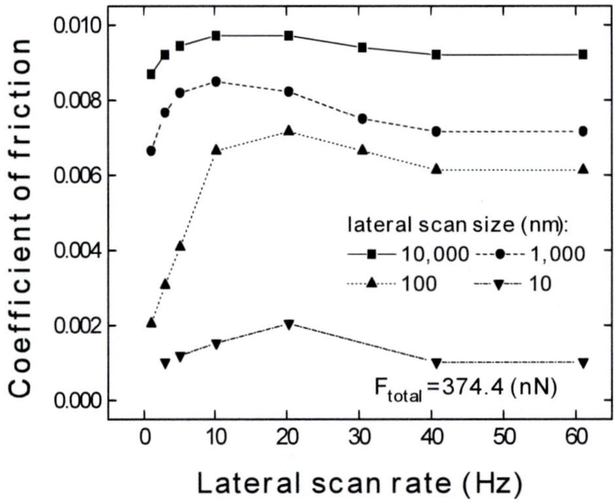

(b)

Figure 22. Friction coefficient vs. lateral scan size (a) and lateral scan rate (b) measured on (0001) graphite with LFM. (Figure b from E. Liu et al. *J. Appl. Phys*. 84 (1998) 4859.)

4. Postscript

In macrotribology, friction is generally accompanied by elastic and plastic deformation, wear and even fracture in the contacting surfaces. In a ball-on-flat reciprocating tribological test, the friction regimes are derived from the relationship between static and kinetic friction.

Nanotribology was stimulated by the fabrication of micro-electro-mechanical systems (MEMS). With the development of scanning force microscopy (SPM), specifically, lateral force microscopy (LFM), an experimental approach of nanotribological regimes can be realized. LFM with a bidirectional position-sensitive detector allows the normal and lateral tangential forces acting on a cantilever tip to be detected simultaneously. In nanotribology, visible wear may not always be the case depending on the contact pressure.

In the case studies, the macrotribological behaviors of different carbon materials, such as bulk diamond, diamond coating, DLC coatings and graphite, were studied by submitting them to a ball-on-flat reciprocating testing, while the nanotribological behavior of the same graphite sample was investigated with LFM.

In the ball-on-flat experiments, it was found that the coefficient of friction at the initial stage can be considerably different from that measured in the steady regime under the predetermined experimental conditions due to elastic and plastic deformation, wear and material transfer, whereas the friction coefficient measured under a low load with LFM can be relatively constant throughout a test.

Further Reading

Adibnazari, S., Hoeppner, D.W. (1992). *Wear*, 159, 43.

Albrecht, T.R., Akmine, S., Carver, T.E., Quate, C.F. (1990). *J. Vac. Sci. Technol. A*, 8, 3386.

Amontons, G. (1699). *Histoire Acad. Roy. Sci*, 12, 206.

Ashby, M.F. (2005). *Materials Selection in Mechanical Design*; 3rd ed.; Butterworth-Heinemann: Oxford, pp 157-160.

Beer, F.P., Johnston, Jr.E.R., Dewolf, J.T. (2002). *Mechanics of Materials*; 3rd ed, McGraw-Hill: New York, NY, pp 150-152, 186-188, 532-533.

Bhushan, B. *Tribology and Mechanics of Magnetic Storage Devices*; 2nd ed.; Springer-Verlag: pp 883-900.

Binnig, G., Quate, C.F., Gerber, C. (1986). *Phys. Rev. Lett*, 56, 930.

Binnig, G., Rohrer, H., Gerber, C., Weibel, E. (1982). *Phys. Rev. Lett*, 49, 57.

Coulomb, E. (1785). *Mem. Math. Phys*, 10, 161.

Hills, D.A. (1994). *Wear*, 175, 107.

Kamo, M., Sato, Y., Matsumoto, S., Setaka, N. (1983). *J. Cryst. Growth*, 62, 642.

Kelly, B.T. (1981). *Physics of Graphite*, Applied Science Publishers, London.

Kroto, H.W., Heath, J.R., O'Brien, S.C., Curl, R.F., Smalley, R.E. (1985). *Nature*, 318 162.

Liu, E., Blanpain, B., Celis, J.P. (1996). *Diam Relat. Mater*, 5, 649.

Liu, E., Blanpain, B., Celis, J.P. (1996). *Wear*, 192, 141.

Liu, E., Blanpain, B., Celis, J.P., Roos, J.R. (1998). *J. Appl. Phys*, 84, 4859.

Liu, E., Blanpain, B., Shi, X., Celis, J.P., Tan, H.S., Tay, B.K., Roos, J.R. (1998). *Surf. Coat. Tech*, 106, 72.

Liu, E., Shi, X., Tay, B.K., Cheah, L.K., Tan, H.S., Shi, J.R., Sun, Z. (1999). *J. Appl. Phys*, *86*, 6078.

Lubkin, J.L. (1962). In *Handbook of Engineering Mechanics*, W. Fliegge, Ed, McGraw-Hill: New York, NY.

Mate, C.M., McClelland, G.M., Erlandsson, R., Chiang, S. (1987). *Phys. Rev. Lett*, 59, 1942.

Robertson, J. In *Diamond and Diamond-Like Films and Coatings, NATO-ASI Series B: Physics*, R. E. Clausing, L. L. Horton, J. C. Angus, P. Koidl, Eds., Plenum Publishing Co.: New York, NY, 1991.

Schermer, J.J., Hogenkamp, J.E.M., Otter, G.C.J., Janssen, G., Van Enckevort, W.J.P., Giling, L.J. (1993). *Diam Relat. Mater*, 2, 1149.

Tabor, D. (1995). *Tribo. Lett*, 1, 1.

Vingsbo, O., Ödfalk, M. (1992). *Wear*, 157, 435.

Vingsbo, O., Schön, J. (1993). *Wear*, 162-164, 347.

In: Tribology Research Trends
Editor: Taisho Hasegawa

ISBN: 978-1-60456-912-4

Chapter 6

HUMIDITY EFFECTS ON DRY SLIDING PERFORMANCE OF SINTERED POLYIMIDE/GRAPHITE COMPOSITES

Pieter Samyn* and Gustaaf Schoukens
Ghent University – Department of Textiles, Technologiepark 907, B-9052 Zwijnaarde, Belgium

ABSTRACT

Polyimides are known as high-performance polymers with extreme thermal and chemical resistance, supposed to function under severe conditions with variable environmental conditions. Graphite can be added as an internal lubricant during sintering for controlling friction and/or wear. The effect of humidity on sliding properties of sintered graphite-polyimide composites cannot be clearly predicted at present. The tribological properties of the polyimide matrix and the graphite lubricant seem to depend on the moisture content in an opposite way: theoretically, water molecules are needed for easy shear of the graphitic structure, while they have detrimental effect on the sliding properties of the polyimide surfaces. The friction and wear performance of pure and graphite-filled polyimides will therefore be experimentally investigated at three humidity levels during unlubricated sliding against a steel counterface. Test results will be discussed in relation to microscopic evaluation of the worn surfaces. A parallelism between the tribological properties during sliding at different humidity and different temperatures will be demonstrated and confirmed by Raman spectroscopy.

Keywords: polyimide, graphite, composite, friction, wear, humidity

---

* Corresponding author: Pieter.Samyn@UGent.be

1. Introduction

Polymers often replace metal parts in sliding applications, because they function under dry sliding conditions and can be used when external oil lubrication is impossible or undesirable. Softening and/or melting of the polymer surfaces allow for self-lubricating properties. Polymers posses visco-elastic mechanical properties, which are weaker than metal parts and strongly dependent on the operating and environmental conditions. Besides discussions on the influence of sliding parameters such as sliding motion [1], normal load [2], sliding velocity [3] or temperature [4], some studies on the influences of humidity [5] are available, but less frequently reported. Kawakame et al. [6] reported that variations in the relative air humidity from 50 to 70% can duplicate the lost wear volume of PTFE composites and, consequently, double the wear rate of self-lubricating seals. Under atmospheric conditions, different testing atmospheres can be applied according to ISO-291: (i) 23°C, 50 % RH as recommended for most applications, or (ii) 27°C, 65 % RH as recommended for tropical regions. Also different environmental humidity during winter time (< 20% RH) or summer time (> 80% RH) may cause different results. Engineering polymers under sliding have different sensitivities to moisture: water vapours directly affect the mechanical properties, in contrast to metals, and indirectly change the transfer mechanism. The absorption of a water layer may cause boundary lubrication under extreme conditions of 100% humidity [7].

For metals or alloys, the effect of humidity is often clearly explained by the formation of oxide layers. E.g., the wear rates of Al–Si alloy decreases by two orders of magnitude as the relative humidity increases from 3% to 100% [8]. At low humidity conditions adhesive wear is predominant, whilst at high humidity conditions a layer of compacted oxide–metal debris film is formed on the slider surface, which reduces the direct metal–metal contact. The friction coefficient for Al-Si is maximum at 3% and 100% relative humidity conditions. For diamond-like carbon coatings, Tanaka et al. [9] observed that coefficients of friction were higher and wear rates were smaller in wet air than those experienced in dry air. These observations suggest that the sliding surface under wet air conditions has ability to deform plastically. This property translates into smaller wear and better film toughness. Liu et al. [10] found that the correlation between tribological properties and humidity for diamond-like carbon coatings strongly depends on the coating composition. The coefficient of friction for hydrogen-free (amorphous a-C) coatings decreased with an increase in relative humidity. A water layer physically adsorbed at the interface between the mating surfaces plays two major roles: acting as a lubricant and increasing the true area of contact. However, it was noticed that the friction coefficient of the hydrogenated DLC (a-C:H) coatings first increased and then decreased with increasing relative humidity in the steady state. There appeared to be a critical relative humidity for the a-C:H coatings, at which the steady-state friction reached the maximum value. According to Yang et al. [11] the coefficient of friction of DLC is lower in air compared to vacuum sliding. Some authors found also and increase in friction at higher relative humidity [12]. The interaction between hydrogen and oxygen at the interface between the a-C:H coating and water layer was mainly responsible for such behaviour.

For engineering polymers, the effect of humidity generally depends on whether the polymer has free hydrogen bonds into its structure. Polymers with no or only few of those hydrogen bonds, such as ultra-high-molecular weight polyethylene (UHMWPE), polyacetal

(POM), polypropylene (PP), polytetrafluouroethylene (PTFE) or polyethylene terephtalate (PET), are less sensitive to atmospheric conditions because the physical material properties are not strongly altered by humidity. For those polymers repelling water, the presence of humidity eventually may have positive effects due to lubrication. Polyacetals are completely insensitive to any variation in humidity due to low water absorption and insolubility. Vale Antunes et al. [13] also showed that a variation in humidity did not greatly influence the behaviour of polypropylene composites. Hammerschmidt et al. [14] indicated that humidity effects for PET are not as strong as for other more hydrophilic polymers: only 0.4% water content is absorbed under equilibrium conditions at 40% relative humidity, causing a very small shift in relaxation temperature. Da Silva et al. [15] demonstrated for high density polyethylene (HDPE), in contrast, that the coefficient of friction increases at higher humidity for low applied loads, while there was no variation in friction observed for high applied load. McNicol et al. [16] found that humidity influenced the wear of both PTFE and HDPE. For worn PTFE, it is known that a band-like structure develops under dry conditions while it generally changes into a fiber-like aspect at higher humidity due to destruction of the sheet-like molecular planes [17]. Tanaka et al. [18] showed that the friction of PTFE decreased with increased humidity independently of crystalline transitions. Morgan et al. [19] performed indentation-recovery tests on composites of PTFE/glass fibres and found that higher humidity was associated with greater resilience and reduced relaxation. Bearings using this liner material exhibited larger torque at increasing humidity. Polymers with active hydrogen bonds in their structure, such as polyamides (PA) or epoxies, are more sensitive to the effect of water absorption on their sliding properties. Polyamides are extensively studied as engineering materials due to easy synthesis and flexibility in composition [20-22]. The coefficient of friction for polyamide may increase from 0.40 to 1.20 at higher humidity, because the polymer becomes tough and the deformation component of friction increases [23]. The wear resistance of polyamide may improve at higher humidity [24]. The formation of additional bonds in between the polymer chains under high humidity generally disfavours the molecular mobility and prevents the formation of an oriented transfer film. The adsorption of moisture causes weakening of the polymer structure and consequently increases the coefficient of friction. Small water molecules can generally permeate into the amorphous regions of semi-crystalline polymers through diffusion. Water is considered to be present in the free volume and becomes active when it is attached to polymer chains by hydrogen bonds. For polyamides, water permeation occurs in a somewhat thicker surface layer compared to polyethylene and this is may be observed as plasticization of the surface [25]. Therefore, polyamides are prone to swelling in moist environment and mechanical properties deteriorate with consequently changing dimensions. For oil-filled polyamide grades, exposure of the clean material to moisture may prevent the absorption of oil into the capillaries [26]. The main modifications in properties of polyamides result from the interaction of the amide groups with polar organic molecules [27]. The degree to which polyamides absorb moisture is related to the number of amide groups per given polymer chain length. At room temperature and 50 % relative humidity (RH), polyamide could absorb up to 2.75 % and every 1 % moisture increase may result in 0.2 to 0.3 % increase in its dimension. The moisture acts as a plasticizer that reduces the entanglement and bonding between molecules, therefore increases their volume and mobility [28]. The moisturized material exhibits lower glass transition temperature (Tg), which makes it easier for further crystallization [29]. Lim et al. [30] studied the water-vapour transmission rates, which were enhanced above 60 to 70 % RH due to a transtion of the polymer from glassy to rubbery states. Based on the frequency shift of FTIR peaks, moisture sorption appeared to reduce the average hydrogen-bond strength of the N - H Group, while an increase was seen for the C = O groups. More in general, Garoff et al. [31] studied the influences of humidity on hydrophilic polymer surfaces, such as cellulose, concluding that the magnitude of friction and adhesion and their dependence on humidity decreased with increasing hydrophobicity of the contacting surfaces. In the limit of water-lubricated sliding, one of the two rubbing surfaces should be hydrophilic. It was suggest that friction of hydrophilic polymer surfaces under ambient conditions is greatly influenced by capillary condensation.

For high-performance polymers, the tribological properties are improved by the formation of molecular superstructures in the bulk polymer or by copolymerisation. They should withstand sliding systems under severe working conditions with high loads, high sliding velocities and/or high temperature, ensuring longer life-expectance and reliability due to their strength, load-carrying capacity and thermal resistance. The chemical resistance of high-performance polymers, however, still depends strongly on their structure and is different for, e.g., polyphenylenesulphide (PPS), polyether-etherketone (PEEK) or liquid crystalline polymers (LCP). Hoa et al. [32] measured significant moisture absorption into PPS composites and a large reduction in tensile strengths from 25 to 46%. Lhymn et al. [33] reported that absorbed moisture reduces the shear strength and plasticization can be observed for PPS and its composites. Wang et al. [34] showed that the fracture toughness of PEEK was unaffected by crystallinity and moisture content. Also polyimides (PI) are a broad polymer family with extremely high mechanical strength, chemical inertness and thermal stability, having a linear or cyclic imide unit connected by aliphatic [35] or aromatic [36] groups. After sintering, they have virtually no melting temperature and behave as semi-thermosetting materials, making them potential candidates for bearing materials at extremely high temperature. Low friction is then desirable, as the conversion of mechanical energy into energy losses should be minimized. The earliest work on polyimide tribology by Buckley [37] and Fusaro [38] was made under vacuum conditions. They concluded that low friction and wear for thermoplastics is caused by a beneficial transfer film developing on the counterface. Transfer was attributed to plastification ability of the polyimide surface in absence of water vapour. It can be predicted that water molecules restrict the formation of a transfer film and cause a transition from low to high friction and/or wear, before achieving a low steady-state friction regime. Water molecules possibly act as anti-plasticizers restricting the molecular mobility (relaxation) and orientation under sliding [39], i.e. they suppress the secondary relaxation processes. The surface then acts brittle and the secondary transition temperature artificially increases. Under full water lubrication, however, Tanaka [40] reported lower friction and higher wear compared to dry sliding of polyimides. The effect of other environments such as cooling liquids was investigated by Sheiretov et al. [41], indicating no chemical degradation and no strong effect of refrigerants on friction and wear behaviour of polyimides, as they are resistant to most common solvents or chemicals.

For internally filled polymer composites, mainly the fillers influenced the sliding behaviour at different atmospheric conditions: organics either improved or deteriorated the wear rates compared to dry atmospheric sliding. Shen and Dumbleton [42] reported on galling mechanisms and three-body abrasive wear under moist conditions while good performance under dry sliding was noticed for polyimide and polyimide(amide) copolymers. This was attributed to low cohesion of fillers and polymer matrix resulting in stress-cracking.

Dislodgements on the wear surface were most likely observed near the fillers, as the matrix may change size in contact with moisture while fillers does not. For glass-filled polymers, Tsenoglu et al. [43] found that water penetration is enhanced through a network of micro-channels formed along the imperfectly bonded polymer-fiber interface. For carbon-fibre reinforced composites, the effect of moisture often becomes worse due to different water-uptake of the constituent phases as demonstrated for PEEK [44]. In order to lower the coefficients of friction for polymers, graphite powder can be added as an internal lubricant. It has a molecular structure of parallel planes with low shear strength along the basal planes and is consequently expected to have low friction. Pure graphite, however, has no intrisic lubricating properties as they strongly depend on the test conditions and interaction with the test environment. Lancaster [45] ascribed some transitions in friction of graphitic and non-graphitic carbons to mechanical and/or thermal effects, being most important for carbons with high graphiticity. The shear resistance of graphite under dry or vacuum conditions is often high and not uniquely related to its lamellar structure. Since the work of Savage [46], it is known that the shear resistance along the graphitic basal planes becomes low only in presence of moisture or other vapours suitable to adsorb on the graphite surface and to desactivate the reactive sites in the graphite structure. On the other hand, transfer of graphite-based composites is hindered under extremely high humudity conditions due to water-lubrication, leading to very large wear rates compared to dry conditions.

From the available literature, the effect of humidity on sliding properties of sintered graphite-polyimide composites cannot be clearly predicted. The tribological properties of polyimide and graphite lubricant seem to depend on the moisture content in an opposite way. Water molecules are needed for easy shear of the graphitic structure, while they have detrimental effect on the sliding properties of the polyimide surfaces. The friction and wear performance at three humidity levels will therefore experimentally be investigated and compared to microscopic evaluation of the worn surfaces. Test results will be discussed in relation to mechanisms known in literature for polyimide and graphite. The tribological performance of the system graphite powder – polyimide – humidity is dubiously influenced by humidity and has not intensively been described before. It will also be demonstrated that the effect of humidity for sintered polyimide is different than expected from the behaviour of thermoplastic polyimides.

2. Experiments

2.1 Test Materials

Polyimides were synthesized from a polycondensation between pyromellitic dianhydride (PMDA) and 4,4' diamino diphenyl ether or oxydianyline (ODA). The reaction scheme is shown in Figure 1. Due to limited solubility of the polyamide acid precursor, it has relatively low molecular weight (M_w = 10000 to 250000 g/mol, M_n = 13000 to 55000) with a polydispersity index 2.0 to 5.0. Pure sintered polyimide (SP-1) resins with initial grain size diameter of 10 to 20 μm were pre-compacted into 100 μm particles before sintering. Sintering happened under pressures of 500 to 2000 bar at a temperature of about 300°C for 10 minutes. Graphite-filled sintered polyimide (SP-2) contained 15 wt.-% synthetic graphite flakes with density 2.25 g/cm^3, surface area 9 m^2/g and 5 to 10 μm diameter. The mechanical properties of graphite sheets have been investigated in terms of the density changes and flake sizes by Leng et al. [47]. An optimum concentration of 15 wt.-% graphite flakes depends on the balance between low shear resistance and sufficient strength. According to Xian et al. [48], an increase in graphite content from 15 wt.-% to 20 wt.-% did not further reduce the wear rate of thermoplastic polyimides. The mechanical properties of pure polyimide and graphite-polyimide composites are given in Table 1.

Table 1. Mechanical properties of sintered polyimide in pure (SP-1) and graphite-filled (SP-2)

Property	Units	SP-1	SP-2
Specific gravity	g/cm^3	1.34	1.42
Water absorption water 24 h @ 23°C air equilibrium 23°C / 50% RH	 % %	 0.22 1.00	 0.19 0.80
Modulus of elasticity	MPa	2480	3170
Tensile stress at break	MPa	72	62
Tensile strain at break saturated in air	%	7.5	5.5
Compressive stress at 1 % strain 2 % strain 10 % strain (compress. strength)	 MPa MPa MPa	 24 46 112	 23 43 104
Charpy impact strength unnotched notched	 kJ/m^2 kJ/m^2	 7.5 3.5	 3.2 4.5
Rockwell Hardness	-	M 100	M 90
Shear strength	MPa	90	77

Steel counterfaces were 40 CrMnMo864 (DIN 1.2738) high-alloy steel with a hardness HV = 320, yield strength R_e = 765 N/mm^2 and tensile strength R_m = 900 to 1100 N/mm^2. They were ground and polished with GRID 600 Si-C paper to an average roughness Ra = 0.05 μm in order to simulate adhesive sliding.

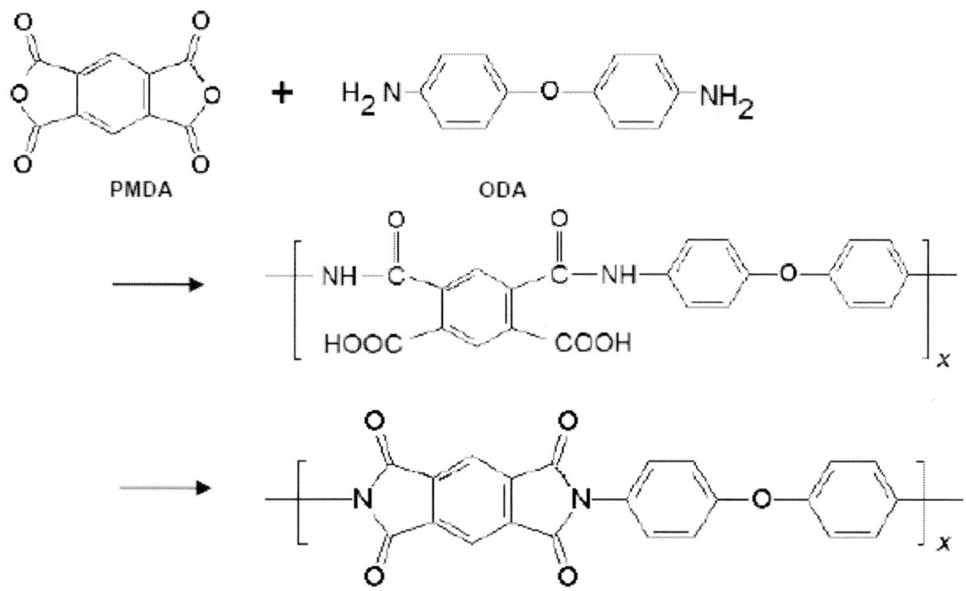

Figure 1. Reaction scheme and molecular structure of PMDA-ODA polyimide

2.2 Material characterisation

The thermal properties of sintered polyimides were evaluated by differential thermal analysis (DTA) in environmental air (60% RH) and argon atmosphere (no moisture), coupled to a mass-spectrometer. Samples were conditioned in the required testing atmosphere for at least 12 hours before starting the measurement. Measurements were done on a Stanton Redcroft 1500 Thermobalance, using 17 mg polymer samples that were heated between 23°C and 700°C at a constant rate of 10°C/min. Calibration was done on an empty Al_2O_3 crucible at the same heating rate. Exothermic reactions are registered as upward peaks and endothermic reactions are registered as downward peaks.

Raman spectra of the worn polyimide surfaces were made on a a Brucker FT spectrometer Equinox 55S (Bruker Optik, Ettlingen, Germany), equipped with a Raman module FRA 106 fitted to a nitrogen cooled (77 K) germanium high sensitivity detector D418-T. The applied laser wavelength during the experiments was the 1.064 μm line from a Diode Laser Pumped Nd:YAG laser.

2.3 Tribological testing

Polyimide cylinders (diameter 5 mm x width 15 mm) were reciprocating slid in a line contact against a steel counterface on a PLINT TE 77 High Frequency sliding tester. The total sliding distance was 15 km, corresponding to 5000 cycles with a single stroke of 15 mm. The effect of normal loads 50, 100, 150 and 200 N was evaluated at a fixed 0.3 m/s sliding velocity. The effect of sliding velocities 0.3, 0.6, 0.9 and 1.2 m/s was evaluated at a fixed 50 N normal load. The test environment was surrounded by a plastic box that is connected to an external climate conditioner. The environmental temperature was fixed at 23 ± 2°C and the relative humidity was varied between 30 ± 2, 40 ± 2 and 60 ± 2 %. Conditioned air was continuously circulated and parameters were PID-controlled with feedback. After mounting the test samples in a mechanical fixation, they were conditioned in the required testing atmosphere at zero load for at least 12 hours before starting the test.

The coefficients of friction were recorded on-line with a piezo-electrical force transducer in contact with the stationary steel counterface. Wear is recorded on-line with a contactless displacement transducer as the diameter reduction of the polyimide sample, representing material loss and deformation. Real wear rates or material losses are calculated from weight measurements of the samples before and after testing. The samples underwent a drying procedure of 12 hours at 60°C before weight measurements at the beginning and end of the test.

3. Material Characterisation

The DTA/TGA curves and mass spectrum during heating of polyimide SP-1 are shown in Figure 2. Sintered polyimides are thermally stable up to 592°C and degrade at higher temperatures without melting. The heat flow curve shows an important transition at around 180°C, with an endothermic peak indicating a transition between hydrolysis reactions (at 100 to 180°C) and imidisation reactions (at 180 to 260°C). The release of water from the polyimide structure was ensured by mass spectroscopy, indicating the water and oxygen fraction.

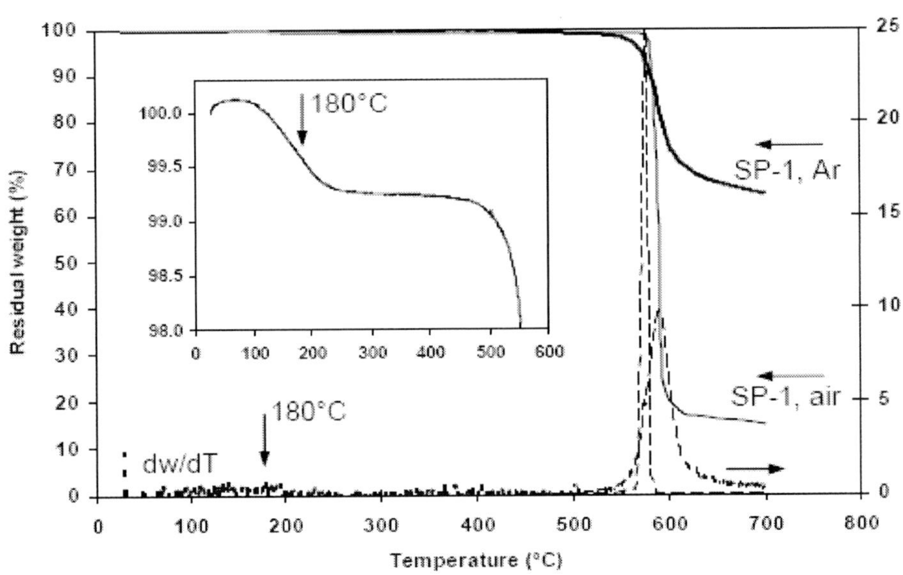

(a)

Figure 2. Continued on next page.

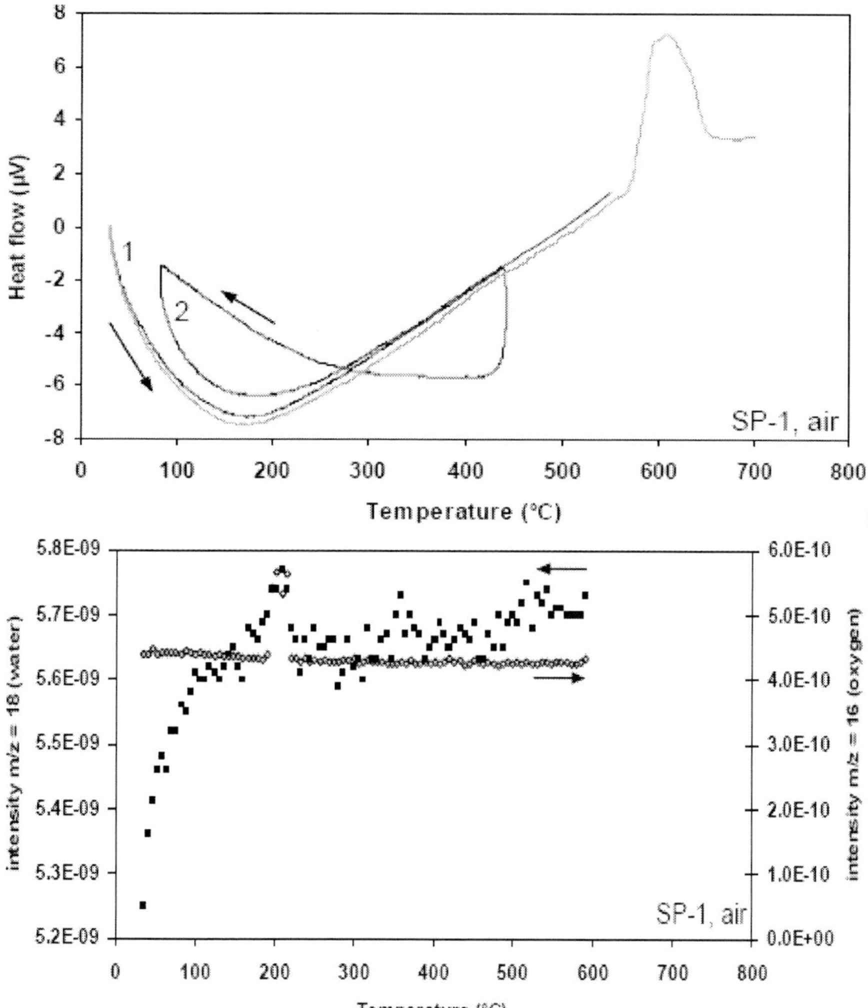

(b)

(c)

Figure 2. Thermal analysis of polyimide: (a) thermogravimetric analysis (TGA), (b) differential thermal analysis (DTA), (c) mass spectroscopy of volatilised products water and oxygen (MS)

Similar trends were measured under air and argon atmosphere, although the endothermic peak seems to have a little lower intensity in argon atmosphere. Also the weight-loss step at 180°C is somewhat larger in air compared to argon atmosphere. The small differences depending on the testing atmospheres are explained by influences of either chemically or physically adsorbed water in the polyimide structure. In argon atmosphere, the water release at 180°C only results from chemically bonds, while in air atmosphere (60 % RH) there is additional water release from physically bonded water. It clearly indicates that during preconditioning of the samples under moist air, water penetrated into the polyimide structure and is additionally released under heating. The water uptake by polyimide in this temperature region will significantly alter its tribological properties.

4. Tribological Test Results

4.1 Influence of sliding distance

On-line measurements (Figure 3) show the variations in coefficients of friction and diameter reductions of SP-1 and SP-2 as a function of the sliding distance. The effect of 40 or 60 % relative humidity at 50 N, 0.3 m/s is illustrated. Low sliding velocities and low normal loads are represented because variations in sliding and transfer are most pronounced at mild sliding conditions. At higher sliding velocities and normal loads, thermal heating and overload interfere with sliding mechanisms. The environmental atmosphere influences importantly the evolutions in friction and diameter reduction with sliding distance.

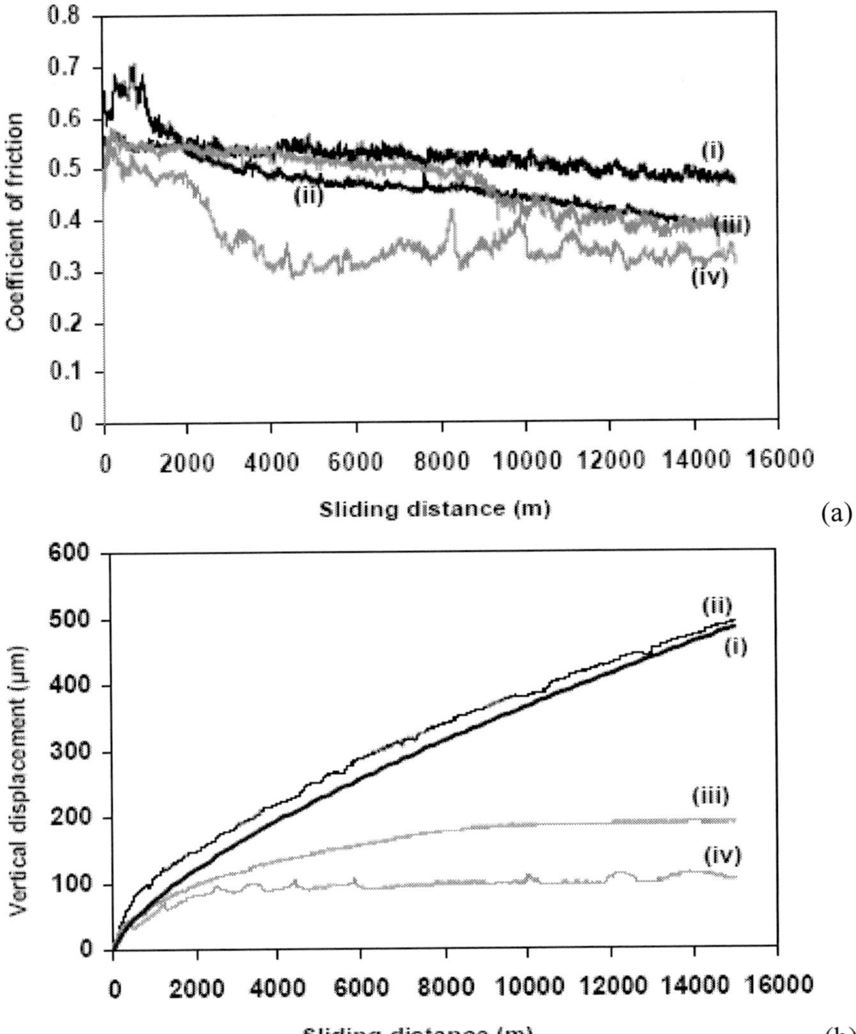

Figure 3. On-line measurements of tribological properties for SP-1 (i, ii) and SP-2 (iii, iv) at 50 N, 0.3 m/s under relative humidity of 60 % RH (i, iii) and 40 % RH (ii, iv), (a) friction as a function of sliding distance, (b) wear depth as a function of sliding distance

Friction for SP-1 at high humidity (60 % RH) has a maximum at running-in and stabilises with ongoing sliding distance. This behaviour is typical for polymers sliding under Hertz contact conditions, and is attributed to a change from line contact (running-in) into flat contact (steady-state). It was previously demonstrated that the stabilisation in coefficient of friction after 2000 m sliding distance coincides with levelling of the contact pressure from initially 34 MPa (line contact) to 2 MPa (full contact) under steady-state [49]. The friction value under steady-state remains constant and is reasonably high, which indicates a constant interaction between the polyimide and the steel counterface without significant alterations by transfer. Friction for SP-1 at low humidity (40 % RH) is more stable at running-in and progressively decreases with higher slope after longer sliding time. The peak-value at running-in disappears at low humidity, which indicates that the interaction between polyimide and steel is altered by transfer from the first meters of sliding on. Friction for SP-2 indicates smooth sliding during running-in and decreasing friction with sliding distance. The initial slope is parallel between pure and graphite-filled polyimide, with a transition after 8000 m at 60 % relative humidity and after 2000 m at 40 % relative humidity. The transition into steady-state sliding happens more rapidly at high loads or velocities. The steady-state friction of SP-2 is however less stable compared to SP-1, reflecting larger bulk inhomogeneities with dispersed graphite powder within the matrix. Therefore, the transfer film is supposed to be built up of graphite layers with low shear strength as reported by Langlade [50], while the debris particles consist of both polyimide and graphite. Instabilities in friction are most pronounced at 40 % relative humidity and indicate strong interactions with interfacial transfer particles while relatively smooth sliding at higher relative humidity suggest the removal of transfer particles outside the interface.

Diameter reductions for SP-1 linearly increase with sliding distance under steady-state conditions. The change in wear depth evolution between running-in (high slope) and steady-state (moderate slope) relates to the Hertz line contact. The wear evolutions are slightly influenced by the test environment. At 40 % relative humidity, running-in wear is somewhat higher and some scatter in the steady-state wear depth curve is noticed. Both observations are due to the periodical formation of a transfer film and removal of polymer particles from the interface. The smooth wear curve at 60 % relative humidty refers to the permanent polyimide/steel interaction without interfacial debris interactions. Diameter reductions for SP-2 have two different slopes with a clear separation between running-in and steady-state sliding: the transition into constant wear depth coincides with the transition into lower friction. The graphite-filled polyimide cylinders wear until the contact pressure is sustained by the graphite transfer film. Wear at 40 % relative humidity is significantly lower than at 60 % relative humidity, but it shows more instabilities in parallel to friction.

4.2 Influence of sliding parameters

The steady-state values for coefficients of friction and wear rates of SP-1 and SP-2 are summarised in Tables 2 to 5, including various normal loads and sliding velocities at 30, 40 and 60 % relative humidities. Test results are averaged from three runs per sliding parameter, showing a statistical variation of 7 % on coefficients of friction and 8 % on wear rates. These are acceptable for tribological data [51] and the trends described below are repeatable under different sliding conditions.

Table 2. Coefficients of friction (μ) and wear rates (w, 10^{-4} mm^3/m) for SP-1 at different normal loads and constant sliding velocity (0.3 m/s)

Relative humidity	50 N		100 N		150 N		200 N	
	μ	w	μ	w	μ	w	μ	w
30	0.20	9.2	0.27	27.4	0.32	65.2	0.35	63.4
40	0.39	8.0	0.45	20.3	0.53	64.7	0.49	58.6
60	0.51	7.7	0.47	19.5	0.45	31.7	0.41	50.9

For pure SP-1, the coefficients of friction are lowest at low relative humidity and gradually increase at higher relative humidity for every sliding velocity. At 30 % relative humidity, coefficients of friction increase with an increase in sliding velocity. At 40 to 60 % relative humidity, in contrast, coefficients of friction decrease with an increase in sliding velocity. A theoretical model of Matsubara [52] considers the orientation of molecules at the sliding surface in relation to the sliding velocity: it predicts higher friction at high sliding velocities because the polymer chains react stiffer at high shear rates and disfavour a conducive surface texture. This trend applies to SP-1 at 30% relative humidity and indicates that molecular mobility is limited at low relative humidity while improving at higher relative humidity. Also at 50 to 100 N normal loads, the coefficients of friction are lower at low humidity compared to moist air, but this trend may reverse at 150 to 200 N due to overload and brittle fracture. The increase in friction with normal loads does not agree with the sliding performance of thermoplastics, where plastification of the sliding surface contributes to lower friction. For sintered polyimide, there is a tendency of brittle fracture at high load levels, characterised by a monotoneous increase in friction. This behaviour is most pronounced at low compared to high humidity. Due to mechanical overload, some tests were stopped prematurely. The wear rates are highest at low humidity (30 %) and monotoneously decrease at higher relative humidity (40 to 60 %) for all test conditions. The decreasing wear rates at higher relative humidity are accompanied with increasing coefficients of friction. A critical normal load of 150 N is due to mechanical overload, in parallel to the transition into brittle fracture noticed for friction. The behaviour of sintered polyimide under variable humidity is clearly different to previous reports on thermoplastic polyimide by Tanaka [53].

For graphite-filled SP-2, the coefficients of friction are lower than SP-1 for most sliding parameters, except at some overload conditions. In contrast to SP-1, graphite-filled SP-2 presents both lower coefficients of friction and lower wear rates at low relative humidity compared to high relative humidity, for all sliding velocities and normal loads. At 30 % relative humidity, coefficients of friction increase at higher sliding velocties. At 40 to 60 % relative humidity, coefficients of friction decrease at higher sliding velocities. These trends are similar to pure SP-1 and indicate that the sliding properties of the graphite-filled composite may be strongly determined by the polyimide matrix. At 50 to 200 N, the coefficients of friction generally increase at higher loads and sometimes become higher than the values of unfilled polyimide, because the graphite fillers somewhat reduce the mechanical strength of the polyimide bulk. It is assumed and visually observed that the sliding behaviour of graphite-filled polyimides is controlled by the formation and the quality of a graphite transfer film. Such films are able to support the load, but have limited strength due to their lamellar structure. From Table 1, it reveals that sintered polyimide with graphite additives has lower strength and is more brittle. The influence of intrinsic mechanical strength mainly manifests at low sliding velocities with a continuous increase in friction as a function of normal load. This mechanical overload is also reflected in high wear rates and lumpy transfer, being most pronounced at high humidity. It was previously noted [54] that low velocity / high load conditions are sometimes unfavourable, as the lubricating mechanisms through 'easy shear' of the graphite structure is not fully exploited. The wear rates for graphite-filled SP-2 are only 10 to 20 % of the values for pure SP-1, and gradually increase at higher normal load and/or sliding velocity. A discontinuous increase in wear rates above 150 N refers to the mechanical overload situation for pure SP-1. It indicates that the bulk properties of pure sintered polyimides – although inferiorly – influence the behaviour of its composites. A transition into brittle fracture does not happen for SP-2 at any humidity condition and all tests have run over the full sliding distance.

4.3 Transfer film formation

The sliding behaviour of smooth surfaces depends on adhesion and formation of an interlayer with low shear strength, consisting of a lubricant or transfer film [55]. Early theories [56] state that the formation of a thin film is favourable for reducing the counterface roughness and therefore lowers friction and wear. For polymers, often a reverse trend is observed depending on the morphology of a polymer transfer film. Polymer/steel contact is reversed into (local) polymer/polymer contact with variations in adhesion and/or deformation.

For pure polyimides SP-1, optical micrographs of the transfer films are shown in Figure 4 after sliding at different normal loads, sliding velocities and variable humidity. Some details of the transfer films at 100 N, 0.3 m/s are investigated by secondary electron microscopy (SEM) in Figure 5 for different humidity. Transfer of sintered polyimides occurs mainly at low humidity and develops more difficult at high humidity. At 30 % relative humidity, transfer orients favourably along the sliding direction and becomes smoother at higher normal loads or sliding velocities. This film type is called 'platelet-like' transfer. A thin film shown at 50 N, 0.3 m/s is most favourable for low friction and is similar to the transfer type observed after sliding against stainless steel [57]. However, the platelet-like film causes higher wear rates compared to smoooth films because it acts more abrasive. It is concluded that mechanical snteraction in the sliding interface is most important above thermal interaction: (i) the contact area is reduced from the entire steel surface to local debris particles, reducing friction and (ii) the platelet particles are brittle and cause deformation and abrasion of the polyimide sliding surface. At 40 % relative humidity, the transfer films look rough with separate flakes at mild sliding conditions and a gradually thicker film at more severe sliding conditions. This film type is called 'lumpy' transfer. It causes higher friction because of large variations in adhesion and deformation in the sliding interface. At 60 % relative humidity, there is no film formation at mild sliding conditions and only some separate debris particles accumulate at severe sliding conditions. This film type is called 'island-like' transfer. It suggests that particles detach from the polyimide bulk and deposit without conglomeration in the interface. The transfer types for sintered polyimide under different atmospheric conditions are different from the continuously thin layers observed in vacuum sliding [58]. A smooth and continuous film is mostly expected at 1.2 m/s due to high shear, but only develops at low humidity.

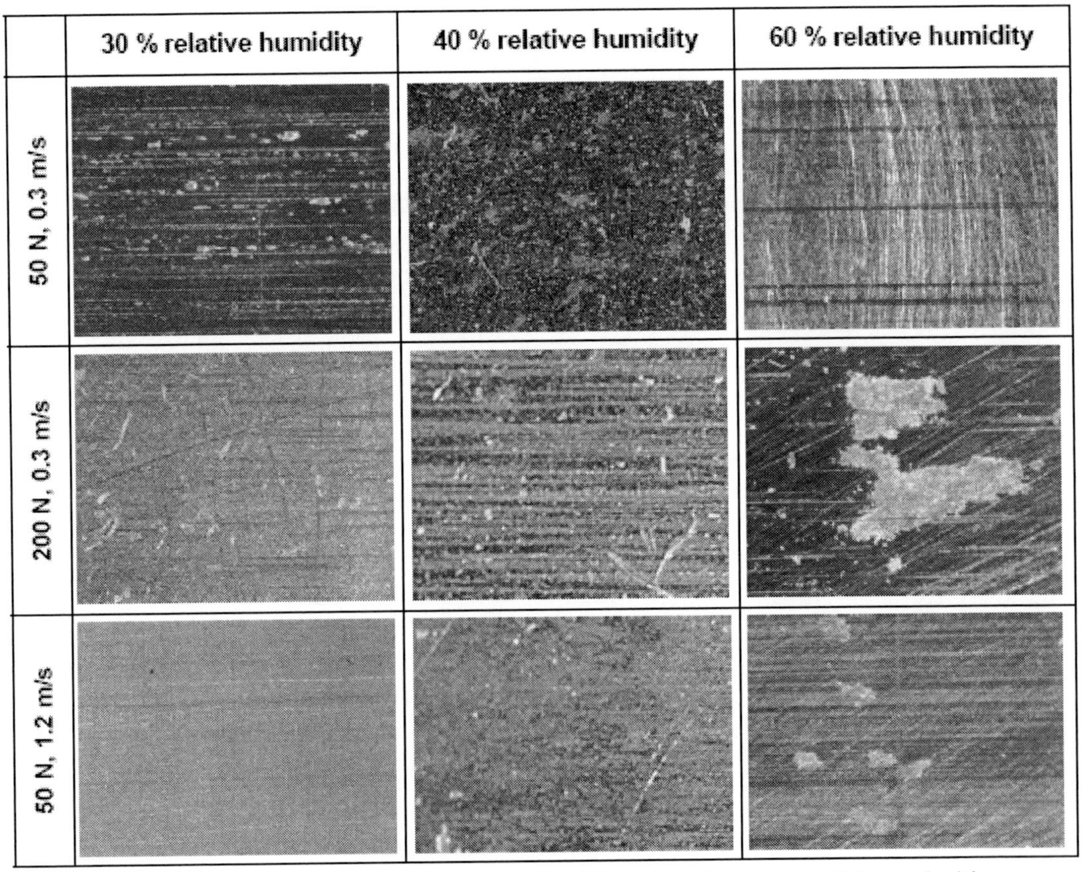

Figure 4. Overview of SP-1 ttransfer at different hummidity, normal loads and sliding velocities

Optical micrographs of the graphite-filled SP-2 transfer films are shown in Figure 6. At 30 % relative humidity, transfer consists of a thin and viscous film with mixed polyimide and graphite particles that are smoothened in the direction of sliding. At 40 % relative humidity, the polyimide and graphite transfer particles demix and form two separate phases on the counterface. At 60 % relative humidity, the transfer only contains graphite particles and no polyimide phase. The latter is in agreement with the observations for pure SP-1 that did not show transfer at high humidity. The lumpy transfer of graphite at 200 N explains the increase in friction and wear due to limited loading capacity of SP-2 composites.

5. Discussion

High performance polymers have good thermal resistance and are supposed to present chemical resistance, but their sliding behaviour is strongly influenced by humidity. The humidity effects become more important for graphite-filled than for pure polyimides. The kinetics of water vapour absorption into a porous polyimide structure was studied by Bertrand et al. [59], using an infiltration model. Depending on the sample thickness, it was measured

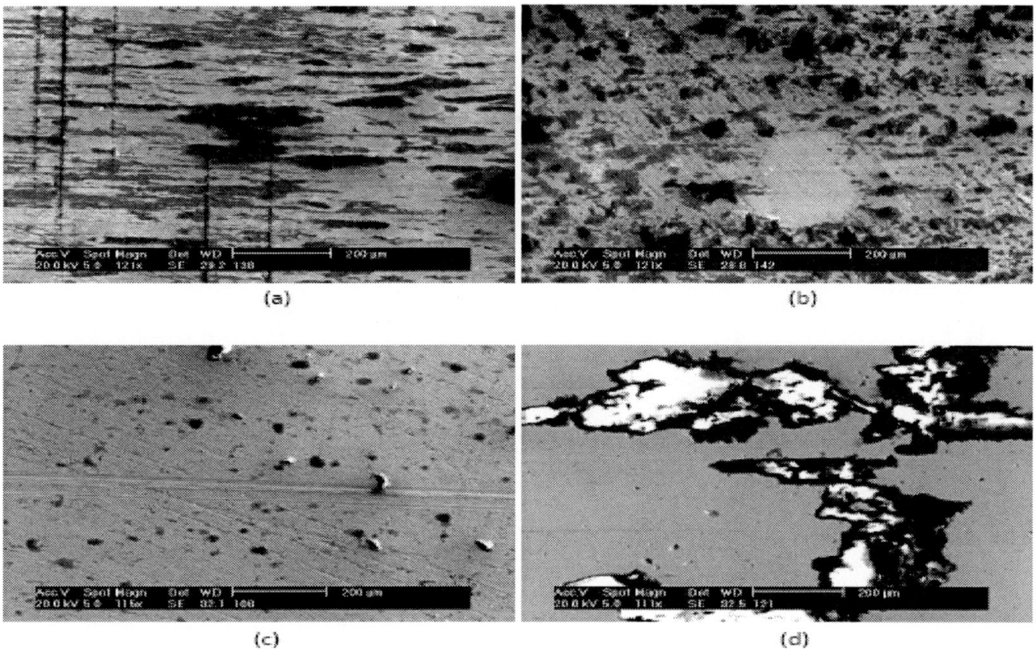

Figure 5. Details of SP-1 transfer at different humidity for 100 N, 0.3 m/s, (a) 30 % centre, (b) 40 % centre, (c) 60 % centre, (d) 60 % border

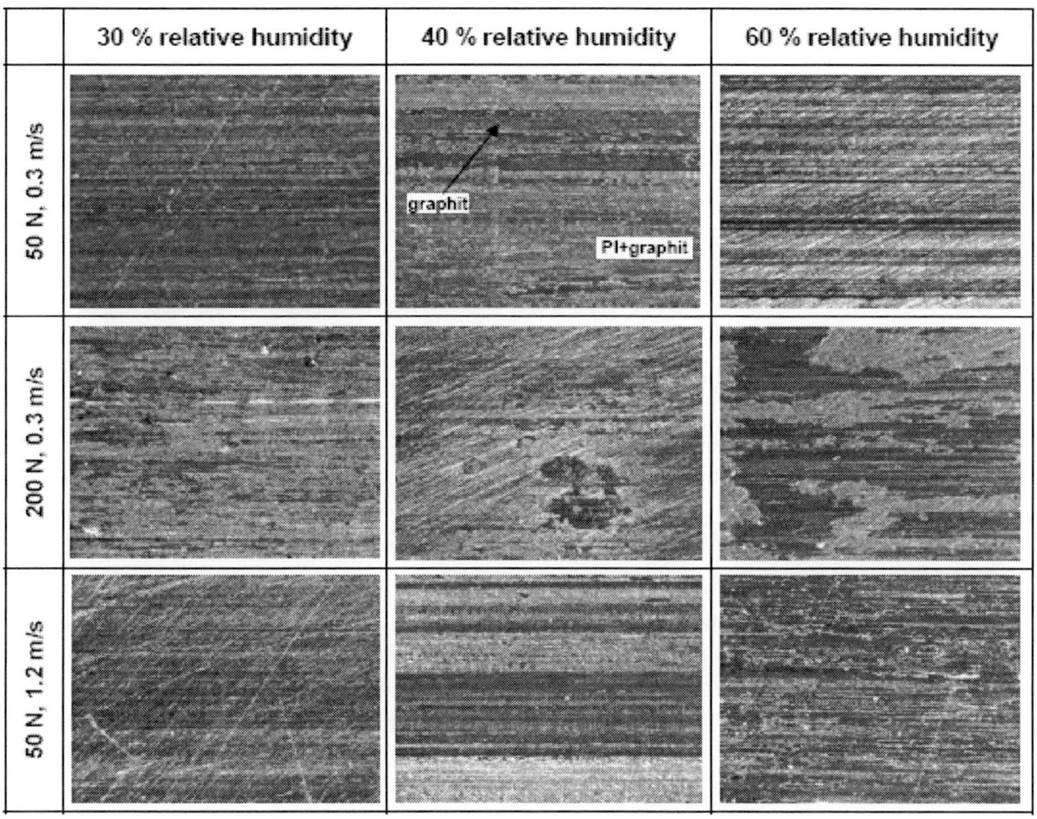

Figuure 6. Overview of SP-2 ttransfer at dgraphifferent humhitmidity, normaPI+graphital loads and ssliding velocities

that 90% of the saturation value is attained after 300 min (thickness 0.25 cm) or 60 min (thickness 0.12 cm), while slower absorption into the smallest 10% of the pores took about two days.

5.1 Influence of humidity on the sliding mechanisms of polyimide

From literature [38-39], the effects of humidity on polyimide are related to absorption of water molecules at the sliding surface and formation of hydrogen bonds between the carbonyl groups in adjacent molecular chains. The effect of humidity on sliding properties, however, strongly depends on the polyimide structure. Kang et al. [60] reported that for thermoplastic polyimides from benzophenone tetracarboxylic dianhydride (BTDA) and bisaniline (Bis-P) with glass transition temperature T_g = 285°C, the wear rates increased at higher humidity due to weakening of the polymer bulk through water absorption. The carbonyl linkage in the BTDA molecule is a site where water can directly attach to the chain through hydrogen bonding. It was postulated that the flexibility of a molecule permits it to reorient under stress and fracture rather than plastic flow occurs when this is restricted. For sintered PMDA polyimides, a different trend is noticed with decreasing wear rates at high humidity suggesting that strengthening effects dominate at high humidity. The absence of clear glass transition temperature and lower concentration of carbonyl groups in the molecular structure makes it less sensitive to weakening in presence of water. For PMDA polyimides, it can be expected that cross-links are formed which give the surface structure better mechanical strength with lower wear rates and lower tendency for overload at high humidity. On the other hand, also the brittleness seems to reduce by hydrolysis as revealed from the smoother surfaces at high humidity, while overload conditions at 150 to 200 N at low humidity cause unstable friction. An increase in wear rates for sintered polyimides at low humidity is in contrast to literature data on thermoplastic polyimides [39], reporting smooth transfer and progressively lower wear as the moisture content in air decreases. Present results on tribological properties of polyimides show some parallel tendencies with adhesion measurements made by Hu et al. [61]. They reported that the peel strength of polyimide films on silicon substrates decreases with increasing relative humidity due to the hydrolysis of polyimide, reaches a minimum, and then increases with increasing relative humidity due to the hydrogen bonding at the weak boundary layer. In a high-humidity environment, peel crack tips are attacked by moisture and result in weak adhesion measurement. Water adsorption by the polyimide films and diffusion into the peel crack tip was the main mode of moisture attack. Sager et al. [62] correlated the humidity absorption to a linear volume exapansion over a wide range of humidity. The observed material behaviour, however, strongly depends on the conditions of the polyimide fabrication process.

The transfer film morphology for pure SP-1 is clearly affected by relative humidity. At high humidity, therefore, transfer lacks and friction increases. At low humidity, transfer establishes and another reason for high wear rates is found in the abrasive action of the smooth-lumpy transfer film. The lack of transfer at 60 % RH and easy transfer at 40 % RH agrees with findings for thermoplastic polyimides by Jia et al. [63], who tested explicitly in dry and water environments. The worn polyimide surface under dry sliding was characterised by severe plastic deformation and micro cracking, while a large amount of transferred debris particles was observed on the counterpart. The worn polyimide surfaces after sliding at high humidity were smoother and no signs of transfer were noted. In other words, water inhibited the transfer of polyimide to the metal counterface. Lower wear rates of polyimide composites under water-lubricated conditions compared to dry sliding were related to the polar imide radicals which were liable to adsorb water and lead to swelling and decreasing shear strength. On the contrary, Kang et al. [60] observed an increase in wear rates of polyimide coatings under fretting conditions at high humidity compared to dry humidity, however more likely attributed to the formation of abrasive iron oxide debris. Earlier investigations by Fusaro [58] on thermoplastic polyimides reveals a friction coefficient $\mu = 0.60$ during a 50 % RH test, dropping to $\mu = 0.10$ in a 0.13 Pa vacuum or $\mu = 0.05$ in a 0.00013 Pa vacuum while wear rates progressively lowered under more severe vacuum conditions. These differences were also ascribed to water vapour concentration. The sensitivity of polyimides to changes in humidity was reported to be lowered for siloxane-modified copolymers, however, they have generally larger wear rates compared to pure polyimides. Also the effects of fluorine were favourably exploited to reduce the effect of moisture on tribological behaviour, because of its hydrophobic properties.

5.2 Influence of humidity on the sliding mechanisms of graphite

From literature, the lubricating action of graphite depends strongly on the test conditions and interaction with the test environment [46]. Fillers successfully used in one typical sliding condition may not perform equally in another. Under dry conditions the shear resistance is high and the associated increase in friction coefficient is often ascribed to the presence of dangling bonds with high chemical reactivity, leading to an increase in adhesive surface interaction [64]. Graphite does therefore not function well under vacuum conditions. Low friction may result from the complete desactivation of dangling bonds, e.g. through chemical reaction with available molecules from the environment. In ambient atmosphere, those bonds are rapidly desactivated by chemisorption of water and result in low friction. Desactivation of dangling bonds may also be created from reaction with wear debris. Hence, the mechanism by which water is influencing the shear resistance is still not fully clear. Referring to Lancaster [65], low friction for graphitic materials prevails as long as the fraction of the graphitic surface area covered by the adsorbents remains larger than a critical value. Possible transitions from low to high friction often occurs abruptly and have been associated to vapour desorption. However, referring to Gardos [66] the adsorbed vapours do not necessarily play the role of boundary lubricants, but rather modify the electronic orbitals within the graphite and thus the shear resistance.

The general believe of low friction for graphite at high humidity atmosphere does not agree with present test results on graphite-filled polyimide SP-2, performing low friction at 30 % RH and high friction at 60 % RH. So far, very few papers are found in literature focussing on sliding applications of sintered polyimide filled with graphite flakes. Most polyimide composites studied include graphite fibre reinforcements and/or PTFE and/or MoS_2 fillers [67]. Only Xian et al. [48] investigated the effect of percentage graphite flakes and the effect of temperature for polyetherimide. For pure graphite, higher friction at high humidity was recently observed in certain sliding conditions of low sliding velocities by Brendlé et al. [68, 69], modelling the tribo-reactions of moisture at the sliding surface of graphite as a triboreactor. He concluded that the real amount of water entering and consumed in the contact depends not only on the relative humidity but also on the sliding temperature and sliding velocity, responsible for the initiation of chain reactions by, e.g. particle detachment.

A hypothesis can be drawn from previous discussion of mechanisms found in literature in relation to present test results. For graphite-filled sintered polyimides the fraction of graphite is only 15 wt% and it can be estimated that the polyimide bulk properties are most important. At low humidity, the improved transfer of polyimide is beneficial for the formation of a mixed graphite-polyimide transfer film with good sliding properties. At low humidity, also the theoretical presence of dangling bonds on graphitic surfaces makes them chemically reactive and may also stimulate interaction with polyimide debris for more homogeneous transfer.

5.3 Relation between the humidity effects versus temperature effects

The influence of temperature on friction and wear under the same contact conditions as used in this chapter was investigated during our previous research for both sintered polyimide [70] and graphite-filled sintered polyimide [71].

For SP-1, high friction occurred at temperatures between 100 to 180°C due to hydrolysis reactions, while low friction was observed at temperatures between 180 to 260°C due to imidisation In a similar way, it was demonstrated in this chapter that polyimides have high friction at high atmospheric humidity and lower friction at low atmospheric humidity. Concerning coefficients of friction, conditions of low humidity seem to correspond to conditions of high temperatures. In an opposite way, it was observed that wear rates for SP-1 stabilise at high temperatures while wear rates increase at low humidity. Concerning wear, conditions of low humidity does not correspond to conditions of high temperatures. It indicates the influence of additional chemical reactions that manifest at high temperatures, such as imidisation, improving the strength of the transfer film. As a function of sliding temperature, no coherent transfer film was observed at low temperatures, while a transfer film only developed above 180°C. In a similar way, most favourable transfer is observed at low humidity. When transfer lacks, separate debris particles collect into the interface and act as third-body abrasive particles that strongly increase the wear rates. Both the counterface and the polymer surface are then characterised by abrasive grooves in the sliding direction.

The relation between humidity and temperature can be further explained by studying the variations in structure on the worn surfaces of sintered polyimides. A Raman spectrum of worn polyimide surfaces at different temperatures is shown in Figure 7. The relation between imide structure and hydrolysed carboxylic acid structure is illustrated by comparing the 1612 cm^{-1} and 1601 cm^{-1} band, respectively (Figure 8). For a constant 1601 cm^{-1} position, there is a downward shift in the 1612 cm^{-1} wavenumber at 100 to 140°C at different normal loads. The lower frequencies represent deterioration of the imide structure, partially changing into polyamic acid that is characterised by the 1601 cm^{-1} band. Enhancement of the imide structure through ring-closing of carboxylic acid is given by an upward shift of 1612 cm^{-1} and establishes at 180 to 200°C. Hydrolysis at 100 to 140°C is also reflected in the relative

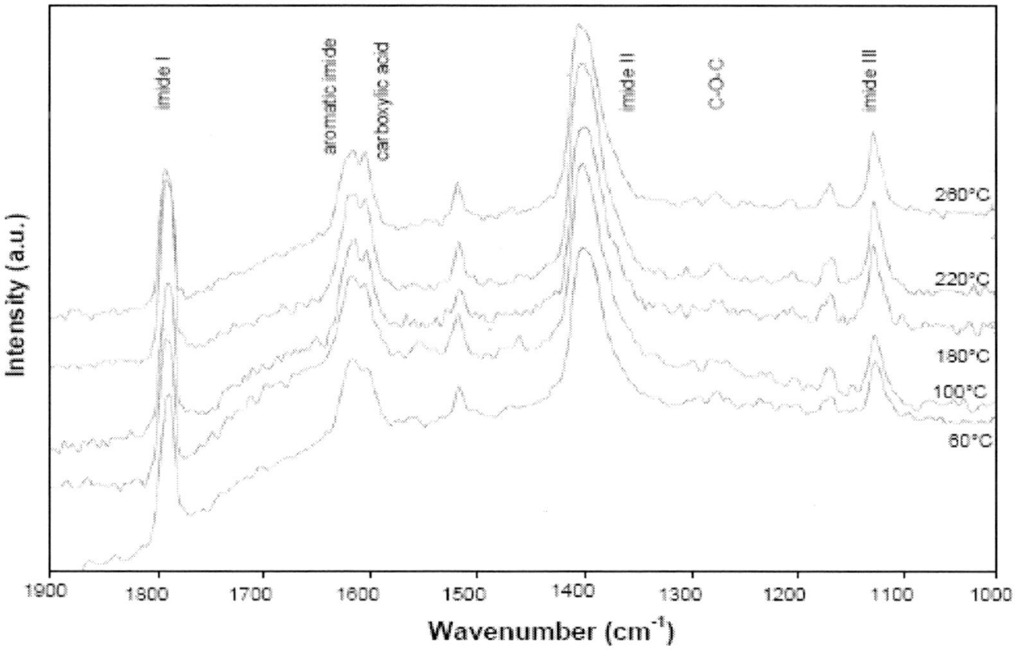

Figure 7. Raman spectrum of polyimide surfaces worn at different temperatures

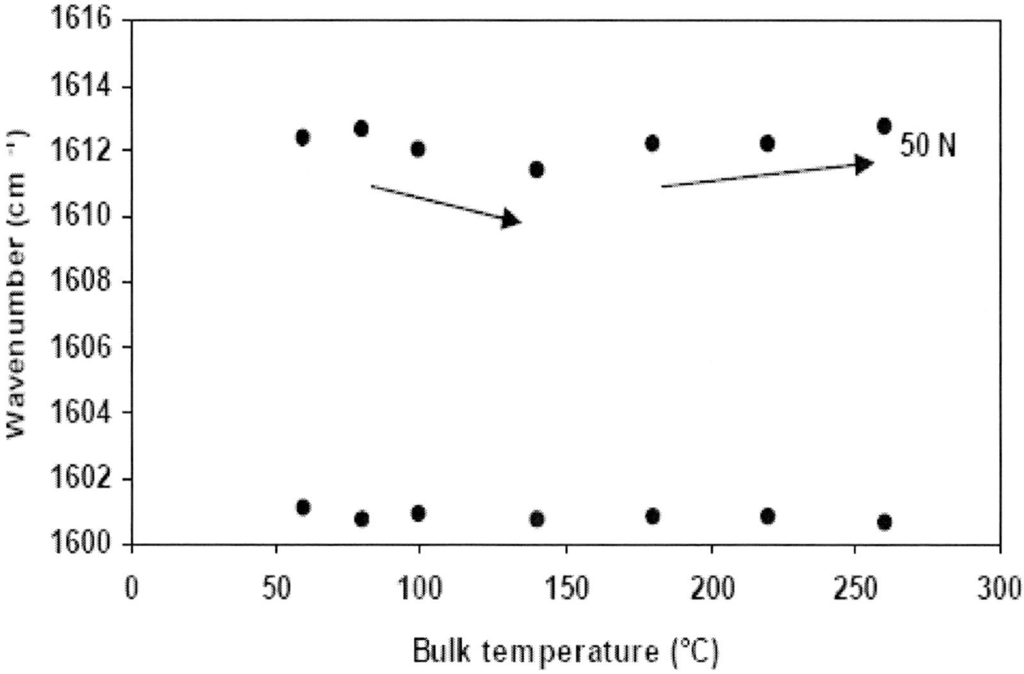

(a)

Figure 8. Continued on next page.

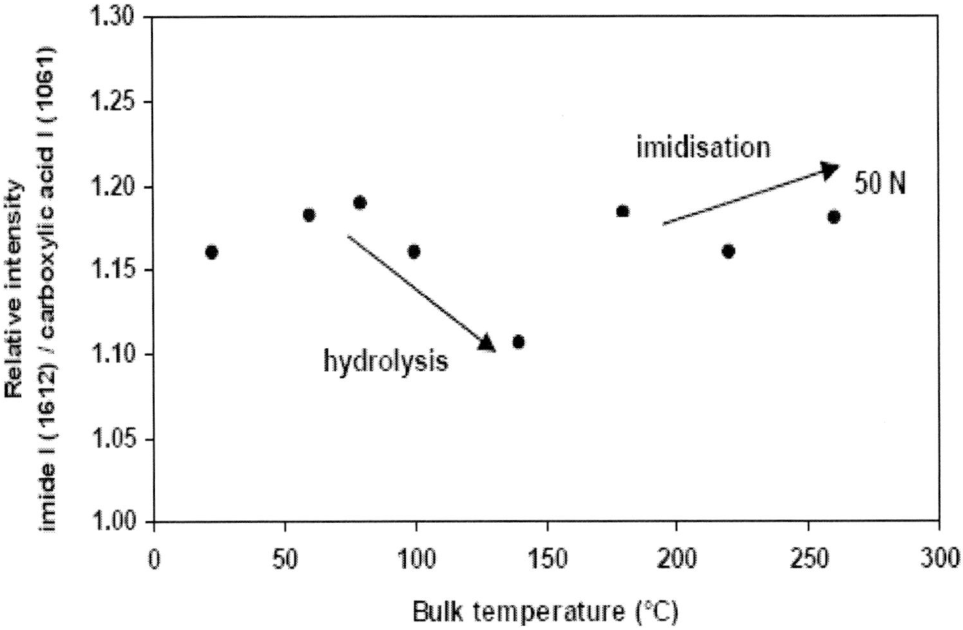

(b)

Figure 8. Raman analysis of imide aromatic band (1612 cm^{-1}) and carboxylic acid band (1601 cm^{-1}) for SP-1, (a) wavenumber position, (b) relative intensity

intensity I (1612) / I (1601), with a maximum in hydrolysis and minimum in imidisation reactions at 140°C for each normal load. Dehydration happens most intensively at 180°C irres-pective of the 50 to 150 N normal loads, represented by a maximum 1612 cm^{-1} band. The 1612 cm^{-1} band only represents the aromatic imide part, but also characteristic C=O imide groups (1788 cm^{-1}) and C-N-C imide groups (1395 cm^{-1}) show similar trends relatively to the carboxylic acid structure (Figure 9). The 1788 cm^{-1} position shifts down from 1787.12 cm^{-1} (60°C) to 1785.19 cm^{-1} (140°C) and it shifts up from 1789.12 cm^{-1} (180°C) to 1791.11 cm^{-1} (260°C) at 50 N normal loads. The minimum imide I intensity through hydrolysis at 140°C is more pronounced for the 1788 cm^{-1} band than observed for the aromatic imide ring at 1612 cm^{-1}, as mainly the C=O groups with high polarity are affected through hydrolysis. Imidisation starts at 180°C and shows clearer evolutions at 180 to 260°C by additional orientation of the functional C=O polyimide groups. The C-N-C imide bonds relatively to the carboxylic acid indicate identical decreasing and increasing tendencies. The minimum in the curve has shifted, however, from 140 to 180°C as the C-N-C structures are not directly hydrolysed. Hydrolysis is mainly affecting the C=O bonds and aromatic imide rings. At high loads, other reactions happen to the C=O and C-N-C structures with a decreasing intensity of imide-related bonds. According to Li et al. [72], a drop in C=O was also found for PEEK in the high load region and not in the low load region, correlated to bond rupture and radical formation by higher mechanical energy input. According to Cong et al. [73], XPS analysis indicated carbonation on the friction surfaces of thermoplastic polyimides, which was presently not revealed for sintered polyimides. Referring to the molecular structure of sintered polyimide in Figure 1, it is clear that imide ring-opening is explained either by hydrolysis (low temperatures) or by water supply (high humidity). Thus, it is demonstrated that besides the environmental humidity, also the sliding temperature induces various tribochemical reactions that may change the effective amount of water in the sliding interface. Therefore, sliding conditions of low temperatures (100-180°C) can be related to sliding conditions of high humidity.

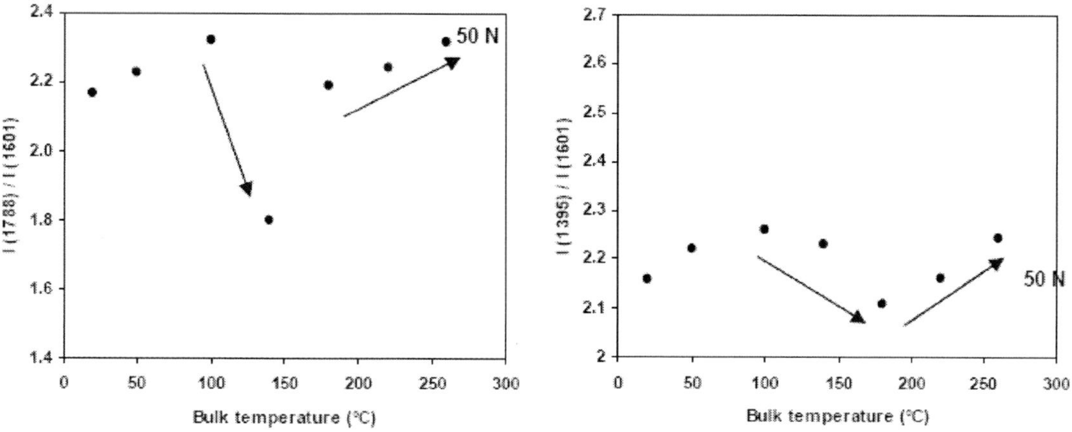

(a) (b)

Figure 9. Raman analysis of imide bands (1788, 1395 cm^{-1}) and carboxylic acid band (1601 cm^{-1}) for SP-1, (a) imide I (C=O) relative intensity, (b) imide II (C-N-C) relative intensity

For SP-2, there was observed high friction at temperatures between 60 to 100°C and lower friction at temperatures between 140 to 260°C, explained by the interaction of graphite lubricant and effective humidity in the sliding interface. In a similar way, it was demonstrated in this chapter that graphite-filled polyimides have high friction at high atmospheric humidity and lower friction at low atmospheric humidity. Concerning coefficients friction, there is a correlation that conditions of low humidity agree to conditions of high temperature in parallel to the observations for pure polyimide SP-1. The wear rates for graphite-filled polyimide are lowest at low humidity, while they are also lowest at low temperatures. Concerning wear, conditions of low humidity does not seem to correspond to conditions of high temperatures. In parallel to sintered polyimide SP-1, the variations in mophology of the polyimide transfer film mainly control wear rates: transfer films for graphite-filled polyimide are smoother under both low-humidity and high-temperature conditions.

6. Conclusions

Although polyimides are highly aromatic and have good chemical resistance, their tribological properties are strongly affected by humidity. In this chapter, the friction and wear observations for pure and graphite-filled sintered polyimides are evaluated at 30, 40 and 60 % relative humidity. According to literature, the sliding properties of polyimide and graphite are affected by water in an opposite way: while water molecules increase the friction of polyimide due to chain constraint, they lower the friction of graphite due to easy shear.

For pure polyimides, coefficients of friction decrease and wear rates increase at low humidity in parallel to the formation of a smooth-lumpy transfer film. For graphite-filled polyimides, both coefficients of friction and wear rates decrease at low humidity due to the formation of a thin and mixed polyimide/graphite transfer film. The reaction of graphite dangling bonds with wear debris is likely at low humidity. At higher humidity, phase separation of the graphite and polyimide transfer is observed. The sliding behaviour under low humidity corresponds to the sliding behaviour under high temperatures.

The role of moisture is confirmed from DTA/TGA and mass spectra, indicating the chemisorption and physisorption of water under air conditions and smaller water content under argon conditions. The release of water above 180°C corresponds to the reversion of hydrolysed into imidised structures, as further confirmed by Raman spectroscopy.

REFERENCES

[1] Abarou, S., Play, D., Kennedy, F.E. (1986). *ASLE Tribol. Trans*, 30, 269-281
[2] Bahadur, S. (1974). *Wear*, 29, 323-336
[3] Liu, Y., Schaefer, J.A. (2006). *Wear*, 261, 568-577
[4] Zhu, J., Schweigen W. (2001). *Kunststoffe Plast. Eur*, 91, 55-56
[5] Uetz, H., Richter, K., Wiedemeyer, J. (1983). *Wear*, 15, 103-114
[6] Kawakame, M., Bressan, J. D. (2006). *J. Mater. Proc. Techn*, 179, 74-80
[7] Lee, S., Spencer, N.D. (2006). *Tribol. Internat*, 11-12, 922-930
[8] Bing, K.Y. (1997). *J. Mater. Sci*, 32, 821-828
[9] Tanaka, A., Nishibori, T., Suzuki, M., Maekawa, K. (2003). *Diamond Rel. Mat*, 12, 2066-2071
[10] Liu, E., Ding, Y.F., Li, L., Blanpain, B., Celis, J.P. (2007). *Tribol. Internat*, 40, 216-219
[11] Yanh, S.H., Kong, H., Lee, K.W., Park, S., Kim, D.E. (2002). *Wear*, 252, 70-79
[12] Donnet, C., Le Mogne, T., Ponsonnet, L. (1998). *Tribol Letters*, 4, 259-265
[13] Vale Antunes, P., Ramalho, A. (2005). *Tribol. Internat*, 38, 363-379
[14] Hammerschmidt, J.A., Moasser, B., Wayne, L., Gladfelter, G., Haugstad, R., Jones, R.(1996). *Macromolecules*, 29, 8996-8998
[15] da Silva, C.H., Tanaka, D.K., Sinatora, A. (1999). *Wear*, 255-229, 339-342
[16] McNicol, A., Dowson, D., Davies, M. (1995). *Wear*, 181-183, 603-612
[17] Tanaka, K. (1986). In: Friction and Wear of Polymer Composites, Elsevier: Amsterdam, 137-174
[18] Tanaka, K., Miyata, T. (1977). *Wear*, 41, 383-392
[19] Morgan, J.E., Plumbridge, W.J., (1987). *Wear*, 119, 215-223
[20] Samyn, P., Schoukens, G., Van Driessche, I., Van Craenenbroeck, J., Verpoort, F. (2006). *Polymer*, 14, 5050-5065
[21] Apichartpattanasiri, S., Hay, N. H., Kukureka, S.N. (2001). *Wear* , 251, 1557-1566
[22] Kukureka, S.N., Hooke, C.J., Rao, M., Liao, P., Chen, Y.K. (1999). *Tribol. Internat,* 32,107-116
[23] Kohan, M. (1995). *Nylon Plastics Handbook,* Hanser Gardner Publications: New York.
[24] Yamahuchi, Y. (1990). *Tribology of plastic materials*, Elsevier: Amsterdam.
[25] Dlubek, G., Redmann, F., Krause-Rehberg, R. (2002). *J. Appl. Pol. Sci*, 84, 244-255
[26] Bertrand, P.A. (1995). Aerospace Report ATR-95(5826)-1, NASA Goddard Space Flight Center: Greenbelt
[27] Dosière, M. (1988). *Microchim Acta*, 94, 251-254

[28] Uddin, A.J., Ohkoshi, Y., Gotoh, Y., Nagura, M., Hara, T. (2003). *J. Pol. Sci. B: Pol. Phys*, 41, 2878-2891
[29] Park, Y., Ko, J., Ahn, T.K., Choe, S. (1998). *J. Pol. Sci. B: Pol. Phys*, 35, 807-815
[30] Lim, L.T., Britt, I.J., Tung, M.A. (1999). *J. Appl. Pol. Sci*, 71, 197-206
[31] Garoff, N., Zauscher, S. (2002). *Langmuir*, 18, 6921-6927
[32] Hoa, S.V., Lin, S., Chen, J.R. (1989). In: *Advances in Thermoplastic Matrix Composite Materials*, Newaz, G.M., Ed, ASTM: Philadelphia, PA, pp 213-230.
[33] Lhymn C. (1986). *J. Mater. Sci. Lett*, 5, 260-262
[34] Wang, G. (1989). *J. Comp. Mater*, 23, 434-447
[35] Mathews, A.S., Kim, I., Ha, C.S. (2006). *J. Appl. Pol. Sci*, 120, 3316-3326
[36] Kumar, D., Gupta, A.D. (1992). *Polym. Advanced Technol, 3,* 1-7
[37] Buckley, D.H. (1966). Research Report TN D 3261, Nasa: Washington DC
[38] Fusaro, R.L. (1988). *Tribol. Trans*, 31, 174-181
[39] Fusaro, R.L. (1978). *ASLE Trans*, 21, 125-133
[40] Tanaka, K. (1980). *ASME Trans*, 102, 526-533
[41] Sheiretov, T., Van Glabbeek, W., Cusano, C. (1995). *Tribol. Trans*, 38, 914-922
[42] Shen, C., Dumbleton, J.H. (1976). *Wear*, 40, 351-360
[43] Tsenoglou, C.T., Pavlidou, S., Papaspyrides, C.D. (2006). *Comp. Sci. Techn*, 66, 2855-2864
[44] Jacobs, O., Jaskulka, R., Yan, C., Wu, W. (2005). *Tribol. Letters*, 19, 319-329
[45] Lancaster, J.K. (1975). *Wear*, 34, 275-290
[46] Savage, R.H. (1948). *J. Appl. Phys*, 19, 1-10
[47] Leng, Y., Gu, J., Cao, W., Zhang, T.Y. (1998). *Carbon*, 36, 875-881
[48] Xian, G., Zhang, Z. (2005). *Wear*, 258, 783-788
[49] Samyn, P., Quintelier, J., Schoukens G. (2007). *Experiment. Mech*, in press
[50] Langlade, C., Fayeulle S., Olier, R. (1994). *Wear*, 172, 85-92
[51] Klaffke, D., Hartelt, M., Koesling, F.P. (2006). *Tribotest*, 7, 281-299
[52] Matsubara M. (1981). In *Friction and Wear of Polymers*; Bartenev, G.M., Lavrentev, V.V., Ed, Elsevier: Amsterdam, 115
[53] Tanaka, K. (1977). *Lubr. Technol*, 99, 408-414
[54] Samyn, P. (2007). *Polym. Polym. Comp*, in press
[55] Bahadur, S. (2000). *Wear*, 245, 92-99
[56] Hollander, A.E., Lancaster, J. K. (1973). *Wear*, 25, 155.
[57] Samyn, P., Schoukens, G., Quintelier, J., De Baets, P. (2006). *Tribol. Internat*, 39, 575-589
[58] Fusaro, R.L. (1990). *Tribol. Internat*, 23, 105-121
[59] Bertrand, P.A., Carré, D.J. (1997). *Tribol. Trans*, *40*, 294-302
[60] Kang, C. *Wear* 1995, *181-183*, 94-100
[61] Hu, J., Chen, H.C. (1992). *J. Mater. Sci*, 27, 5262-5268
[62] Sager, K., Schroth, A., Gerlach, G. (1996). *J. Intell. Mater. Syst. Struct*, 7, 264-266
[63] Jia, J.H., Zhou, H.D., Gao, S.Q., Chen, J.M. (2003). *Mat. Sci. Eng. A*, 356, 48-53
[64] Lancaster, J.K. (1975). *ASLE Trans*, 18, 187-201
[65] Lancaster, J.K. (1981). *J. Appl. Phys. D*, 14, 747-762
[66] Gardos, M. (1997). In *World Tribology Congress: New directions in tribology*, Hutchings, I.M., Ed, Mechanical Engineering Publications: London.
[67] Bijwe, J., Indumathi, J. (2004). *Wear*, 257, 562-572

[68] Brendlé, M., Stempflé, P. (2003). *Wear*, 257, 818-826
[69] Brendlé, M., Fatkin, J., Gilmore, R. (1990). *Tribol. Trans*, 22, 471-480
[70] Samyn, P., De Baets, P., Schoukens, G. (2007). *Macromol. Mater. Eng*, 292, 523-556
[71] Samyn, P., Schoukens, G. (2007). *Carbon*, submitted
[72] Li, T.Q., Zhang, M.Q., Song, L., Zheng, H.M. (1999). *Polymer*, 40, 4451-4458
[73] Cong, P.H., Li, T.S., Liu, X.J., Zhang, X.S., Xue, Q.J. (1998). *Acta Polym Sinica*, 5, 556-561

INDEX

A

B

C

D

E

F

G

H

I

J

K

L

M

N

R

S

T

U

V

W

X

Y

Z